CAMBRIDGESHIRE RE
(formerly Cambridge Antiquarian Records Society)

VOLUME 8

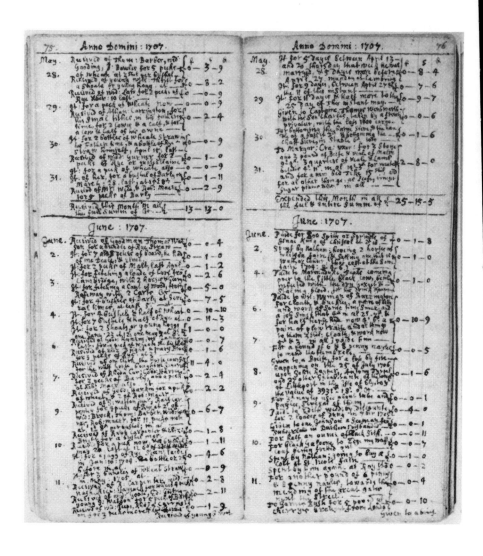

Crakanthorp's household account book for May–June 1707, transcribed below, pp. 209, 212 (receipts) and 211, 214–15 (disbursements)

ACCOUNTS OF THE REVEREND JOHN CRAKANTHORP OF FOWLMERE

1682–1710

EDITED BY
PAUL BRASSLEY
ANTHONY LAMBERT
AND
PHILIP SAUNDERS

CAMBRIDGE
1988

Published by the Cambridgeshire Records Society,
County Record Office, Shire Hall, Cambridge, CB3 0AP

© Cambridgeshire Records Society 1988

This volume has been published with the help of generous grants from Lord Walston, The University of Cambridge Department of Land Economy and the Twenty-Seven Foundation

British Library Cataloguing in Publication Data
Crakanthorp, John, *1642-1719*
Accounts of the Reverend John Crakanthorp of Fowlmere 1682-1710
1. Farms — England — Fowlmere (Cambridgeshire) — Finance — History
2. Agriculture — England — Fowlmere (Cambridgeshire) — Accountancy — History
I. Brassley, Paul II. Lambert, Anthony
III. Saunders, Philip, *1950-*   IV. Cambridgeshire Records Society
338.1'3'0942657   HD1930.F6/

ISBN 0 904323 08 0

*Printed and bound in Great Britain by
E. & E. Plumridge Ltd., Linton, Cambridge*

CONTENTS

	Page
Abbreviations	vi
List of figures and tables	vi
Old Measures	vi
Preface	vii

INTRODUCTION

The Manuscripts and Editorial Method	1
Authorship and Provenance	3
John Crakanthorp	5
Farming in Fowlmere	10
The Harvest Accounts	19
The Household Accounts	32

TEXTS

The Harvest Accounts

I 1682–1684	35
II 1685–1689	61
The Household Accounts 1705–1710	121

GLOSSARY	274
INDEX OF PERSONS AND PLACES	278
INDEX OF SUBJECTS	287

ABBREVIATIONS

C.R.O.	County Record Office, Cambridge
C.U.L.	Cambridge University Library
P.R.O.	Public Record Office
V.C.H.	*Victoria History of the Counties of England: Cambridgeshire and the Isle of Ely*

FIGURES

	Page
1 Fowlmere: The Open Fields	13
2 Market Locations	23
3 Wheat Prices	26
4 Barley Prices	27

TABLES

1 Fowlmere and Cambridgeshire Hearth Tax	11
2 Fowlmere Inventories	16
3 Summary of Harvest Accounts	22
4 Acreages sown to cereal crops	31
5 Apparent Yields	31
6 Summary of Household Accounts	32

OLD MEASURES

4 pecks = 1 bushel   8 bushels = 1 quarter
40 perches = 1 rood   4 roods = 1 acre
12 pence = 1 shilling   20 shillings = 1 pound sterling

PREFACE

This volume was originally suggested by the late Geoffrey Vinter (1900–1981), a founder member of the Cambridge Antiquarian Records Society and treasurer of the society until his death. Vinter, as his notes and collections bequeathed to the County Record Office amply testify, did much to preserve the documentary heritage of his village of Thriplow and its larger neighbour, Fowlmere, and to illuminate their history. In initiating the deposit of the parish records of Fowlmere, which include one of Crakanthorp's account books, he also played a part in bringing together in the County Record Office the three volumes of accounts edited here, which in the course of two and a half centuries had been separated one from the other and lain unconnected and virtually lost, their authorship even a matter for speculation. Much of his historical activities he shared with his secretary, Marie Swann of Fowlmere, who as assistant treasurer managed the day-to-day running of the society's finances to within a year of her untimely death in 1985. It is hoped that the appearance of this volume will be a fitting tribute to their service to the society, the preservation of archives and the promotion of local history in Cambridgeshire.

The present edition is the product of the efforts of several persons. Paul Brassley transcribed the harvest accounts with some assistance from Nicholas Wells and also wrote the introduction, with the exception of the section on the manuscripts and editorial method, contributed by Philip Saunders. Anthony Lambert made a transcript of the household accounts, which has been revised for publication by Philip Saunders, who has also compiled the glossary and (with Brian Lessware) the indexes. For permission to publish these accounts the society is grateful to the rector and parochial church council of Fowlmere, Commander C. Scott-Fox, R.N., and the County Record Office, Cambridge. Thanks are also due to Michael Farrar, County Archivist and a former general editor of the society, the staffs of Cambridge University Library and the library of Seale-Hayne College, Newton Abbot, to Audrey Grant, Joan Thirsk, Pat Taylor and Sue Wheeler. Finally, but most importantly, an incalculable debt of gratitude is owed to our wives for assistance, patience, understanding and encouragement.

P.W.B.
P.C.S.

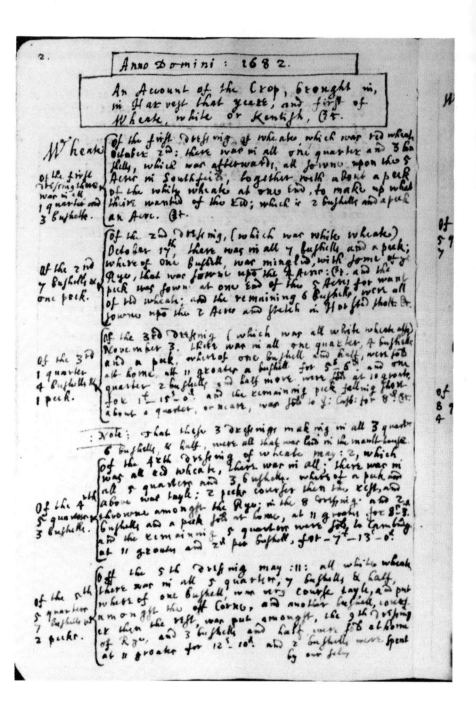

The first page of the earlier volume of Crakanthorp's harvest accounts, transcribed below, pp. 35–6

INTRODUCTION

> 'The historian is like a man visiting a distant country, trying to understand a society vastly different from his own in which people speak a strange language and think and act in unfamiliar ways. His task is even more difficult than that of the traveller, for he cannot speak to the dead; he can only interpret the remains which they have left.'
> G. Holmes, *The Later Middle Ages 1272-1485* (1967), p. 5.

This book contains in printed form the surviving notebooks of the Reverend John Crakanthorp, Rector of Fowlmere in Cambridgeshire from 1667 to 1719. The first two books are of harvest accounts, in which he noted the produce of his harvest and details of how it was used. The third is of household accounts. The time when Crakanthorp lived is now recognized as one in which important changes were affecting the English countryside. These accounts provide a glimpse of life in a fairly typical open field parish in eastern England.

The Manuscripts and Editorial Method

Each of the three account books is of pocket-book size. The two books of harvest accounts are approximately 4 x 6 inches, bound only in a stiffer paper and saddle-stitched. The 1682-4 book[1] is of 58 pages and that of 1685-92[2] of 110 pages, each neatly numbered with a protective blank sheet at the front. The headings at the beginning and for each crop are contained within ruled boxes, and the year is given as a running head at the top of each page, also in a box. A brace on the left of the entry for each dressing directs the reader to the marginal summary. For the sake of economy these are omitted in this edition.

The household account book[3] is taller and narrower (3 x 7 inches) than the harvest accounts, and when deposited in the County Record Office was disbound and its original covers and end-papers lost. It has recently been repaired and re-bound. It was bought for Crakanthorp in London and cost 1s. 2d.[4] It is of 134 pages with the year at the head of each page separated by a rule from the accounts themselves; the month, and below it

1. C.R.O., P72/3/1.
2. C.R.O., 691/A1.
3. C.R.O., 603/A1.
4. Below, p. 126 (household accounts, 9 May 1705, disbursements).

the days, are written in the left-hand margin. As the accounts are heavily compressed and the days in the margin consequently often fall between two lines, it is sometimes uncertain to which entries they refer. These accounts are written in parallel in the manuscript, the receipts on the left and disbursements on the right of each opening. They are not balanced, however, and for economy they are printed here in sequence for each month, first receipts then disbursements.

All three books are written in a minute, neat, hand with many abbreviations. Where the intention of the abbreviation is quite certain it has been extended in this edition; where it is not, the abbreviation, with its sign, most commonly a colon, has been printed as in the manuscript. Partly for this reason, partly for economy, proper names have not been extended, except in a few cases where identification would be otherwise difficult, and an extension is shown in square brackets. Colons, etc., were used prolifically by Crakanthorp (e.g. between anno and domini). They have been omitted here, and both punctuation and the use of capitals conservatively adapted to modern conventions. Crakanthorp's running heads, catchwords at the foot of pages and occasional minor scribal errors such as interlineations of omissions and accidental repetitions have also been suppressed. Numbers have been left in figures or words as the case may be in the manuscript, except in the harvest accounts where, in the interests of economy and readability, sums of money have been reduced to figures only, but without altering denominations: 'one pound 8 shill. & 9 pence' is thus printed 1*l*. 8*s*. 9*d*., '3 pounds' as 3*l*., and '20 pence' as 20*d*. 'Ob.' for *obolus*, a halfpenny, and 'qu.' for *quadrans*, a farthing, have been retained, except in the columns of household accounts, where they are shown as fractions.

The year shown in the running heads of the harvest accounts is that of the harvest, though many 'dressings' took place in the next calendar year. The household accounts follow the normal practice of the time in beginning each year on Lady Day, 25 March. For ease of reference in this edition the modern historical year beginning on 1 January has been used throughout; where in the text a year has been modified from that given in the manuscript it is printed in italics. The internal dating of the household accounts presents a further problem as it seems that Crakanthorp often laid out a month's accounts by first writing the days in the margin. The entries correspond only very approximately to these, so that days occasionally fall between entries and sometimes two or more days stand against a single receipt or disbursement. The fault is aggravated where Crakanthorp, in order to confine the month's accounts to equal lengths for receipts and disbursements and keep them in parallel, obliged himself to compress whichever side produced the longer or greater number of entries. Elsewhere the dating is sparser and less mechanical, and yet still fails to inspire confidence; but as it is used by the author as a means of cross-reference and is at least a rough guide, it cannot be abandoned. In this edition, therefore, where inevitably lines of entries do not correspond exactly to those of the manuscript and it is not practical to place dates between lines, the days have been located against the line beginning with or containing the word that follows it, whether adjacent or just below.

Authorship and Provenance

In discussing the provenance of the notebooks, it is first necessary to distinguish between the two volumes of harvest accounts, which relate to the years 1682-4 and 1685-92, and the household accounts for the years 1705-1709. What happened to the first volume of harvest accounts after Crakanthorp's death is not known, but it seems likely that the second volume, and the household accounts, remained in the possession of Crakanthorp's descendants.

The harvest accounts re-appeared in the early part of this century. In 1905 *The East Anglian* printed a note from Mr. A.H. Arkle, who lived near Birkenhead, which began 'Some little time ago I became possessed of a small manuscript book, giving an account of the harvests in the years 1682, 1683 and 1684...'[5] From internal evidence Arkle deduced that the accounts related to a farm in the neighbourhood of Royston, but he could be no more precise than that. Neither could he identify the author. Following his note, *The East Anglian* printed part of the manuscript, which Arkle had transcribed, and in the following eight issues the rest of the transcription appeared. With the third instalment there appeared a note from the Reverend A.C. Yorke, then Rector of Fowlmere, who demonstrated that the field names in the harvest accounts were the same as those appearing in the seventeenth-century surveys of Fowlmere, and many of the personal names also occurred in the parish registers. He therefore concluded that the manuscript was written at Fowlmere. He was less certain about its authorship, but suggested reasons for believing that it could be a member of the Wedd family.[6] In 1909 the notebook was presented by Arkle to the church at Fowlmere, and remained in the parish chest until it was transferred to the County Record Office in 1969.[7]

The second book of harvest accounts, covering the period 1685-1692, was found in 1937, among some family papers which had been inherited by Sir Cyril Fox. In 1974 his son, Commander C. Scott-Fox R.N., deposited it in the County Record Office. The discovery of this second volume led to renewed interest in the authorship of the first volume. From a comparison of the contents and handwriting of the two volumes it was obvious that the second was the sequel to the first. A note on the cover of the second volume ascribed it to Crakanthorp, who was an ancestor of Sir Cyril Fox. But Yorke, thirty years earlier, had suggested that it was written by Benjamin

---

5. A.H. Arkle, 'Old Fashioned Harvests', *The East Anglian,* xi (1905-6), 161.
6. A.C. Yorke, 'Reply', *The East Anglian,* xi (1905-6), 223.
7. W.P. Baker, 'The Authorship of a seventeenth-century Harvests' Account Book from Fowlmere', *Proceedings of the Cambridge Antiquarian Society,* xlii (1948), 27-32.

Wedd, who was probably also an ancestor of Fox.[8] The problem was taken up by W.P. Baker, who, in a paper read to the Cambridge Antiquarian Society in 1940, demonstrated that Crakanthorp was the author.[9] However, it is interesting to note that on the first page of the accounts Crakanthorp records that he is about to write an account 'of my Seventh Crop, gathered in, in Harvest Anno 1682:'. It would follow from this that his first harvest was in 1676, yet Crakanthorp was instituted to the rectory of Fowlmere in 1667.[10] This point was noted by Baker, whose only explanation of the discrepancy was a suggestion that Crakanthorp was a very young man when he came to Fowlmere.[11] He was in fact 25 years old (see below) which does not seem too young to begin farming. An alternative explanation (but equally, pure guesswork) is that his first harvest relates to corn harvested from the land he held in his own personal capacity, as opposed to his glebe. It is possible that he did not acquire this until 1676. Conversely, he may have leased the glebe in the years 1667 to 1675. There is no evidence for or against any of these hypotheses.

The household accounts were found in 1970 in a deed box which had been recovered from a bombed London office block in 1940. The box contained papers belonging to the Reverend Frederick Fox Lambert, Rector of Clothall (Herts.), 1879-91, and Vicar of Cheshunt (Herts.) 1891-1911. Lambert was the great uncle of Sir Anthony Lambert and his sister, Mrs. Audrey Grant, into whose possession the papers eventually passed. They transcribed the household accounts and deduced that they must have been written by a parson living somewhere between Royston and Cambridge. In 1974 Sir Anthony wrote an article in *Blackwood's Magazine* explaining how they traced the authorship to Crakanthrop.[12] In 1976 he and his sister presented the book to the County Record Office, thus bringing together again the three remarkable account books.

Crakanthorp entitles the surviving book of household accounts 'Tabulae vel Codex Sextus... Book the sixth'. The previous volumes, and any subsequent volumes, have yet to be found.

8. Ms. pedigree in the possession of Cdr. Scott-Fox.
9. W.P. Baker, *op. cit.*
10. British Library, Add. MS. 5823, fo. 33 (William Cole's Parochial Antiquities of Cambridgeshire; microfilm also in C.R.O.). The earliest entry in the parish register by Crakanthorp is a baptism, 24 November 1667 (C.R.O., P72/1/2).
11. W.P. Baker, *op. cit.* p. 31.
12. Anthony Lambert, 'The Sixth Codex', *Blackwood's Magazine* cccxvi, no. 1908, October 1974 pp. 289-301; Audrey Grant, 'Hertfordshire References in the Diaries of John Crakanthorp', *Hertfordshire's Past* no. 1 (Autumn 1976), pp. 14-15; Audrey Grant, 'Sickness and Health; references in the account book of John Crakanthorp, 1705-10', *Hertfordshire's Past*, no. 5 (Autumn 1978), pp. 19-20.

John Crakanthorp

John Crakanthorp was born at Stanford Dingley, near Reading in Berkshire, on 26 January 1642. His father was also a clergyman, and had been introduced to the living of Stanford Dingley in the last year of his life.[13] A pedigree of the family prepared by Samuel Crakanthorp of Norwich in 1795 stated that Crakanthorp's father, who was also called John, died at Markshall in Essex 'leaving his wife in a state of Pregnancy who was delivered of a son'. This pedigree also states that Crakanthorp's grandfather Richard Crakanthorp, who was born in 1567 and also a clergyman, married a daughter of Lady Honywood of Markshall.[14] This seems unlikely, since the youngest child of the only Lady Honywood to live at Markshall was born in 1647, so that she was unlikely to have been born before 1600 herself.[15] It is more probable that it was Crakanthorp's father who married Lady Honywood's daughter. A number of pieces of evidence point to this conclusion. Lady Hester Honywood was the daughter of John Lamott, a London merchant, who in 1635 bought the manor of Fowlmere.[16] She married Sir Thomas Honywood of Markshall in 1634, having previously been married to John Manning.[17] If she had had a daughter by Manning (say around 1620), the daughter would have been old enough to have been Crakanthorp's mother in 1641. This might also explain why Crakanthorp's father, who lived near Reading, should have died at Markshall: he was presumably visiting his mother-in-law. Lady Honywood appears to have owned the manor and advowson of Fowlmere between 1655 and 1681. It therefore seems that it was her own 25-year old grandson whom she presented to the living of Fowlmere in 1667. Crakanthorp owed her a debt of gratitude for this, which he acknowledged when he christened his only daughter Hester.[18]

Nothing is known of Crakanthorp's early life until he entered Trinity College, Cambridge, in 1658. He took his B.A. in 1661 and M.A. in 1665. Two years later, as we have seen, he was presented to the living of Fowlmere.[19] His immediate predecessors in the Rectory had not enjoyed settled possession of the living. John Morden, rector until 1646, was ejected, several of his parishioners having complained of his Laudian sympathies. His successor, Ezekias King, was ejected in 1660. King was followed by

---

13. I owe this information to the kindness of Mrs. Theo Ziolkowska (née Crakanthorp).
14. Samuel Crakanthorp's pedigree is in the possession of Cdr. Scott-Fox of Ayshford House near Tiverton. Richard Crakanthorp (1567-1624) held the rectories of Black Notley near Braintree and Paglesham in Essex. He was noted as a puritanical disputant, according to the *Dictionary of National Biography*.
15. Alan Macfarlane, (ed.), *The Diary of Ralph Josselin 1616-1683*, British Academy, Records of Social and Economic History, new series iii (1976) p. 666.
16. *V.C.H.*, viii, 157.
17. Macfarlane, *op. cit.*, p. 666.
18. *V.C.H.*, viii, 157 and 161; Macfarlane, *op. cit.*, p. 445.
19. J. Venn, *Alumni Cantabrigienses*, Part 1: *To 1750*, i, 410.

Simon Potter, who held the rectory in plurality until Crakanthorp arrived in 1667. He then remained in possession for fifty-three years.[20]

In 1670 Crakanthorp married Margaret Sherwin, who was then living at Bennington, near Stevenage. She was the daughter of William Sherwin (1607–87?), a nonconformist minister who had been ejected from the living of Wallington (Herts) in 1660 or 1662. Sherwin was living with his daughter and son-in-law at the time of his death, but whether he had done so since their marriage is not known.[21] The new rector and his bride lived in a medieval house which had just been altered and repaired. In 1674 it had eight hearths, more than any other house in the village. We can get some idea of what the house was like from the inventory taken when Crakanthorp died, although it must be remembered that this was more than fifty years after he first moved in:[22]

An inventory of all the goods and chattels of Mr. John Crakanthorp, late of Fowlmire, deceased, taken and appraised by us whose names are underwritten, 26 August 1719

	l.	s.	d.
In the hall chamber: 1 box, 1 trunk, 5 napkins	0	10	0
In chamber over ye kitching: 1 bed and bedsted, one rug, two pillows, 2 blanks, one trunk and other lumber	1	10	0
In the best chamber: one bed and bedstead and bedding, 7 chairs, 1 chest drawers, 1 table, fire shovel and tongs	3	10	0
Over the buttery: 1 hanging press, 2 chairs, one close stool	0	7	6
In the kitching: one jack, 3 spits, one grate, five pewter dishes, other od things	2	0	0
In the hall: 1 long table, tongs and fire shovel	0	8	0
In the brew house: one copper, tubs and other lumber	2	5	0
In the garner: one quern to grind malt	0	10	0
Wearing apparrel and money in his purse	2	0	0
Severall books	6	0	0
	19	0	6

Appraised by us: James Miles, Willm. Brooks

The hall, kitchen, buttery, brewhouse and garner were presumably on the ground floor, with the best chamber, hall chamber, and chambers over the

20. *V.C.H.*, viii, 162.
21. Audrey Grant, 'Hertfordshire References in the Diaries of John Crakanthorp', *Hertfordshire's Past*, no. 1 (Autumn 1976), pp. 14–15; *Dictionary of National Biography*, sub Sherwin.
22. *V.C.H.*, viii, 162; P.R.O., E179/244/23 (Hearth tax, 1674); C.U.L., Ely Consistory probate records, inventory of John Crakanthorp, 1719. The house was demolished in the nineteenth century, but there is a water-colour of it in Fowlmere church.

kitchen and buttery above. Although the inventory does not specifically mention it, it is clear from the harvest accounts that there was also a study.[23]

In 1672 John and Margaret had their first child, who was also called John. He later followed his father, going to Cambridge University, where he was admitted to Sidney Sussex College in 1689, and being ordained into the Church of England, although when older and living at Royston he was said to be 'a Quaker, and Preacher or speaker among them.'[24] In 1674 the second son, William, was born, in January 1676 Samuel, and in 1678 Benjamin. In 1707 Samuel was living in Essex and Benjamin in London.[25] Their only daughter, Hester, was born in 1680. She married Benjamin Wedd, a dissenter and farmer who lived in Fowlmere, in 1705.[26] His youngest son, Nathaniel, was born in 1684, and by 1705 was farming land which he had hired for himself. Later he took over his father's glebe.[27]

By the time that harvest accounts begin, therefore, Crakanthorp had a wife, five young children, and the biggest house in the village to support. His income varied from year to year (for a more detailed discussion see below pp. 32–3), but was normally between about £150 and £200. Does this mean that he was rich, or poor? This question can only be answered by comparison with the incomes of others. The wages paid to day labourers shown in the household accounts vary, depending upon the job done, between 10d. and 1s. 3d. per day. If a labourer worked six days a week for fifty-one weeks in the year at 1s. 3d. per day (all of which are optimistic assumptions) he would earn £19 2s. 6d. But this is hardly a fair comparison, for the income figures which can be calculated from the harvest and household accounts are gross rather than net figures. Crakanthorp had numerous expenses which a labourer would not have had. Indeed, in four out of the five years for which figures are available his expenditure exceeded his income.[28] It is therefore more useful to compare his income to that of others in the same position. A survey of clergy incomes at the beginning of the eighteenth century indicates that over half of all livings were worth less than £50 a year, a figure which presumably refers to gross income. Gregory King's data, based on the year 1688, estimated the incomes of eminent clergymen at £72 and those of lesser clergymen at £50 per family per year. Freeholders of the better sort had an income of £91 per family per year on average, persons in the law £154, and gentlemen £280. These figures suggest that, at the very least, Crakanthorp was reasonably well off, in so far as that phrase has any meaning. Yet there is little evidence of conspicuous consumption in the household accounts; comparing them

23. Below, pp. 51, 52, 117.
24. C.R.O., P72/1/2 (Fowlmere Parish Register), 30 April 1672; Venn, *op. cit,* i, 410; British Library Add. MS. 5808 (William Cole's Parochial Antiquities of Cambs.), fo. 172.
25. C.R.O., P72/1/2, 31 March 1674, 6 January 1675/6, 30 May 1678; Grant, *op. cit.*
26. C.R.O., P72/1/2, 20 April 1680; below, p. 132 (household accounts, 23 June 1705, disbursements); C.U.L., Add. 6586 ('Chronicles of Fowlmere' by A.C. Yorke), p. 22.
27. C.R.O., P72/1/2, 27 May 1684; below p. 142 (household accounts, 7, 10–11 September 1705, disbursements).
28. See below, pp. 32–3.

with the diary of James Woodforde suggests plainer living in the Crakanthorp household.[29]

Towards the end of the first decade of the eighteenth century Crakanthorp was still active as a clergyman, but his career as a farmer was drawing to an end. In May 1708 his youngest son, Nathaniel, was married, and at the beginning of July 1708 leased the glebe from his father, who thus retained only a few acres of copyhold land in his occupation. Just before Christmas 1708 Nathaniel 'entered into housekeeping' and thereafter Crakanthorp paid for his and his wife's board. He had not retired suddenly from economic life (Nathaniel had been closely involved in his marketing activities since at least 1705) but his inventory (see above) indicates that he had transferred his live and dead stock to Nathaniel by the time of his death.[30] In May 1713 Margaret Crakanthorp died. Crakanthorp survived his wife by only six years, although to judge from his handwriting in the parish register he was increasingly frail in the last year of his life. He was buried at Fowlmere on 19 July 1719, 'under a grey handsome stone near the door as you enter the church'.[31]

He had made his will on 12 July 1714. In it he left his copyhold land in Fowlmere to his daughter Hester and her husband Benjamin Wedd, who were also appointed his executors, on condition that they paid £50 to his son Samuel one year after his death, and another £50 to his son Benjamin two years after his death. To his eldest son John he left 'my best bed bedsted and furniture as it now stands and two pair of good sheets.' He left the malt quern to Nathaniel, and the rest of his possessions, except his books, were to be divided equally among Hester, Samuel, Benjamin and Nathaniel. He did not say who should receive his books, which was perhaps an oversight, considering that they formed nearly one-third of the value of his moveable goods.[32]

29. G.R. Cragg, *The Church and the Age of Reason 1648–1789* (Harmondsworth, 1970), pp. 125-6; King's figures are given by J.F.C. Harrison, *The Common People* (1984), pp. 113-118. It is interesting to note that King's figure for the labourer's family income was £15, which is what would be earned by working for 6 days a week for 50 weeks at one shilling per day. James Woodforde, *The Diary of a Country Parson*, (5 vols., 1924-31). See also A. Macfarlane, *The Family Life of Ralph Josselin* (Cambridge, 1970), p. 77.
30. Below, pp. 245 (household accounts, 5 May 1708, disbursements), 271 (17 January 1710, receipts), 258 (16 February 1709, disbursements), 260 (30 March 1709, disbursements). There is an interesting discussion of retirements and inheritance customs in M. Spufford, 'Peasant inheritance customs and land distribution in Cambridgeshire from the sixteenth to the eighteenth centuries', in Jack Goody, Joan Thirsk and E.P. Thompson (eds.), *Family and Inheritance: Rural Society in Western Europe 1200-1800* (1979), pp. 173-176.
31. C.R.O., P72/1/2, 11 May 1713, 19 July 1719; Grant, *op. cit.* The antiquary William Cole described the stone as '...a grey handsome Slab, which had formerly an Inscription at the Feet of a Priest dressed in his Cope, & by a Projection on the right Side of his Face, it looks as if there had been a Crosier in his Hand: I took a Draft of it on that Account: his Head appears shaven: but all the Brass is stolen. ...The Clark said the Stone belonged to Mr. Crakanthorp, who was buried under it:' (William Cole, British Library Add MS. 5824, partly published in *Monumental Inscriptions and Coats of Arms from Cambridgeshire*, ed. W.M. Palmer (Cambridge Antiquarian Society, 1932) p. 60).
32. C.U.L. Ely Consistory probate records, original wills, 1719, and WR C38, fo. 44.

Thus it is possible to produce an outline of the life of John Crakanthorp. Yet there are still two important questions which remain unanswered: what kind of man was he, and why did he keep such enormously detailed harvest and household accounts?

The most easily accessible source of information about his character is the account books themselves. They suggest a neat and tidy man with an organised mind, easily moved to pity by the misfortunes of others. But any such impression must be a personal one, since Crakanthorp did not set out to record his thoughts or his reactions to the world in which he lived. It is known that he voted for a whig candidate in the 1705 Cambridgeshire election, as did both other Fowlmere electors, but otherwise the only independent testimony as to his character is the remark by the antiquary William Cole that he was 'a Puritanical kind of man' and that '... by his Principles he detested anything tending to Popery'. This is not surprising, given that his grandfather, his father-in-law and his children were all described, at one time or another, as puritanical, nonconformist, or dissenters, and that he ministered to a village which Dr. Spufford classified as one with a large number of dissenters in 1676.[33]

It is also impossible to say for certain why he kept the account books. The obvious answer is that he needed to be able to refer to the information contained in them, for business purposes. But if so, did he really need records in the minute detail in which he compiled them? And did he need such precise details of his charitable activities as he kept in the household accounts? He also kept the parish registers in the same detailed fashion. He recorded dates of birth as well as of baptism, which was unusual, although not unique. He even noted the hours of birth of his grandchildren. He wrote the bishop's transcripts of the registers himself from 1672 to 1717, and in a way that was much more uniform and neat than was common at the time.[34] All this may simply indicate that when he did a job he preferred to do it well. Or does it indicate an obsession with detailed record keeping? He may simply have felt that some of the details were worth recording to remind himself of curious events, but there may also have been deeper reasons. Macfarlane has pointed out that there was a rapid growth in diary keeping in the second half of the seventeenth century, and suggested religious reasons for it. He quotes Riesman's view that, among protestants, '... diary keeping... may be viewed as a kind of inner time-and-motion study by which the individual records and judges his output day by day.' Crakanthorp may have been influenced by such considerations. It is also worth remembering that he plainly knew Lady Hester Honywood, who in turn was well acquainted with two writers of substantial diaries: Ralph Josselin and Samuel Pepys. A degree of emulation may have been among Crakanthorp's motives for keeping his accounts with the precision that he

---

33. C.R.O., *A Poll for the Election of Two Knights of the Shire... 1705.* British Library, Add. MS. 5808, fo. 172 and Add. MS. 5854; *V.C.H.,* viii, 162; M. Spufford, *Contrasting Communities: English Villagers in the sixteenth and seventeenth centuries* (1974), pp. 312-8; and see notes 14, 21 and 24 above.
34. I am grateful to Dr. P. Saunders for this information.

did, and it is quite possible that he kept a diary which has not survived, just as several volumes of his accounts have not survived. Certainly the accounts fall a long way short of being a diary. They contain only a few details of family events, very few of local and none of national events. He mentions more than one visit to London, but most of his life seems to have been concerned with the affairs of his family, his farm and his parish.[35]

Farming in Fowlmere

Fowlmere lies on what was, in the seventeenth century, the main road from London to Cambridge, about nine miles south of Cambridge. Royston lies about five miles to the south-west. In the north of the parish, which contains 2,272 acres, the land is about 75 feet above sea level, rising gently over about 4 miles to about 150 feet above sea level at the southern end. In the north of the parish were the Great and North moors. The Great Moor was an area of marshy ground about 200 acres in extent, which was still undrained in the seventeenth century. The rest of the land was in cultivation or used for grazing. The whole parish is underlain by chalk, over which the soil is a chalky loam.[36]

During the period that Crakanthorp lived in Fowlmere the population probably fell, but only slightly. In 1665 there were 45 families, and in 1674 50 families. The 'Compton census' of 1676 noted 139 adults, but in 1728 there were only 40 families, comprising 183 people.[37] Fowlmere was therefore a medium sized village both in numbers of people and density of population, according to Dr. Spufford's criteria.[38] In 1674 most of the Hearth Tax payers had three hearths or less, some had more. Crakanthorp himself and Mr. Bowles paid tax on 8 hearths, Abraham Killingbeck on 7 and Thomas Nash and Henry Thompson on 6 apiece.

The population is categorized according to the number of hearths in table 1 and compared with the total for the county.

35. A. Macfarlane, *The Family Life of Ralph Josselin* (1970), pp. 3–8; A. Macfarlane (ed.) *The Diary of Ralph Josselin 1616–83*, British Academy, Records of Social and Economic History, new series iii (1976), p. 15 n.2; below, p. 88, 269.
36. *V.C.H.*, viii, 155; R.M.S. Perrin and C.A.M. Hodge, 'Soils', in J.A. Steere (ed.), *The Cambridge Region* (1965), pp. 68–85.
37. *V.C.H.*, viii, 155–6; microfilm of the hearth taxes there cited is also available in the County Record Office.
38. M. Spufford, *Contrasting Communities: English Villagers in the sixteenth and seventeenth centuries* (1974), pp. 313–5.

Table 1: Fowlmere and Cambridgeshire Hearth Tax

	In Fowlmere 1674		In Cambridgeshire 1665
	Number of payers & exempt	% of total number	% of total number
1 hearth	27	54	42.0
2 hearths	9	18	27.8
3 hearths	7	14	13.3
Over 3 hearths	7	14	16.6

It can be seen that the proportion of those with only one hearth was a little greater than the average for the county, while fewer than average had two hearths. On the other hand the proportion of those with three hearths or more was remarkably close to the average. On balance it may be suggested that although it had a few more of the very poor than normal, Fowlmere was not an untypical Cambridgeshire village.

Dr. Spufford has argued that there was a definite relationship between wealth and house size, but this does not necessarily mean that Crakanthorp was one of the wealthiest men in the village, since he was paying for the parsonage house.[39] One of the other larger payers may have been an innkeeper. There were certainly two inns, the Chequers and the Swan, in the village in the latter part of the seventeenth century, and there may have been three more. The existence of such a large number of inns was a result of the village's position on the main London to Cambridge road. It was perhaps this ease of communication which prompted the lord of the manor of Fowlmere in the early part of the seventeenth century to establish a market in the village; it was no great success, presumably as a result of the competition from larger well-established markets at Cambridge, Royston and Saffron Walden, and soon fell into disuse.[40]

The principal business of Fowlmere was farming, and most of the inhabitants worked on the land. When Crakanthorp arrived in Fowlmere the bulk of the land was owned by Lady Hester Honywood, and descended to her children until it was sold by her grandson in 1703 to James Mitchell, a merchant who lived in London. Thus during Crakanthorp's lifetime there was never a resident lord of the manor. During the sixteenth century there was a second manor, known as Heslartons, but in the early seventeenth century it was bought by Edward Aldred, who owned the manor of Fowlmere, and thereafter descended with the main manor. Some land in the southern part of the parish was also in separate ownership and Christ's College, Cambridge, held about 54 acres in 1663 and continued to hold land in the village until the present century. The only other estate of

---

39. M. Spufford, *op. cit.*, pp. 39–41.
40. *V.C.H.*, viii, 156 and 160; the ease of communication may be exaggerated: Defoe referred to 'a village very justly called Foul Mire' (Daniel Defoe, *A Tour through the Whole Island of Great Britain* (1971 edn.), p. 434).

any significance was the Rectory, the extent of which will be discussed later. This land was treated as a separate manor from the fifteenth century.[41]

The majority of the land in the village was laid out in open fields until an act was passed for its enclosure in 1845. There is no comprehensive surviving survey which is contemporary with Crakanthorp, but it is possible to build up a reasonable picture of the open field village by extrapolating earlier and later surveys, although most of these deal with individual estates or manors rather than with the village as a whole. In the harvest accounts Crakanthorp refers to three fields: the North, the Barr, and the South or Heath Fields. These are sometimes referred to by other names in earlier or later sources. The North Field is usually so called, but is occasionally referred to as the Moor Field. Barr Field is also called Barr Lane, Fawdon, Farden and West Field. In each case the different names seem to refer to the same piece of land. The case of the South or Heath Field seems more complex. Crakanthorp uses the two words interchangeably, but later surveys refer to Waterdrain Field as a synonym for the South Field and distinguish a separate Heath Field lying to the south of it.[42] Postgate states that in 1794 Fowlmere had a three-course rotation on four fields, a system which was quite common in the county. It is quite clear from the harvest accounts that a three-course rotation was in operation during Crakanthorp's time. The year after the fallow was called the tilth, and wheat, rye and sometimes barley were grown in the tilth field. The following year was called the broke crop, which was mainly barley, and in the following year the field was left fallow again. It would therefore seem that in the eighteenth century, probably, as will be seen later, in the early part of the century, the Heath Field came to be recognized as a separate field, even though it went with the South Field in the rotation. Again, this was not uncommon in Cambridgeshire. However, it should be noted that in the preamble to the harvest accounts for 1688 Crakanthorp states that 'there was very good barly upon the heath, that half of it that lyeth next Norwich roade'. This was in a year in which the North Field was the tilth and the Barr Field broke, so that presumably the South Field was fallow. It would seem that the association of the South and Heath Fields was not as strict in Crakanthorp's time as it may have been later. The enclosure map shows a boundary between the South and Heath Fields, and this has been reproduced in figure 1 (page 13), which is drawn from the enclosure map. For the sake of clarity the fields will be called North, Barr and South hereafter when appropriate.[43]

41. *V.C.H.*, viii, 157-8; Christ's College Record Office, Fowlmere terrier, 1663.
42. Private Acts, 8 Victoria cap. 2 and C.R.O., Q/RDc70 (Fowlmere enclosure award); below, harvest accounts, *passim;* C.R.O., R82/8 (Vinter Collection), Fowlmere notes, fo. 104; M.R. Postgate, *The Open Fields of Cambridgeshire,* unpublished Cambridge University Ph.D. thesis, 1964, appendix 1; *V.C.H.,* viii, 159; Christ's College Record Office, Fowlmere terrier, 1663.
43. Below, harvest accounts, *passim;* C.R.O., Q/RDc70 (enclosure map); M.R. Postgate, *op. cit,* p. 224.

INTRODUCTION

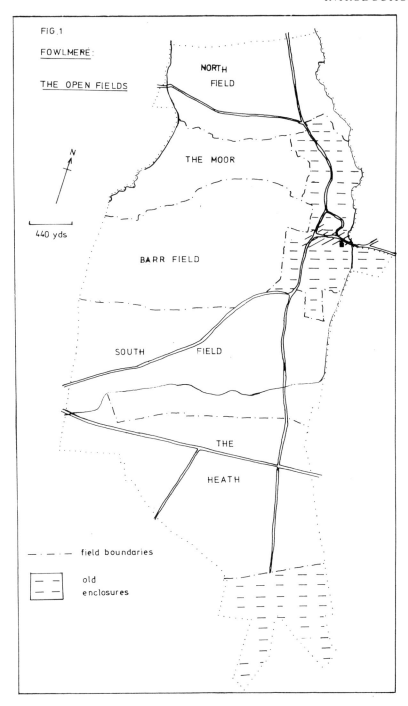

Fig. 1 Fowlmere: The Open Fields

A field book of Fowlmere, compiled about 1615, shows a total of 1268a. 2r. 30p. 'good land acres' in the open fields, and an acreage of 367a. 1r. 0p. in the Heath.[44] It also lists the acreage of each of the fields, and this list may be compared with the abstract which survives of a terrier of the parish of 1763.[45]

	1615			1763		
	a.	r.	p.	a.	r.	p.
North Field	254	1	20	226	1	22
Barr Field	449	1	10	385	3	0
South Field	565	0	0	454	2	39
Heath	367	1	0	515	0	5
Enclosures				195	3	16
Grange, Moor, Roads				363	3	30
Croft ground				65	3	9
TOTAL	1635	3	30	2207	2	1

There is no certain explanation of the discrepancies in these two surveys. As far as the North, Barr, and South Fields are concerned, it may be that the earlier survey used field measure, while the later used chain measure. In each case the later acreage is between 80 and 90 per cent of the earlier. It might also be that pasture enclosures encroached on the open fields near the village, and an examination of the enclosure map indicates that this is possible in Barr Field, although it does not seen to have happened in North Field. Neither of these explanations would account for the apparent increase in the size of the heath between 1615 and 1763. However, this may be explained by differences in the definition of the heath on each occasion. The earlier figure could apply only to land which was cultivated as arable (open field furlongs and the Chrishall Grange enclosures). The later figure is not very different from that of about 510 acres which are distinguishable as heath on the enclosure map, which suggests that it includes common pasture, but not the Chrishall Grange fields which are old enclosures.[46] Within each of the open fields further sub-divisions were recognized. In the harvest accounts for example, Crakanthorp notes in 1687 that wheat was sown 'upon the 4 Acres in North Field'; but he also refers to sub-divisions of a field, such as 'Branditch Shott' and 'Short Shott'. Furlongs and shotts were blocks of arable in the open fields, many of which were individually named, as in the examples above. Their size varied considerably. The field book of 1615 shows that Branditch Shott contained 21a 1r 20p, but some, such as 'Cross Shott abutting on Sheperheath way'

44. C.R.O., P72/3/9; M.R. Postgate, 'Field Systems of East Anglia', in A.R.H. Baker and R.A. Butlin (eds.) *Studies of Field Systems in the British Isles* (Cambridge, 1973), pp. 294, 296–7; *V.C.H.*, viii, 159.
45. C.R.O., P72/3/13 and C.U.L., Add. 6586, p. 116.
46. M.R. Postgate, thesis, *op. cit.*, p. 33; *V.C.H.*, viii, 159; C.R.O., Q/RDc70 (map). The enclosure map covers a total of 2,212a. 1r. 6p. comprising 1,881a. 0r. 30p. former open fields and 331a. 0r. 16p. old enclosures. The author is grateful to Dr. P. Saunders for a number of suggestions on this point.

(5a 3r 0p), were much smaller and others, such as Waterdrain Shott (107a 1r 20p), much larger. There were in all sixteen shotts in South Field, seven in Barr Field and seven in North Field. Several different tenants might cultivate the land in each shott. The holding of one individual in a shott is usually referred to now as a strip, although the word strip is found more often in the work of modern historians than in contemporary documents. A strip is therefore a unit of tenure rather than cultivation.[47]

The unit of cultivation is usually referred to as a selion. A ploughman would begin a piece of ploughing by making a ridge, after which he would plough round and round the ridge. Thus all the furrows on one side of the ridge would be turned one way, and all those on the other side the other way. When he had ploughed the whole selion he could then go to the next selion and again begin by making a ridge. Where two selions joined, the last furrows of each would be turned away from each other, thus forming a small depression which would divide one selion from another. The size of the selion might vary considerably: conventionally it was 220 yards long and 22 yards wide, giving an area of one acre, but it might be more than this and was very often less. In the eighteenth-century transcript of the 1615 field book, which lists each selion, measurements were added according to a convention that a selion was one rood and one perch. But this appears to be a later addition, it is not maintained throughout the book, and it produces some inexplicable totals for the acreages of shotts. It therefore seems probable that the size of a selion was variable within certain limits, perhaps between one and two roods. Postgate found that in Cambridgeshire as a whole it was often between one quarter and one half of an acre (there are four roods in an acre). The number of selions in a strip was also variable. There might be several selions in a strip, or only one. For Cambridgeshire, he demonstrated that strips were commonly between one and two roods in area, and therefore suggested that single selion strips were common.[48]

Postgate also suggested that the term 'stitch' or 'stetch' was used to denote strips of meadow in the open fields in Fowlmere. The only time that Crakanthorp uses the word 'selion' he refers to rye being sown 'upon 10 stetches or selions, in Short-furlong, and Duck Acre Shott in Southfield', which suggests that 'stetch' and 'selion' were synonymous. This appears to be confirmed by the 1663 survey of Christ's College land in Fowlmere, where the description of the arable land in North Field included 'a stetch butting on ye moore on ye south, on ye Mill path on ye North, on Mr. Morden on ye east, and John Rand on ye west'. Since the term stetch is so frequently used in this survey and in the harvest accounts, and there are many examples of corn being sown upon stetches, it seems fair to conclude that the stetch was the basic unit of cultivation. Its size was variable and

---

47. Below, pp. 76, 70; C.R.O., P72/3/9; and cf M.R. Postgate, thesis, *op. cit.*, pp. 51, 55; I.H. Adams, *Agrarian Landscape Terms: a glossary for historical geography*, Institute of British Geographers, Special Publication no. 9 (1977), p. 90.
48. The process is described in detail in C.S. and C.S. Orwin, *The Open Fields* (3rd edn., Oxford, 1967), pp. 32-3; C.R.O., P72/3/9. M.R. Postgate, thesis, *op. cit.*, pp. 55, 64-6.

was not always specified in the harvest accounts: in 1687 Crakanthorp 'had 7 stetches of my owne barly this year and 2 acres rye'. In 1689, however, he was more specific, having 'of my owne in South Field, 6 stetches, and 2 in the North field, in all 3 acres,...'[49]

Crakanthorp's harvest accounts are mainly concerned with the open fields and their products, but it should be remembered that the land of the parish provided many other products. Sheep could be grazed on the heath and cattle on the Great and North moors. Cattle could also be fed on the balks in the open fields, and cattle, pigs and poultry were allowed to graze over the open fields for a month after harvest. The moors also provided game, reeds for thatching and clay. The old enclosed land lay around the village on the eastern side of the parish, and to the north of these closes, between the brook and the Shepreth road, lay the meadows.[50]

Of the probate inventories which exist for residents of Fowlmere during the period 1660–1720, only seven relate to people whose main activity was farming, which is too few for any significant statistical analysis, but enough to demonstrate the main trends which are summarised in table 2.

Table 2: Fowlmere Inventories

	Total value of inventory			Percentage of total value derived from livestock	Percentage of total value derived from crops
	£	s	d		
W. Finckel, 1671	158	4	2	20.8	51.6
T. Eversden, 1688	336	13	0	12.8	67.4
J. Welch, 1702	169	0	4	19.4	35.6
T. Nash, 1703	230	0	10	15.0	31.3
W. Newling, 1709	415	8	6	45.7	32.5
J. Rayner, 1713	270	3	8	14.0	48.8
W. Fordham, 1714	49	19	0	24.0	23.0

Only two of the seven farmers had a greater proportion of their wealth in livestock than in crops, and the other five derived less than twenty per cent of their wealth from livestock. All except one derived more than thirty per cent of their wealth from crops. Only three of them had sheep: one of these was William Newling, who in 1709 had a flock of 297 worth £121. All of them had cattle, horses and pigs, but in no case was the number large. Thomas Eversden, who died in 1688, had goods to the value of £336 13s. 0d., yet in addition to his 27 sheep he had only 17 cows and bullocks, four horses and nine swine in the yard. John Welch in 1702 had only 5 cows, 4 horses and three shoats in the yard, although he had 55 acres of land either sown to corn or waiting to be sown. This pattern is a typical one for the

---

49. *Ibid.,* p. 56 n.1; below pp. 85, 76, 91; Christ's College Record Office, Fowlmere terrier, 1663.
50. *V.C.H.,* viii, 159; M.R. Postgate, *thesis, op. cit.,* appendix 1.

county: Dr. Spufford found that the median stock numbers in Cambridgeshire inventories for 1661-1670 were 6 cattle, 4 horses, 2 pigs and 15.5 sheep.[51]

Arable cropping was clearly the mainstay of agriculture in Fowlmere. John Rayner, who died in 1713 appears to be fairly typical of those who farmed there. His inventory shows that he had 4 cows, 3 horses and swine to the value of £5. His inventory was presumably taken just before harvest, since all his crops are listed in terms of acres. He had 6 acres of wheat and 11 of rye, together worth £25 15s. 0d., 42 acres of barley worth £73 10s. 0d., 15 acres of oats, 8 of peas, and 5 acres of lintels and tares, together worth £32 0s. 0d. His hay and 'sanfoile' (sainfoin), which had presumably already been cut, were valued at £4 5s. 0d. The cultivations which he had carried out in the fallow field were carefully valued: 40 acres had been ploughed twice, and 20 acres thrice, and the value of this work was estimated at £18 0s. 0d. He had also spread 50 loads of dung, which was worth £6 11s. 4d., and the dung still waiting to be spread was worth £5 10s. 0d. The grass in his closes was valued at £3 0s. 0d. For his horses there were 'collars, cart saddles, plough gear etc'., and he had one waggon and four carts. In his house was a hall, parlour, brewhouse, 'darey and cheese-chamber' (containing a cheesepress and 60 cheeses) and buttery downstairs, with the hall chamber, chamber over the buttery and another chamber upstairs. The house contained goods worth £32 16s. 8d., and his purse and apparel were valued at £10 0s. 0d.[52]

Rayner grew all the crops which are commonly mentioned in the inventories for Fowlmere residents. He may have been unusual in growing sainfoin, since in only one other inventory, that of Thomas Nash, who died in 1703 is 'saint fine' noted, although in 1705 Crakanthorp was collecting tithes of sainfoin. Later, in 1727, John Fairchild had 7 quarters of 'sankfoylseed' and grew turnips. These references to fodder crops such as sainfoin, cinquefoil and turnips are the only evidence that exists for the practice of convertible husbandry at Fowlmere, (if the argument on p. 15 about the meaning of the term stetch is correct), and even then these crops may have been grown in the enclosed land near the village rather than in the open fields. There is no reference in the harvest accounts to their cultivation in the open fields.[53]

Barley was the most important of the cereal crops in Fowlmere as in other parishes in upland Cambridgeshire. An early seventeenth-century edition of Camden's *Britannia* noted that this area 'yeeldeth plentifully the best barly; of which they make store of mault: By venting and sending out

---

51. C.U.L., Ely probate records, inventories; M. Spufford, *Rural Cambridgeshire 1520-1680*, unpublished Leicester University M.A. thesis, 1962, p. 25-6; the median livestock numbers for Cambridgeshire may be compared with those for a county where the emphasis was on livestock: in the lowland areas of Durham in the 1680s the figures were 17.5 cattle, 20 sheep and 4 horses (P.W. Brassley, *The Agricultural Economy of Northumberland and Durham in the period 1640-1750*, (New York, 1985), p. 214).
52. C.U.L., Ely Consistory probate records, inventory of John Rayner, 1713.
53. C.U.L., probate inventories, *passim;* below, p. 129 (18 and 19 June 1705, receipts).

whereof into the neighbor-counties, the Inhabitants raise very great gaine'. In 1700 it was said of the district that it was 'chiefly Barley that has the reputation of being very good'. At about the same time Defoe found that it was 'almost wholly a corn country; and of that corn five parts in six of all they sow, is barley...' Defoe's estimate of the proportion of barley was perhaps an overestimate. Dr. Spufford found (admittedly in a small sample of inventories) that barley formed about half of the sown acreage, but Defoe's sin is clearly one of overemphasis rather than complete distortion. The barley malt went to satisfy the demands of the London market. As Defoe again pointed out, it was 'generally sold to Ware and Royston, and other great malting towns in Hertfordshire, and is the fund from whence that vast quantity of malt, call'd Hertfordshire malt is made, which is esteem'd the best in England'.[54]

The other crop which was grown for consumption outside the district was saffron. James Douglas, writing in 1727, reported that it was grown extensively in the district between Saffron Walden and Cambridge. Although there are no references to it in the Fowlmere inventories, Christopher Sell, who died in neighbouring Thriplow in 1705, had two roods of saffron ground worth £2 0s. 0d. and 2½ pounds of saffron in the house valued at £3 5s. 0d.[55]

Crakanthorp lived at a time when, in other parts of England, many farmers were agreeing amongst themselves to the enclosure of their open fields. It is pertinent at this point to ask why this did not happen in Fowlmere. To begin with, this was not unusual in Cambridgeshire. Vancouver in his *General View* estimated that two thirds of the county was still unenclosed in 1793. It has been argued that the persistence of the open fields reflected unenterprising land management policies on the part of the Cambridge Colleges which owned large amounts of land in the county, or a lack of market stimulus. Clearly neither of these explanations will do in the case of Fowlmere. Only a small part of the land in the village was owned by a college, and local corn markets were stimulated by the demand from London. Postgate's suggestion that the process of enclosure was made more difficult by the shortage of common grazing may be true, but doubts remain about the significance of pasture in an agricultural economy which was so dominated by arable farming. The explanation which seems most satisfactory is that put forward by Dr. Spufford, who suggested that enclosure might have been delayed until the nineteenth century simply because open field arable farming was reasonably profitable. It was also flexible enough in this area to allow the pattern of cropping to respond to market forces. If there was a demand for barley, for example, it might not matter that barley was traditionally sown on the broke field while the tilth was reserved for wheat and rye; so in 1683 Crakanthorp sowed his barley on 21 acres and a rood in the South Field,

54. Camden and Defoe are quoted by H.C. Darby, 'Agriculture', *V.C.H.*, ii, 74; C. Taylor, *The Cambridgeshire Landscape* (1973), p. 174; M. Spufford, M.A. thesis *op. cit.*, p. 28.
55. Douglas is quoted by Darby, *op. cit.*, p. 74; C.U.L., Ely Consistory probate records, inventory of Christopher Sell, 1705.

which was the broke field that year, and on 19 acres and a rood in the Barr Field, which was the tilth field that year. In areas where farmers might wish to respond to market forces by increasing their acreage of fodder crops for grazing, such flexibility would have been difficult in the open fields, because of the necessity for fencing (although this did not prevent it happening elsewhere, as in Oxfordshire), but in this area the problem was much simpler since it was the acreage of a cereal crop or another arable crop such as saffron which the tenants wished to increase. Thus they could maximise their farm incomes in the open fields without the expense of enclosing, and without the necessity of dispossessing the smaller landowners who might have been left without viable farms after enclosure.[56]

The Harvest Accounts

The examination of the harvest accounts should begin with the process of threshing. At harvest the corn was cut and then left in the field for a few days until it was ready for carting. In the preamble to his account of the 1688 harvest, for example, Crakanthorp notes that '... Harvest began generally on Thursday July 26. (some began to reape on the Munday before, but no corne was carried till that day, because they reaped some corne greener than they commonly doe in this Towne, and so left it the longer in the field)...'

Sometimes the corn was bound into sheaves immediately, but sometimes it was left loose and raked into rows, known as gavels (see the preamble to 1689). When it was dry enough to be stored without heating or going mouldy it was carted home, and either put into the barn or made into a rick or stack which was subsequently thatched to keep it watertight. When the grain was required for use it would have to be threshed. This was done with a flail, which consisted of a swingel made of some hard wood, such as holly or blackthorn, coupled to a handle made of ash. Ewart Evans has described how 'the thresher swung the handle over his shoulder and brought down the swingel across the straw just below the ears so that the grains of corn were shaken out without being bruised'. The corn was usually threshed in the barn where there would be a hard threshing floor

56. R.H.P. Watson, *The Agrarian Revolution in Cambridgeshire 1770-1850*, unpublished London University M.A. thesis 1951, pp. 12-13, 27; M.R. Postgate, thesis, *op. cit*, p. 259; M. Spufford, M.A. thesis, *op. cit.*, p. 43; below, p. 49; M.A. Havinden, 'Agricultural Progress in Open Field Oxfordshire', reprinted in E.L. Jones (ed.) *Agriculture and Economic Growth in England 1650-1815* (1967). The persistence of open fields has recently been discussed using economic concepts such as risk aversion and transactions costs, in D.N. McCloskey. 'The Persistence of English Common fields', in W.N. Parker and E.L. Jones (eds.), *European Peasants and their Markets* (Princeton, 1975) pp. 73-119, and in C. Dahlman, *The Open Field System and Beyond: a property rights analysis of an economic institution* (Cambridge, 1980). These ideas are reviewed in J.A. Yelling. 'Rationality in the Common Fields', *Economic History Review*, 2nd series, xxxv (1982-3), 409-15.

REV. JOHN CRAKANTHORP

made of wooden planks or beaten earth between the barn doors. After threshing the grain and chaff were sieved to remove the cavings (small pieces of straw and refuse) and then winnowed, i.e. thrown into the air with a shovel. The heavier grains would carry furthest and the lighter or shrivelled grains less far, forming a tail against the pile of good grain. Thus these inferior grains were called tailings, and Crakanthorp either used them as animal feed or sold them at a lower price. The chaff would fall first, and so the whole crop was separated into its constituent parts.[57]

The account of each kind of grain in the harvest accounts is arranged by dressings. The precise meaning of the term dressing is not immediately obvious. Crakanthorp notes, for example, that the 18th dressing of barley in 1690 was 'threshed by Will: Thrift', suggesting that dressing is equivalent to threshing, but also notes that one quarter and five bushels of the first dressing of oats in the same year 'were partly fanned up before hand', suggesting that it is equivalent to winnowing. Each dressing is given a specific date, suggesting at first sight that it was the work of one day. But many dressings are said to be the work of one man, as, for example, the second and third dressings of oats in 1689, which were the work of William Thrift. The fourth dressing, on the other hand, was quite clearly the work of two men. The problem here is that the quantity of corn involved in each of these dressings was more than ten, eight and seventeen quarters respectively; yet the usual rate of work by a man threshing with a flail would produce between five and six bushels of wheat, nine and eleven of barley, or between eleven and thirteen bushels of oats. Each dressing must therefore have taken more than one day to complete. This conclusion is confirmed by the evidence of the household accounts, which show that the payment for threshing one quarter of barley was roughly equivalent to the normal payment for one day's work. In January 1706, for example, Thomas Watson was paid either 9d. or 10d. per day for ordinary farm work, and in the same month William Thrift was paid 9s. 3d. for threshing 12 quarters and 3 bushels of barley at 9d. per quarter. This clearly took more than one day. As would be expected, the rate of payment for threshing wheat, which took longer than barley, was greater. In March 1706 Crakanthorp paid 1s. 3d. per quarter for threshing wheat and rye. Finally, it is worth noting that the household accounts contain no payments for winnowing. It may therefore be concluded that the term dressing refers to a discrete quantity of corn threshed, winnowed and ready for use or for market.[58]

The total output from all these dressings is summarised in table 3, which reveals a number of interesting features. The output of each individual crop might vary significantly from year to year: the worst wheat and rye crops, for example, were less than half the size of the best, and although the oat and barley crops do not exhibit quite the same variation the difference was nevertheless significant. A bad year for the winter sown crops, wheat

57. G. Ewart Evans, *Ask the Fellows Who Cut the Hay* (1965), pp. 87–95; W. Fream, *Elements of Agriculture* (10th edn., 1918), p. 80.
58. E.J.T. Collins, 'The Diffusion of the Threshing Machine in Britain 1790–1880', *Tools and Tillage*, ii (1972), 16–33.

and rye, might not be so bad for the spring sown crops, and vice versa, so the variation in the total output of all the crops was not so great as the variation in any single crop; nonetheless, total output in the worst year (1686) was only 62% of that in the best year (1683). As might perhaps be expected, there was much greater variation in the receipts from corn sales, where the receipts in the worst year (1688) were only about one third of those in the best year (1692), although it should be remembered that in the years 1688, 1689 and 1690 Crakanthorp had a much larger proportion of his barley malted than was usual. He almost certainly produced more malt than he could use himself, but there is no record of its sale, as once it had been sent to the maltster it was no longer the concern of the harvest accounts. It is clear from the table that barley was Crakanthorp's major crop, normally accounting for more than half his output and half his cash receipts. The other major cash crops were wheat and rye. The bulk of the oats produced were fed to his horses, and the peas were consumed both by his family and his livestock. The rectory of Fowlmere was said to be worth £168 in 1650 and £200 in 1728. The variation in the receipts from corn sales indicate that these figures are an assessment of the mean returns: in practice, when the great tithes were taken in kind, as they were by Crakanthorp, the value of the living could vary significantly from year to year.[59]

Marketing

It has already been mentioned that Royston was a major market centre for the barley crop of north Hertfordshire, south Cambridgeshire and north Essex. It is plain that the pull of the Royston market had a significant effect on Crakanthorp's marketing decisions. Although it is not possible to make a precise quantitative analysis of the relative importance of the markets he used, because he does not invariably state the buyer or the location, it is clear that Royston was the main centre at which he sold his main cash crop, barley. In 1683, for example, he sold at least 16 loads of more than 5 quarters to various buyers in Royston. The next most common destination, Ashwell, received only 3 loads.

Deliveries of 5 or 6 quarters (or multiples thereof) at a time were most common for sales outside Fowlmere or the immediate locality, which suggests that this was the size of the normal cart or wagon load.[60] Local sales, in Fowlmere (those which Crakanthorp describes as 'sold at home' or words to that effect), were normally of only a few pecks or a few bushels, but even consignments to Foxton and Thriplow, the neighbouring villages, would be made up into decent-sized loads containing a few quarters at least.

Crakanthorp sold only a small proportion of the peas and oats that he

59. *V.C.H.*, viii, 161, and see below, pp. 32–3.
60. This suggestion is also made by J.C. Wilkerson in his introduction to *John Norden's Survey of Barley, Herts., 1593–1603* (Cambridge Antiquarian Records Society, 1974), p. 11.

Table 3: Summary of Harvest Accounts

	WHEAT		RYE		BARLEY		OATS		PEAS		Total Payment Received			% of total from barley
	Quantity harvested	% Sold	Quantity harvested	% Sold	Quantity harvested	% Sold	Quantity harvested	% Sold	Quantity harvested	% Sold	£	s	d	
	q-b-p		q-b-p		q-b-p		q-b-p		q-b-p					
1682	29-3-3	86.5	37-6-2½	75.2	171-5-2	88.5	44-7-0	19.6	24-4-1	8.7	239.	2.	2	69.0
1683	33-6-3	67.5	52-1-3	78.3	220-2-2	75.8	72-1-0	26.1	22-1-0	30.8	237.	19.	4	63.8
1684	16-1-0	35.2	38-3-1	69.8	217-0-1	81.4	59-5-2	29.8	5-0-2	22.2	268.	16.	10	76.6
1685	27-6-1	59.2	72-5-1	71.0	139-0-0	58.5	60-3-2	2.4	2-4-2	0	134.	13.	11½	57.1
1686	15-6-3	44.2	43-6-2	83.2	130-5-2	79.3	43-1-0	0.7	14-5-3	12.1	139.	4.	2½	70.8
1687	30-7-1	62.2	81-2-1	78.6	214-6-0	69.2	43-3-2	9.7	16-4-0	21.0	196.	5.	2½	67.6
1688	29-3-3	60.8	52-1-3	84.4	184-4-3	37.5	33-2-1½	9.4	22-7-0	28.6	99.	5.	4½	47.4
1689	35-2-0	77.3	41-7-3	68.6	201-3-1	37.8	41-0-2	0	13-1-1	26.1	111.	12.	10½	44.2
1690	45-6-3	83.6	59-2-2	88.7	179-7-0	58.0	60-5-2	10.4	15-5-0	6.8	134.	2.	1½	42.6
1691	27-6-0	69.9	56-1-2	82.9	161-7-1	66.5	38-3-1	27.4	28-1-3	36.7	168.	11.	10	46.6
1692	37-0-1½	72.1	77-7-3	84.9	198-6-1	83.2	33-5-3	0	11-3-0	0	336.	8.	9½	53.4

had threshed. The consignments were small, and the buyers were normally from Fowlmere itself or, occasionally, from Foxton or Thriplow. His main cash crops, sold to non-local buyers, were wheat, rye and barley. Taking a wagon load as one of 5 quarters or more, the greatest sales of wheat in any one year amounted to 4 wagon loads, and in some years (e.g. 1684 and 1686) no loads at all were sold, sales being confined to local buyers. The same pattern emerges for rye, with perhaps 5 or 6 wagon loads being sold outside Fowlmere in most years, although in 1685, with the exception of two small consignments of 4 and 6 bushels, all the rye was sold in the village. Barley sales, in contrast, were on a much bigger scale; only in 1688, when a greater proportion than normal was malted, did sales outside Fowlmere amount to less than 12 wagon loads, and in 1687 they were almost double that.

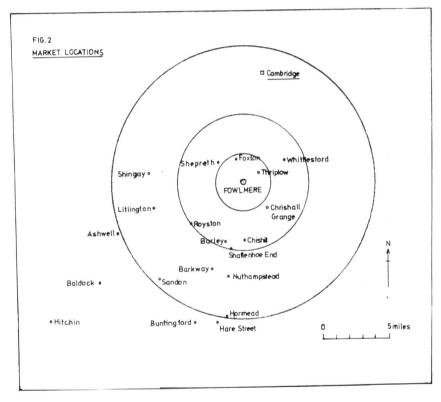

FIG. 2
MARKET LOCATIONS

Although Royston was the most important market for Crakanthorp's barley it was not the only one. Figure 2 shows the places to which he sent barley, wheat, and rye. Its most striking feature is the preponderance of markets to the south and west of Fowlmere. Leaving aside Shepreth, Foxton and Thriplow, the neighbouring villages, and excepting Cambridge and Whittlesford, which received a few loads of wheat and rye, the bulk of Crakanthorp's trade was directed to the south and west. The importance of

Royston partially accounts for this; it was a major market and malting centre and it was only 5 miles along the London Road from Fowlmere. It was already a market centre, specialising in corn and malt, by 1640.[61] Baldock, Barkway, Buntingford and Hitchin also had markets by 1640, and Ashwell, Baldock and Hitchin were established in the malt trade by then.[62] This trade demonstrates the effect of the London market in drawing raw materials from the agricultural areas of eastern England. Whilst grain from northern Cambridgeshire was carried north and via the coasting trade to London, barley and malt, in particular from the southern part of the county, went overland to the head waters of the rivers Stort and Lea. Mathias points out that 'the carts of maltsters formed the bulk of the heavy land traffic between Royston and Ware in the mid-seventeenth century'. At Ware their loads could be carried to London down the Lea, which had been improved in the 1570s and formed an important waterway by the early eighteenth century. Thus Crakanthorp was linked with a national market. He also sold to smaller centres such as the maltster at Shaftenhoe End (to which he referred as Shafnoe End).[63]

It can be seen from figure 2 that most of the places to which Crakanthorp sold his corn were within ten miles' radius of Fowlmere. This was not unusual. After the middle of the sixteenth century the network of markets in eastern England had developed to the extent that, on average, only 15% of people had to travel more than ten miles to a market.[64] He was also fortunate in that Fowlmere was on one of the main roads from London to Norwich, which ran along the line of the modern B1368. Going south from Fowlmere, the road passed through Barley and close to Chishill and Shaftenhoe End, then on through Barkway to Hare Street, where there was a turn off for Buntingford. Even given the notoriously poor state of seventeenth-century roads, particularly in the winter, this was presumably one reason why it was worthwhile for Crakanthorp to sell corn to these smaller market centres, although their importance should not be exaggerated, since they were used infrequently.[65]

Not surprisingly, Crakanthorp names only a few buyers in the centres to which he sold infrequently. Thus at Buntingford Mr. Turpentine bought

61. J. Thirsk (ed.), *The Agrarian History of England and Wales: volume 4, 1500–1640* (1967), pp. 508, 589–90.
62. J. Thirsk, *op. cit.*, p. 474; P. Mathias, *The Brewing Industry in England 1700–1830* (Cambridge, 1959), pp. 428, 436, 438–9, 442; J.A. Chartres, *Internal Trade in England* (1977), pp. 18, 42.
63. Wilkerson, *op. cit.*, p. 11: Wilkerson states that Uriah Bowes carried on a malting business at Shaftenhoe End. This may be so, but although Crakanthorp undoubtedly sold barley to Bowes, he nowhere states that he lived at Shaftenhoe End. Crakanthorp writes 3 times of 'Shafnoe End': in two instances the buyer there is not specified, in the third he is identified as 'J:N:' (below, pp. 101, 108, 72). The only buyer of any consequence with these initials is John Norris (not Morris, as Wilkerson identifies him), who bought several consignments of barley in 1682. Shaftenhoe End is a hamlet in Barley about half a mile from the village.
64. J. Thirsk, *op. cit.*, pp. 498–9.
65. C. Taylor, *Roads and Tracks of Britain* (1979), p. 196; L.M. Munby, *The Hertfordshire Landscape* (1977), p. 203.

his barley and George the miller his rye, whilst the buyer of his wheat is not specified. It is interesting to note that the barley was not sent direct to Turpentine, but 'delivered at Barly to his trustee'.[66] The greatest numbers of buyers are mentioned in Royston and Ashwell, the markets he used most regularly. At Royston he usually sold his barley either to Mr. Glenister or to Miles Thurgood, but another eight buyers are mentioned, while at Ashwell his barley was sold to seven different buyers. Altogether 48 different buyers (in addition to local buyers) are mentioned by name in the harvest accounts, which suggests the existence of a fairly sophisticated and competitive market, particularly for barley, in the area around Fowlmere.

This conclusion is also suggested by an examination of the prices received by Crakanthorp for his wheat and barley. These are summarised in figures 3 and 4. A comparison of the range of wheat prices with the range of barley prices shows that in most years the variation in barley prices was much less than the variation of wheat prices. Wheat prices might double during the course of the year, and the highest price was often 50% higher than the lowest price. Barley prices, on the other hand, might change by no more than two shillings per quarter over the course of the year. This is perhaps to be expected, given that the barley and malt market in the area around Royston was closely linked to the London market, so that price movements would reflect demand and supply trends in the largest market in the country. Moreover, although Crakanthorp sold most of his barley to Glenister and Thurgood, there were many alternative buyers available to him in Royston and its neighbouring towns and villages. It would therefore be likely that the forces of competition would act to minimise price differences, and there do not appear to have been any significant differences between Royston prices and, for example, Ashwell prices. In contrast wheat prices were much more variable, presumably reflecting the fact that Crakanthorp was selling to a much smaller number of buyers in a much more local market.

A less competitive wheat market is also suggested when the prices received by Crakanthorp are compared with Cambridge prices. Figures 3 and 4 show that the variations from year to year in Crakanthorp's prices are roughly similar in magnitude and direction to the Cambridge price variations. These indicate that the second half of the 1680s was marked by a fall in corn prices, which is seen not only in the prices reported from Cambridge, but also those from Winchester and Wiltshire. Wheat prices did not regain the levels of the early 1680s until 1692. But Cambridge wheat prices were normally at the upper end of the range of prices received by Crakanthorp, while Cambridge barley prices were, as often as not, less than Crakanthorp's price. Again, this presumably reflects a slightly greater level of competition in a barley market more closely linked to the London

---

66. Below, p. 102.

REV. JOHN CRAKANTHORP

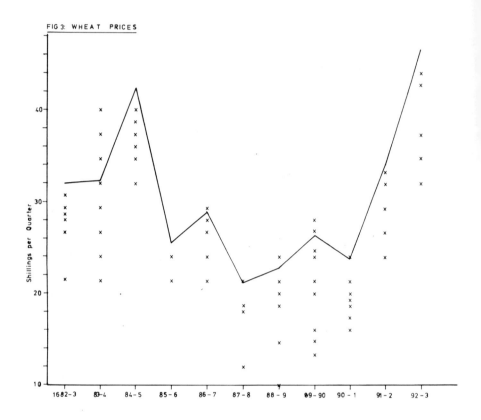

The crosses show prices received by Crakanthorp for loads of wheat sold during the marketing year. The solid line shows Cambridge average wheat rent prices for the corresponding years, taken from J.E.T. Rogers, A History of Agriculture and Prices in England *(1887), v, 226–30.*

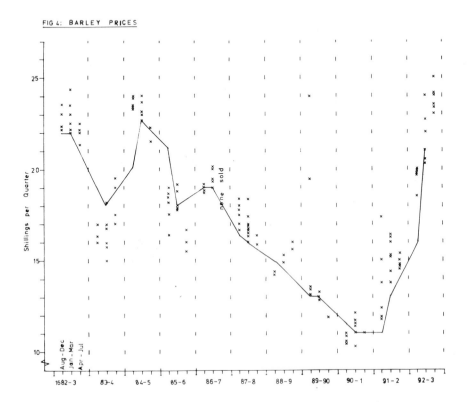

The crosses show prices received by Crakanthorp for loads of barley sold during August to December, January to March, and April to July in each marketing year. The solid line shows St. John's College, Cambridge, prices for Michaelmas and Candlemas, taken from J.E.T. Rogers, A History of Agriculture and Prices in England *(1887), vi, 84–93.*

trade. It also provides further evidence of the extent to which Crakanthorp's business was, in effect, integrated with the national barley market.[67]

Yields

There is no clear information on the quantity of land farmed by Crakanthorp. Although there are several surveys and terriers of Fowlmere in general and the glebe in particular, none of them date from the period of Crakanthorp's incumbency. The problem is further complicated by the fact that he farmed land on his own account, in addition to the glebe: in the preamble to the harvest accounts for 1689 he notes 'I had of my owne in South feild, 6 stetches, and 2 in the North feild, in all 3 Acres, besides Gleabe...' and among the land sown with the rye threshed in the 2nd dressing in 1689 was 'my owne 2 Acres of Coppy Land' in Barr Field.[68]

In estimating the area of land available to Crakanthorp it is therefore necessary to distinguish between his glebe and his own land. Estimates of the size of the glebe vary, but not wildly. The field book of c.1615 gives the size of the glebe as 120 acres, made up of 94 acres in the open fields, 22 acres in the heath, and a four acre close. A terrier of the glebe in 1639 showed that it amounted to 25 pieces of land containing 56 acres, 4 pieces of which, amounting to 16 acres, were in the heath, together with a further 141 stetches whose total area was not given. Early nineteenth-century measurements recorded by Crakanthorp's successor, William Metcalfe, give the area of the glebe as 111 acres and 2 roods by field measure, or 97 acres, 2 roods and 36 perches by chain measure.[69]

The evidence for the amount of land he held in addition to the glebe is less clear. On 9 November 1719, in Fowlmere manor court, Hester Wedd paid an admission fine of £21 for 'Three Cottages or tenements and twelve acres of land in severall peeces in the Comon feilds on the death of Mr. Crakenthorpe Minister her father'.[70] However, an examination of the land which he specifically notes as 'my owne temporal estate' suggests that he normally had the 6 stetches in North Field and 2 stetches in South Field, extending to 3 acres, as noted above, together with 2 acres and a stetch in

67. The Cambridge prices are taken from J.E.T. Rogers, *A History of Agriculture and Prices in England*, v, 226–230, vi, 82–93, Winchester prices from W.H. Beveridge, *Prices and Wages in England from the twelfth to the nineteenth century* (2nd impression, 1965), i, 81, and Wiltshire prices from J.M. Stratton, *Agricultural Records A.D. 220–1968* (1969), pp. 56–60. The Cambridge prices are those in the St. John's College series.
68. Crakanthorp was not the only rector of Fowlmere to farm land of his own in addition to his glebe. John Morden had enough of his own land to enable him to remain in the village for several years after he was ejected from the living in 1644 (G.O. V[inter] and M. S[wann]., *Some Historical Notes on Fowlmere*, no date.)
69. C.R.O., P72/3/9 (field book), C.U.L., Ely Diocesan Records, H1 and C.R.O., P72/3/7–8 (glebe terrier), and P72/3/13 (notebook with extract from lost terrier of 1763). Notes on all these surveys by A.C. Yorke are in his manuscript 'Chronicles of Fowlmere' (C.U.L., Add. 6586), the field book analysed on pp. 157–60, the glebe terrier transcribed on pp. 77–108, and the summary of the 1763 terrier transcribed on page 116.
70. C.R.O., R84/39, Manor of Fowlmere entry book of fines, 1695–1725 (formerly *penes* J. Ruston).

Barr Field. He was not always meticulous in noting the area of his own land (as opposed to glebe land) which was sown, so it is possible that there was more. It is also conceivable that he sublet some of his own land, although there is no evidence for this.

These figures suggest an alternative method of discovering the amount of land farmed by Crakanthorp: from the details given in most years of the harvest accounts of the land sown to each of the various crops. It is not always possible to be precise about the acreage involved, since the figures are sometimes given in stetches, and a stetch did not have a fixed size. From the information given in the first dressing of rye in 1682 it can be calculated that the average size of the stetches mentioned was 0.38 acres, while the information given in the second dressing of rye in 1682 suggests a figure of 0.75 acres. Nevertheless, it is possible to calculate the areas sown with sufficient accuracy to say that in the years when Barr Field was the tilth field, he had 75 acres or more, and when it was not, he had 60 acres or so.

There is a further piece of evidence which confirms that these figures are of the right order of magnitude. When Crakanthorp died he was succeeded as rector of Fowlmere by Henry Hall. Hall did not farm the glebe himself but leased it to John and Richard Fairchild. When John died, the stock, crop and tithe of the rectory had to be valued for the purpose of compiling the inventory, which shows that the glebe land under cultivation extended to 59½ acres in 1727. Allowing one third of the glebe and Crakanthorp's own land to be in fallow in any one year, it therefore appears that the total size of his holding was between about 90 and 112 acres, which is consistent with the findings of the surveys made before and after Crakanthorp's incumbency.[71]

Using the evidence of the harvest accounts it is possible, for most years, to break down the total acreage into the acreage sown to each crop (see table 4).

Since the quantity of each crop which was threshed is also known (see Table 3 above) it appears at first sight that it should be possible to calculate the yields achieved, by dividing the quantity harvested by the appropriate acreage. The results of this calculation are shown in table 5.[72]

It is immediately apparent that if these were the yields obtained by Crakanthorp, they were much higher than those produced by other farmers in this period. Using data from Norfolk and Suffolk, Overton calculated that for the period 1675-95 the range of wheat yields was between 11.5 and 15 bushels per acre, of rye yields between 7 and 12 bushels per acre, of barley yields between 12 and 16 bushels per acre, of oat

---

71. C.U.L., Add. 6586, fo. 125; C.U.L., Ely Consistory probate records, inventories, 1727. Fairchild was described as a yeoman. His estate was valued at £744 10s. 1d., together with half of the value of the stock, crop and tithe held by lease from the rector, which was valued at £216 16s. 1d.
72. There are 8 bushels in a quarter, and 4 pecks in a bushel. Approximate bushel weights are: wheat, 63*lbs.;* barley, 56*lbs.;* rye, 57*lbs.;* oats, 42*lbs.;* peas, 63*lbs.* See H.I. Moore (ed.), *Primrose McConnell's Agricultural Notebook* (14th edn. 1962), p. 821.

yields between 11 and 14 bushels per acre and of pea yields between 6.5 and 10 bushels per acre. On a national basis, Chartres suggested yields of wheat of 16 bushels per acre, of rye, 17 bushels per acre, of barley 23 bushels per acre, and of oats, 24 bushels per acre, all lower than those apparently produced by Crakanthorp. The overall impression is that if these figures are an accurate estimate of his yields, he was producing about twice as much as would normally be expected.[73]

There are two possible explanations for this discrepancy: either the area sown is understated or the production is overstated. It has already been pointed out that the total area, as it is derived from the harvest accounts, is consistent with the area derived from other sources. An alternative possibility is that the acre used at Fowlmere differed from the statute acre. There is no evidence of this, and the quantity of seed sown per acre (for wheat, 2 or 2½ bushels per acre, and for barley 3 bushels per acre) is what would be expected for a statute acre.[74]

It therefore appears that the level of production must be overstated, in that some of the corn which Crakanthorp had threshed was not produced on his glebe or his copyhold land. This is confirmed by numerous references to tithes. In 1683, there is a note that the 13th and 14th dressings of barley 'were all tithe and almost all the tithe, of the North field', while the second dressing of peas was 'the tithe of Dods Close'. In 1686 all the peas were tithe, 'for I had none sowne, of my owne'. The first dressing of wheat in 1691 was the tithe of a piece of land farmed by Goodman Rayner.[75] Crakanthorp appears to have employed 'tithe men' specifically for the purpose of gathering his tithes.[76] It is interesting to note that John Fairchild's inventory shows that in 1727 the value of the corn produced by the glebe was £85 8s. 9d., while the corn tithe was valued at £94 0s. 0d. Since it appears that Crakanthorp's copyhold land extended to roughly one tenth the size of his glebe, it would seem likely that about half of his corn would be the product of the land he farmed directly, and half collected in tithe. This would explain why the estimates of yields calculated above are roughly twice the size they would be expected to be.[77] In short, the harvest accounts can provide no reliable estimate of yields.

73. M. Overton, 'Estimating Crop Yields from Probate Inventories: an Example from East Anglia 1585-1735', *Journal of Economic History* xxxix (1979), 363-78; P. Goldsworthy, *Cereal Yields; the extent of and reasons for changes from the medieval period to the present day*, unpublished H.N.D. dissertation, Seale-Hayne College, 1981; recent work in this field is summarised in J. Chartres, 'The Marketing of Agricultural Produce', in J. Thirsk (ed.), *The Agrarian History of England and Wales: volume 5.2, 1640-1750* (1985), p. 444.
74. Primrose McConnell, *Notebook of Agricultural Facts and Figures for Farmers and Farm Students* (9th edn., 1919), pp. 194-200; J.C. Wilkerson, *op. cit.*, p. 11.
75. There are further references to tithe in 1686 (peas), 1689 (preamble, peas and oats), 1691 (peas) and 1692 (oats). Tithing practices in this period are discussed in E. Evans, 'Tithes', in J. Thirsk (ed.), *The Agrarian History of England and Wales: volume 5, 1640-1750* (1985), and E. Evans, *The Contentious Tithe: the Tithe Problem and English Agriculture 1750-1850* (1976).
76. Below, p. 69 (1686 preamble).
77. See note 71 above.

Table 4: Acreages sown to cereal crops

	Wheat			Rye			Barley			Oats			Peas		
	a	r	p	a	r	p	a	r	p	a	r	p	a	r	p
1683	7 +2 stetch	0	0	'neare' 13	0	0	'neare' 37	0	0	—			3	3	20
1684	4	0	0	15 +1 stetch	0	0	40	2	0	13	1	0	3	1	20
1685	5	3	0	15	0	0	30	3	0	12	3	0	3	0	0
1686	—			*30 or 21	0 0	0 0	30	2	0	—			—		
1687	6	2	0	18	3	20	—			7	0	7	3	3	20
1688	—			14	2	0	—			—			5	0	20
1689	—			12	2	0	35 +4 stetches	0	0	12	0	0	—		
1690	4	2	0	17	2	20	43	0	0	*8 or 10	0 0	0 0	—		
1691	—			13	2	0	31	0	0	—			—		
1692	7	2	0	17	0	0	32	2	0	4	0	0	—		

The acreage is listed under the year in which the crop sown on those acres was harvested. For example, 1683 includes wheat sown in the autumn of 1682 and barley sown in the spring of 1683. A dash denotes no information. * *Acreage unclear.*

Table 5: Apparent Yields (bushels per acre)

	Wheat	Rye	Barley	Oats	Peas
1683	36.1	32.1	47.6	—	45.7
1684	32.25	19.8	42.9	36.0	7.5
1685	38.6	38.8	36.2	37.9	7.3
1686	—	11.7/16.7*	34.3	—	—
1687	38.0	34.5	—	49.6	33.9
1688	—	28.8	—	—	37.1
1689	—	26.9	43.5	27.4	—
1690	81.5	26.9	33.5	60.7/48.5*	—
1691	—	33.3	41.8	—	—
1692	34.2	36.7	48.9	67.4	—

A dash denotes insufficient information to calculate yields. * *Acreage unclear.*

REV. JOHN CRAKANTHORP

The Household Accounts

Although it is convenient to refer to the small notebook which Crakanthorp labelled 'An Account of my Receipts and Disbursements' as household accounts, they in fact cover all of his monetary transactions from April 1705 to March 1710. He made no distinction between receipts or payments for household and farm purposes. Indeed he had no need to. The major part of his farming activities ceased in July 1708 when he leased his glebe land to his son, Nathaniel,[78] (although he appears to have retained control over his few acres of copyhold land[79]), but he continued to sell small quantities of grain after that. He calculated the total of his receipts and disbursements at the end of September and March in most years (where the totals are not stated they have been calculated from the monthly totals) and these figures are shown in table 6.

It is interesting to note that receipts exceed disbursements in only one of these years. If the balances in table 6 are added together it appears that between 1705 and 1710 Crakanthorp spent £399 4s. 10½d. more than he received. In practice this would overestimate the position, because in June 1709 he lists as disbursements the gift of corn, dung and farm implements to his son Nathaniel. Nevertheless he still spent more than he received, and it is not clear where this money came from.

Table 6: Summary of Household Accounts

	Receipts			Disbursements			Balance		
	£	s	d	£	s	d	£	s	d
April – Sept 1705	105	8	5	164	10	2	– 59	1	9
Oct 1705 – Mar 1706	119	7	3½	85	18	7	+ 33	8	8½
April – Sept 1706	95	19	0	*256	3	1	–160	4	1
Oct 1706 – Mar 1707	112	16	2	67	1	2	+ 45	15	0
April – Sept 1707	70	9	3	123	11	2	– 53	1	11
Oct 1707 – Mar 1708	123	13	0	79	13	5	+ 43	19	7
April – Sept 1708	50	0	2	63	5	3	– 13	5	1
Oct 1708 – Mar 1709	128	7	7	49	0	8	+ 79	3	8
Oct 1709 – Mar 1710	40	16	9	37	15	4	+ 3	1	5

* Includes a loan of £100 to Crakanthorp's daughter Hester.

† Includes a gift of farming stock and implements to Nathaniel Crakanthorp.

The items listed in the receipts side of the account confirm what is indicated by the harvest accounts: that the bulk of Crakanthorp's income came from corn sales. In the twelve months from April 1705, for example,

78. Below, p. 249 (1 July 1708, disbursements).
79. Below, p. 271 (17 January 1710, receipts).

only 11% of his receipts were not derived from sales of corn and straw. He received small sums for rents, small tithes of hay, wool and sainfoin, using his wagon to carry coal, turves, hay, straw and wood for other people, he wintered cows at 6d. per week each, and sold willow setts. These miscellaneous receipts shed a little more light on his farming. They show (as the harvest accounts do not) that he kept a few cows, and sold their calves from time to time. He also had a bull, which in January 1706 strayed 'almost as far as Kneesworth or Bassingbourn' (about 6 miles from Fowlmere). Richard Carter was sent after it, and had to pay a man 2d. to help to bring the bull home.[80] The harvest accounts show that he kept pigs and chickens, but there are few references to the sale of their products, so they must have been retained for consumption within the household. Another farming accident provides evidence that he had his own boar: on 24 January 1709 he paid 2s. 0d. 'to Ralph Carpenter, having his cow killed (as supposed) by our boar'.

The lists of disbursements show a mixture of payments for both farming and household purposes. There were many payments of wages. Most men were paid by the day, but the herdsman was paid quarterly at so much per cow,[81] and some servants were paid at Michaelmas for the whole year.[82] There are also numerous payments for the services of tradesmen such as thatchers, sadlers and horse-collar makers, and pig killers. Clearly the farming operations provided part of the food needed by the household, but additional meat and cheese were regularly bought. Other items, such as nails, cloth, medicines and the book in which he was writing, which cost 1s. 2d., had to be bought, and taxes and rates had to be paid. Finally, Crakanthorp gave regularly for various charitable purposes, and noted his donations in great detail.[83]

The household accounts do not answer all the questions that they provoke – it would be interesting to know why Crakanthorp decided to buy meat when he appears to have been capable of producing his own, and how many people were living in the household – but they provide an interesting glimpse of the domestic economy of a clergyman's household at the turn of the eighteenth century.

---

80. Below, p. 155 (8 January 1706, disbursements).
81. Below, p. 131 (14 June 1705, disbursements).
82. E.g. Jane Gurner and George Carpenter. See below, p. 253 (30 September 1708, disbursements).
83. Crakanthorp's charitable activities are discussed in Anthony Lambert, 'The Sixth Codex', *Blackwood's Magazine*, cccxvi, no. 1908 (October 1974), pp. 289-301. Medical aspects are discussed in Audrey Grant, 'Sickness and Health: references in the account book of John Crakanthorp, 1705-10', *Hertfordshire's Past*, no. 5 (Autumn 1978), pp. 19-20.

THE HARVEST ACCOUNTS

I 1682–1684

[County Record Office, Cambridge, P72/3/1]

An account of the quarters of corne of all sorts, sold or spent, that there was of my seventh crop gathered in in harvest anno 1682, when harvest began generally on Tuesday, August 1rst, and ended on Wednesday, August 30, lasting a compleat month and 2 dayes. The harvest was good, only the raine hindered us 3 dayes entirely at the latter end. The North-feilde was this yeare tilthe and the Barr-feilde the broake cropp. [*p.1*]

[In the manuscript there follows an abbreviated form of the preamble to the eighth and ninth harvests]

An account of the crop brought in in harvest that yeare, and first of wheate, white or Kentish [*p.2*]

Of the first dressing of wheate, which was red wheate, October 2nd, there was in all one quarter and 3 bushells, which was afterwards all sowne upon the 5 acres in Southfeild, together with about a peck of the white wheate at one end, to make up what theire wanted of the red; which is 2 bushells and a peck an acre.

Of the 2nd dressing, which was white wheate, October 17th, there was in all 7 bushells and a peck, whereof one bushell was mingled with some of ye rye that was sowne upon the 4 acres; and the peck was sown at one end of the 5 acres for want of red wheate; and the remaining 6 bushells were all sowne upon the 2 acres and stetch in Horsted Shott.

Of the 3rd dressing, which was all white wheate also, November 3, there was in all one quarter 4 bushells and a peck, whereof one bushell and half were sold att home att 11 groates a bushell for 5*s*. 6*d*.; and one quarter 2 bushells and half more were sold at 10 groates for 1*l*. 15*s*. 0*d*.; and the remaining peck falling short about a quarter, or neare, was sold to J: Casb: for 8*d*.

Note: That these 3 dressings, making in all 3 quarters 6 bushells and half, were all that was laid in the maulthouse.

Of the 4rth dressing of wheate, May 2, which was all red wheate, there was in all 5 quarters and 3 bushells, whereof a peck and above was tayle; 2 pecks courser than the rest and throwne amongst the rye in the 8 dressing, and 2 bushells and a peck sold at home at 11 groates, for 8s. 3d., and the remaining 5 quarters were sold to Cambridge at 11 groates and 2d. per bushell, for 7l. 13s. 0d.

[p.3] Off the 5th dressing, May 11, all white wheate, there was in all 5 quarters 7 bushells and half, whereof one bushell was very course tayle, and put amongst the off corne, and another bushell courser than the rest was put amongst the 9th dressing of rye, and 3 bushells and half more sold at home at 11 groates, for 12s. 10d., and 2 bushells were spent by [p.3] our selvs, being sent to mill on May 16, and the remaining 5 quarters were sold to one Mr. Brian at Cambridge, baker, at 3s. 6d. per bushell, for 7l. 0s. 0d.

Off the 6th dressing, May 19, which was also white wheate, there was in all 5 quarters and 7 bushells, whereof one bushell was very course againe and put amongst the off corne, and one bushell more, courser than the rest, put amongst the rye, and 5 bushells sold att home at 3s. 6d. per bushell, for 17s. 6d., and the remaining 5 quarters sold to one Cook, a meale man at Cambridge, at the same prise, for 7l. 0s. 0d.

Of the 7th dressing of wheate, which was all red wheate, May 31, there was in all 8 quarters and half, whereof there was 5 pecks of ye coursest tayle left for off corne, and 2 pecks courser than the rest put amongst the rye; and out of the best a peck was given to Widdow Preston, and 5 pecks more were sold before hand at 3s. 6d. per bushell, for 4s. 4d. ob., 2 quarters 3 bushells and 3 pecks more sold also at home for 3s. 6d. also per bushell, for 3l. 9s. 1d. ob.; sold to Cambridge 5 quarters at 3s. 7d. per bushell, for 7l. 3s. 4d.; and 3 bushells and an half spent by our selvs; so there should be a bushell and half more which is either lost or forgotten to be set downe.

Note: That the whole cropp of wheate this yeare is but 29 quarters and 4 bushells, where of 2 quarters and 2 bushells and a peck were sowne and 5 bushells and half spent, and all the rest besides tayle sold.

[p.5] **An account of the rye or miscellaine brought in that harvest**

Of the first dressing of rye, in the maulthouse, which was on September 18th, there was in all 2 quarters and 7 bushells sowne upon 14 stetches and half in Short-furlong, Duck-Acre Shott and the first knap at London Way, and upon the half acre there and the 3 half acres — in all upon 7 acres and half wanting a small matter.

Of the 2nd dressing, that was on September 23rd, there was 7 bushells and about half a peck, which with a bushell that was borrowed and 5 bushells more that were bought at Royston were sowne upon the 4 acres and 2 stetches in the Rowze (those 2 stetches on each side of the 4 acres are for seed another yeare). So that there is sowne in all very neare 13 acres of rye with 4 quarters 5 bushells of seed, which wants at least 6 pecks in the whole of 3 bushells an acre. It is most of it small measure.

Of the 3rd dressing, October 26, there was in all 6 bushells and a peck, whereof one bushell was paide to Mr. Killenbeck for that which was borrowed for seede, one bushell was spent by ourselves, and 4 bushells were sold at home at 2s. 8d. per bushell, for 10s. 8d., about half a peck remaining for the little piggs and the other half squandered and lost.

Memorandum: That these 3 dressings, making in all 4 quarters 4 bushells and one peck, were all that lay in the malt-house this yeare.

Of the 4rth dressing, February 7, there was in all 4 quarters and 5 bushells, whereof half a bushell was tayle saved for the little piggs, and the rest (being made up even 5 quarters by the addition of 3 bushells and 2 pecks of another dressing) was sold at 2s. 6d. per bushell to one Whitehead, a miller at Royston, for 5l. 0s. 0d.

Of the 5th dressing, February 26, there was in all 4 quarters and one bushell, whereof one bushell and half was tayle and fanned up before hand for piggs together, and 3 bushells and half were added to the former dressing (to make up even 5 quarters) and 2 quarters one bushell and 3 pecks were sold at home at 2s. 6d. per bushell, for 2l. 4s. 4d. ob., and 7 bushells were given to the poor, and 3 bushells spent by our selvs; and about a peck squandered.

Of the 6th dressing, March 21, there was in all one quarter and 3 [p.6] bushells, whereof half a bushell was tayle, saved for little piggs, and 9 bushells and half were sold at home, at 2s. 6d. per bushell, for 1l. 3s. 9d. The remaining bushell was spent by our selvs.

Of the 7th dressing, March 30th, there was in all 6 quarters one bushell and 3 pecks, whereof 2 pecks and something better were tayle, 3 bushells and a peck were sold at home at 2s. 6d. per bushell, for 8s. 1d. ob., and 5 quarters and half were sold to G: Whitehead, a baker at Royston, for 2s. 7d. per bushell, for 5l. 13s. 8d., and the remaining bushell was spent by our selvs and sent to mill Aprill 3.

Of the 8th dressing of rye, Aprill the 7th, there was in all 6 quarters and 7 bushells, where of one peck was tayle and 3 bushells spent by our selves, and the remaining 6 quarters 3 bushells and 3 pecks sold at home at 2s. 6d. per bushell, for 6l. 9s. 4d. ob., and about half a bushell lost, for there was half a bushell of course wheate cast amongst it of the 4rth dressing, otherwise it would have been half a bushell more; but that half bushell is reckoned amongst the wheate and so cannot be reckoned here to increase the dressing.

Of the 9th dressing of rye, Aprill 21, there was in all 5 quarters, where of 3 pecks were tayle; and 3 quarters were sold at home at 2s. 6d. per bushell, for 3l. 0s. 0d., and one quarter 4 bushells and one peck were also sold at home, at 7 groats per bushell, for 1l. 8s. 7d.; and 3 bushells were spent by our selves, the last of which was sent to mill June 27.

Of the tenth dressing, Aprill 27, there was in all 5 quarters and two pecks, where of half a bushell was given [p.7] to Goodman Will: Fordham, and [p.7] one quarter and 7 bushells was sold at home at 7 groats per bushell, for 1l. 15s. 0d., and 7 bushells and half were sold at 2s. 2d. per bushell, for 16s. 3d., and 2 bushells and half at 2s. per bushell, for 5s. 0d.; and were spent by our selves 1 quarter one bushell and 3 pecks, which in all doth make but 4 quarters 3 bushells and a peck; so that this dressing falls short 5 bushells and peck; and besides there was 2 bushells and half of tayle wheat put to it; so that in all it falls short 7 bushells and 3 pecks. There was some thrown out divers times spoyld by the water, but I wonder at so much odds nevertheless — the bins were scarce empty in 9 months time.

Note: That the whole cropp of rye this yeare is but 37 quarters 6 bushells and an half. It falls short of the crop of rye the yeare before 20 quarters wanting but 5 bushells and a peck; and of the crop in 1680 it falls short above 10 quarters, which was the worst crop of rye I have had till this.

[p.8] ## An account of the barly brought in and threshed that yeare

Of the first dressing, November 20th, there was in all 6 quarters and 4 bushells, whereof a bushell was tayle put into garner and 6 quarters and 2 bushells were sold to Miles Thurgood at Royston at 1*l*. 3*s*. per quarter, for 7*l*. 3*s*. 0*d*. A bushell remains to be disposed of, added afterwards to the next dressing.

Of the 2nd dressing, November 24, there was in all 6 quarters 5 bushells and an half, whereof a peck was tayle, and a peck sold to W: Thrift at 8*d*. ob., and the 6 quarters and 5 bushells, together with one bushell of the former dressing and 3 bushells of the 3rd dressing (making in all 7 quarters and one bushell) were sold to J: Norris of Barly at 22*s*. 4*d*. per quarter, for 7*l*. 19*s*. 1*d*.

Of the 3rd dressing, November 29th, there was in all 7 quarters 7 bushells and half, whereof the half bushell was tayle; and 3 bushells were sold with the former dressing, and 6 quarters and 2 bushells were sold to Thom: Chapman of Ashwell at 1*l*. 3*s*. 6*d*. per quarter, for 7*l*. 6*s*. 10*d*.; and the remaining ten bushells were sold with the next dressing.

Of the 4rth dressing, December 4rth, there was in all 7 quarters and 2 bushells, where of one bushell heaped was tayle, and 6 quarters more, together with the ten bushells of the former dressing (making in all 7 quarters and 2 bushells) were sold to J: Norris of Barly for 8*l*. 1*s*. 11*d*. at 22*s*. 4*d*. per quarter. The 9 bushells remaining were added to the next dressing.

Of the 5th dressing, December 7th, there was in all 6 quarters and one bushell, whereof the odd bushell was tayle, and the 6 quarters, together with one quarter and one bushell of the former dressing (making in all 7 quarters and one bushell), were sold to J: Norris also, at 1*l*. 2*s*. 4*d*. per quarter, for 7*l*. 19*s*. 1*d*. ob.

Memorandum: That this and the 2 first dressings, makeing in all 19 quarters 2 bushells and half, came out of the middlesteade in the new barne.

Of the 6th dressing, December 14th, there was in all 5 quarters and 6 bushells, whereof one bushell was tayle laid in the garner. The remaining 5 quarters and 5 bushells were sold to J: Norris att Barly at 1*l*. 2*s*. 2*d*. per quarter, for 6*l*. 4*s*. 8*d*.

[p.9] Of the 7th dressing, December 15th, there was in all 6 quarters and 2 bushells, whereof one bushell was tayle, and the other 6 quarters and one bushell were sold also to John Norris of Barly att 1*l*. 2*s*. 2*d*. per quarter, for 6*l*. 15*s*. 9*d*.

Of the 8th dressing, December 21, there was in all five quarters and five bushells, whereof a bushell was tayle laide in the garner, and the remaining 5 quarters and 4 bushells were sold to J: Norris of Barly at 1*l*. 2*s*. 2*d*. per quarter, for 6*l*. 1*s*. 11*d*.

Of the 9th dressing, December 22nd, there was in all 6 quarters and a peck, whereof a bushell was tayle and laide in the garner, and the peck was sold to W: Thrift at 8*d*., and the remaining 5 quarters and 7 bushells were sold to J: Norris of Barly also, at 1*l*. 2*s*. 2*d*. per quarter, for 6*l*. 10*s*. 2*d*.

Of the 10th, January 2nd, there was in all 6 quarters 2 bushells and 3 pecks, whereof 2 pecks were tayle, and one peck sold to Widdow Dovy att home for 8*d*., and the remaining 6 quarters and 2 bushells were sold to J: Norris of Barly at 1*l*. 2*s*. 0*d*. per quarter, for 6*l*. 17*s*. 5*d*.

Of the 11th dressing, January 4, there was in all 5 quarters 5 bushells and an half, whereof 2 pecks were tayle; and the 5 quarters and 5 bushells were sold to J: Norris of Barly at 1*l*. 2*s*. 0*d*. per quarter, for 6*l*. 3*s*. 9*d*. This dressing properly was 2 parts of dressings whereof 2 quarters 5 bushells and a peck threshed by Thrift and 3 quarters and a peck by Casbourne.

Of the 12th dressing, January 8, there was in all 5 quarters 6 bushells and an half, whereof half a bushell was tayle, and a bushell was sent to mill for the sow, and 5 quarters and 5 bushells were sold to J: Norris of Barly at 1*l*. 2*s*. 0*d*. per quarter also, for 6*l*. 3*s*. 9*d*.

Of the 13rth dressing of barly, January 12th, there was in all 6 quarters and one bushell, whereof the bushell was tayle laide into garner, and the remaining 6 quarters were sold to J: Norris of Barly at 1*l*. 2*s*. 2*d*. per quarter, for 6*l*. 7*s*. 0*d*.

Note: That the 7th, 9th, 12th and this 13rth dressings, making in all 24 quarters one bushell and 3 pecks, came all out of the first mow on the right hand in the new barne, and the iland was full of rye besides. The 3 quarters mentioned in the 11 dressing came as it justly supposed, partly of from the rick and partly of from the other mow. And the 3rd, 4th, 6th, 8th, 10th and 2q. 5b. 1p. of the 11th dressing came out of the middlestead and first mow on the left hand in the old barne, making in all 35q. 4b. 2p.

[*p.10*]

Of the 14th dressing, January 19th, there was in all 7 quarters and one bushell, whereof one bushell was sent to mill for the sow, foules, etc., and something above 3 pecks was tayle, and half a bushell sold at hom to Good: Hil: and Gar: for 16*d*., and 6 quarters and 2 bushells were sold to M: Thurgood of Royston at 1*l*. 2*s*. 6*d*. per quarter, for 7*l*. 10*s*. 7*d*., and 4 bushells and 3 pecks remaine added to the next dressing and sold therewith.

Of the 15th dressing, January 26, there was in al 6 quarters 5 bushells and an half and something more, whereof 3 pecks and something above was tayle and one bushell and half were sent to mill for the sow and piggs; the 6 quarters 3 bushells and one peck remaining (being made up even 7 quarters by the addition of the 4 bushells and 3 pecks remaining of the last dressing) were sold to J: Norris of Barly at 1*l*. 2*s*. 6*d*. per quarter, for 7*l*. 17*s*. 6*d*.

Note: That this dressing came most of it of from the topp of the wheate.

Of the 16th dressing, February 3rd, there was in all 6 quarters and 7 bushells, whereof one bushell was tayle, heaped measure; and the remaining 6 quarters and 6 bushells were sold to M: Thurgood of Royston at 1*l*. 2*s*. 11*d*. per quarter, for 7*l*. 14*s*. 8*d*.

Of the 17th dressing, February 14, there was in all 6 quarters and 4 bushells, whereof one bushell was tayle, and one bushell was sent to mill for sow and foules etc., and 2 pecks sold to Mr. Maulden for 1*s*. 6*d*., and 4 quarters and 7 bushells (being added to 7 bushells that were taken out of 3 quarters of lintells, so making in all 5 quarters and 6 bushells) were sold to M: Thurgood at Royston at 1*l*. 4*s*. 4*d*. per quarter, for 6*l*. 19*s*. 10*d*., and the remaining one quarter 2 bushels and half were added to the next dressing.

Note: That the lintelly barly being reckoned here make this dressing 7 quarters and 3 bushells.

[*p.11*]   Of the 18 dressing, February 17th, there was in all 4 quarters 2 bushels and an half, which (with the addition of one quarter 2 bushells and half of the former dressing, both threshed by W. Thrift, being made up 5 quarters and 5 bushells) were sold to J: Norris of Barly at 1*l*. 4*s*. per quarter, for 6*l*. 15*s*. 0*d*.

Of the 19th dressing (threshed also by Thrift), February 26th, there was in all 4 quarters 2 bushels and an half, whereof the half bushell was tayle; and one bushell was sent to mill for piggs etc., and 4 quarters and very neare one bushell remaining were made up even 5 quarters and half (by the addition of one quarter 3 bushells and neare half a peck of the 22nd dressing) and carryed in to Good: Sell of Triplow to be maulted for our owne use on March 16.

Of the 20th dressing, threshed by J: Casbourne, February 28th there was in all 6 quarters, which was all sold, as it was to J: Norris of Barly, at 1*l*. 4*s*. 0*d*. per quarter, for 7*l*. 4*s*. 0*d*.

Of the 21rst dressing, threshed also by J: C:, there was in all 6 quarters and five bushells, March 6, whereof one peck was sold to Good: Hitch for 9*d*. and the rest was all sowne, most of it, excepting a bushell or two, in the South-feild, being tilth this year, and that also sowne, in the North Feild.

Of the 22nd dressing, March 14, there was in all 8 quarters 5 bushells and half, whereof one quarter 3 pecks and half were sold at home to Athan: Carr:, Henry Fordam, Good: Hitch, Thrift, Casbourn, at 23*s*. 6*d*. per quarter, for 1*l*. 5*s*. 6*d*. *ob*., and one quarter 3 bushells and half a peck were added to the 19 dressing (to make up 5 quarters and half for our owne use), and 4 bushells were added to the next dressing; and all ye rest (viz 5 quarters 5 bushells and half) were all sowne in the North Feild; so that of this and the former dressing there are sowne in all 12 quarters 2 bushells and a peck upon 26 stetches and a single rood in the Southfeild and 3 single acres and 2 acres in Waterden Shot and the Rowze — that is upon 15 acres in Southfeild — and upon 31 stetches and the rood next the mill hedge, and the 4 acres in More Shott, North More Shott, etc., in the Northfeild — that is about 15 acres and 3 roods — in both fields neare 32 acres — which is not 3 bushells and a peck per acre, yet Waterden Shott is large measure.

Note: That 14, 16, 20, 21, 22, dressings, making in all 35 quarters 2 bushells and half, came, at least 33 quarters of it, out of the first mow on the left hand in the new barne. [p.12]

Of the 23rd dressing, March 20th, threshed by Thrift, there was in all 9 quarters, whereof 5 quarters and 4 bushells were carryed in March 21 to J: Sells of Triplow (which with so much carryed in to him of the 19 dressing make in all 11 quarters for that purpose) to be maulted for our owne use; and half a bushell was tayle; and the remaining 3 quarters 3 bushells and half (being made up 4 quarters and 2 bushells by the addition of 4 bushells of the former dressing and 2 bushells and half of the next) were sold to M. Thurgood of Royston att 1*l.* 3*s.* per quarter, for 4*l.* 17*s.* 9*d.*

Of the 24rth dressing, Aprill 4, there was in all 5 quarters one bushell and half, whereof 2 bushells and half were added to the former dressing as before and the remaining 4 quarters and 7 bushells were sold to J: Norris of Barly at 1*l.* 2*s.* 6*d.* per quarter, for 5*l.* 9*s.* 9*d.*, and carryed in Aprill 5.

Note: That this was all the barly that lay in my barnes. All that is behind came of from the ricke.

Of the 25th dressing, Aprill 14th, there was in all 6 quarters and 5 bushells and half, whereof half a bushell tayle; and the 6 quarters and 5 bushells were all sold to J: Norris of Barly at 1*l.* 2*s.* 2*d.* per quarter, for 7*l.* 5*s.* 10*d.*

Of the 26th dressing, Aprill 20th, there was in all 6 quarters and half a bushell, where of the half bushell was tayle. The remaining 6 quarters were sold to M: Thurgood at Royston at 1*l.* 2*s.* 0*d.* per quarter, for 6*l.* 12*s.* 0*d.*

Of the 27th dressing of barly, Aprill 26, there was in all 5 quarters and 2 bushells, whereof above 2 pecks were tayle, a peck sold to Thrift for 8*d.*, neare another peck saft for our owne use, and the remaining 5 quarters and one bushell were sold to J: Norris of Barly at 1*l.* 1*s.* 4*d.* per quarter, for 5*l.* 9*s.* 4*d.*

Note: That these three last dressings of barly, making in all even 18 quarters, came all of the ricke. [p.13]

Note: There is of the whole crop of barly this yeare 172 quarters 2 bushells and half, of which were sowne 12 quarters 2 bushells 1 peck

Item maulted for our expenc	11	0	0
Item otherwise spent or ground	0	5	3
Item were made of tayle	2	0	1
Spent all wayes, in all	26	0	1
The rest sold, viz about	146	2	1

[p.14] **An account of the oates or bullimong brought in in that harvest**

Of the first dressing of oates, on January 12th, there was in all six quarters and 3 bushells, and about a peck of sorry tayle that was sold to W: T: for 2d. The whole dressing was put into the granary for use, having very few left for horses.

Of the 2nd dressing, January 25, there was in all one quarter 2 bushells and half, whereof 2 pecks were tayle, mingled with pease, etc., and the fatting hoggs had them; and the rest were laide in the garner for our owne use likewise.

Of the 3rd dressing, March 6, there was in all 2 quarters 2 bushells and half, very neare 3 pecks; laid all into garner for our owne use, as well as the rest.

Of the 4rth dressing, March 29, there was in all 9 quarters 2 bushells, whereof 5 bushells and half were sold to George Spilman at 22d. per bushel, for 10s. 0d., and 5 quarters more were sold to Mr. Killenbeck at the same prise, for 3l. 13s. 4d., and the rest spent by our selvs.

Of the 5th dressing, May 11, there was in all 8 quarters and 3 pecks, of which the 3 pecks were tayle and sold to Will: Thrift for 4d., and 3 quarters more were sold to Good: Nash, Good. Thompson, Good: Godfrey and others about 1s. 11d. per bushell, for 2l. 4s. 2d., and the rest remains to our owne use.

Of the 6th dressing, May 19th, there was in all 8 quarters and 3 bushells all laide into the granary for our owne use.

Of the 7th dressing, May 26, there was in all 9 quarters and one bushell, which were also laide into granary for our owne use.

[p.15] Note: That of the whole cropp of oates this yeare there was in all 44 quarters 6 bushells and 3 pecks, whereof sold and tayle about 9 quarters, so that there remaines for our expence 35 quarters and 6 bushells.

Lintells threshed this yeare

There was but one dressing, which was on February 9th, and there was in all of them 4 quarters; but 7 bushells of barly were taken out of them and are already reckoned to the 17th dressing of barly. There remaines therefore 3 quarters and one bushell of the lintells, of which 2 bushells were sold at Royston, for 5s. 0d., and 2 bushells more the next day at 4s. 6d., and 4 bushells more to Richard Fairchild for 9s. 6d., and 3 bushells to Good: Law at 7s. 0d., and one bushell to J: Bonnis at 2s. 3d., and 4 bushells more to Good: Hitch, Good: Nash, Good: P:, Good: Carr: and others at 2s. per bushell, for 8s. 0d., so that there is sold in all 2 quarters; and some have been spent by sowe and piggs; and still remaine about 4 bushells when there is occasion of expence.

[p.16] **An account of the pease brought in and threshed that yeare**

Of the first dressing, January 16th, there was in all but one quarter and one bushell — fanned up, as the hoggs had need of them — which were of the bottome of the ricke, and for want of straw and good wining in harvest were course and growne, and therefore spent at first.

Of the 2nd dressing, January 22, there was in all 5 quarters and 3 bushells, whereof one bushell was tayle, they being dressed for seed; and one bushell was sold to Wil: Thrift at 2s. 3d. The rest were laid up in the garner and out of them was sold one bushell more to Will: Thrift for 2s. 2d., and 17 bushells were sowne upon 5 stetches and the 2 acre peice in Brook Shott in the Northfield, in all upon 4 acres wanting half a rood large measure, and it is 4 bushells and about a peck upon every acre, and ye remaining 2 quarters and 7 bushells were all spent by fat hogs.

Of the 3rd dressing, March 7, there was in all 3 quarters 4 bushells and a peck, some of which were fanned up before hand for fatting hoggs, and the rest all laide in the garner for our owne use; one bushell sold to Good: Boreham for 2s. 4d.

Of the 4rth dressing, March 16th, there was in all 4 quarters and 4 bushells and half, whereof one bushell was tayle, the half bushell sold to Good: Watson for 1s. 2d., one quarter sold to Good. Symonds at 7 groates, for 18s. 8d., the remaining 3 quarters and 3 bushells laid in the garner, and out of them sold to Good: Sell 3 bushells at 7 groats, for 7s., 2 bushells to Mr. Killenbeck at the same rate, 3 bushells to Good: Bush: for 7s. also.

Of the 5th dressing, May 2nd, there was in all 3 quarters 2 bushells and half, whereof one bushell and half were tayle and 3 quarters and one bushell were of the best, all laide into the granary.

Of the 6th dressing, June 2nd, there was in all 6 quarters and five bushells, out of which were sold a bushell and 3 pecks at 7 groats, for 4s. 1d. The rest remaine for our owne. [p.17]

Of the whole cropp of pease there was in all 24 quarters 4 bushells one peck, whereof there was tayle 0 quarters 3 bushels 2 pecks

And sowne in Northfeild	2	1	0	
And sold already	2	5	1	at 2l. 9s. 4d.
And spent already	5	4	1	
In all	10	6	0	

Remaining about 13 quarters and 6 bushells, which if they be spent, are spent with horses

Finis huius anni 1682

An account of the quarters of corne of all sorts, sold or spent, that there was of my eighth crop, gathered in in harvest anno 1683, when harvest began generally on Tuesday July 24th (though some began the Tuesday and Thirsday before, that finished no sooner than the rest) and ended on Saturday August 25, lasting five weekes compleate. The sheafe corne was brought in very well generally, [p.19] excepting a loade or two at the last, and then there was about 14 dayes very wett — scarce 2 good dayes all the time — which comeing in the beginning of barley harvest caused some of the barly in the swath to grow, and much of it to be badly brought in, excepting some at the beginning and some at the last. The raine and great bulk withall made the harvest long. The Southfeild was tilthe and the Northfeild broake crop this yeare.

[p.20] **An account of the cropp brought in in harvest that yeare, and first of the wheate, white or Kentish**

Of the first dressing of red wheate, October 17, there was in all ten bushells and half, which was all sowne upon Lamps Acre and stetch and 3 stetches besides, and upon the 3 half acres in Short Spots, in all upon 4 acres in the Barr-feild; which is but 2 bushells and half an acre and 2 pecks over in all, though most of it be large measure.

Of the 2nd dressing, of red wheate also, October 20, there was in all 7 bushells, of which one bushell was courser than the rest, yet spent by ourselves, and the remaining 6 bushells were sold at home at 2s. 8d. per bushell, for 16s.

Note: That these two dressings, making in all 2 quarters one bushell and half, were all that lay in the malthouse.

Of the 3rd dressing, December 22nd, there was in all but 7 bushells and a little tayle, to be put to the rye, whereof one peck was given to Mr. Bourne and 3 pecks were sold at 8 groates, for 2s., and the remaining 6 bushells were spent by our selvs.

Of the 4rth dressing, January 17, of red wheate (as the former was) there was in all 8 quarters 5 bushells and half, whereof there was of tayle one bushell and a peck, sold for 2s.; and there was sold att home at 2s. 8d. per bushell 1 quarter 2 bushells and 1 peck, for 1l. 7s. 4d., and 6 quarters were sold to the Widdow Whitehead of Royston at the same rate, for 6l. 8s. 0d., and sold at home at 3s. per bushel 3 bushels for 9s. 0d., and the remaining 7 bushells spent by our selvs — the last went to mill February 26.

Of the 5th dressing of red wheate, February 26, there was in all 1 quarter 6 bushells and 3 pecks, whereof one peck was tayle; a peck course put to the 7th dressing of rye, 2 bushells and 3 pecks sold at 3s. a bushell, for 8s. 3d., 4 bushells and a peck sold at ten groats, for 14s. 2d., and 2 bushells and a peck sold at 11 groats, for 8s. 3d.; the remaining 5 bushells spent by our selvs.

[p.21] Of the 6th dressing, Aprill 17, which was red wheate, there was in all 5 quarters 4 bushells and 3 pecks, whereof 2 pecks were tayle reserved for ye little piggs; and 5 pecks were a 2nd sort of tayle sold nevertheless at 3s. 1d., 2 pecks were given to Thrift and Casbourne, 2 pecks sold at 11 groats, for 1s. 10d., 2 bushells spent by our selvs, and the remaining 5 quarters were sold to Widdow Whitehead of Royston at 4s. per bushell, for 8l. 0s. 0d.

Of the 7th dressing, May 12, which was all white wheate, there was in all 7 quarters one bushell and a peck, whereof 2 bushells and a peck were tayle spent by the sowes, and amongst the off corne. One bushell and half was sold at home at 5s. per bushell, for 7s. 6d., one bushell and 3 pecks more were sold at 4s. 8d. per bushell, for 8s. 2d., and 2 bushells more were sold at 4s. 4d. per bushel, for 8s. 8d.; and there were sold at 4s. per bushel one quarter three bushels 3 pecks and half, for 2l. 7s. 6d.; and there was spent of it by our selvs 4 quarters and 5 bushells, so that there is 3 pecks and half lost, liing in the garner from May 12 till October 13, that is 5 months.

Of the 8th dressing, June 7, which was all red wheate, there was in all 7 quarters and 4 bushells, whereof 3 bushells were tayle, yet such as were sold, at 2s. 4d. per bushell, for 6s. 11d., and 3 pecks were sold at home at 4s. per bushell, for 3s.; 5 quarters were sold to Royston at the same price, for 8l. 0s. 0d., and a bushell and half more were sold at home at 4s. 4d. per bushell, for 6s. 6d.; one quarter and 6 bushells were spent by our selvs, and 3 pecks are lost.

Notandum: Of the whole cropp of wheat this yeare there was in all 33 quarters 6 bushells and 3 pecks.

An account of the rye or miscellaine brought in in that harvest [p.23]

Of the first dressing of rye or miscellaine, September 14th anno 1683, there was in all two quarters and five bushells, whereof neare a bushell and half were the shalings that were scattered in harvest, and the rest was threshed; and it was all sowne upon Branditch Shot in Barr-feild and the furthermost land in Finch-way and Short Furlong, being not so good miscellaine as that that was sowne afterwards.

Of the 2nd dressing, September 22nd, there was in all 3 quarters and 2 bushells, which with ye former dressing was all sowne (excepting about half a peck) upon the 5 furthest stetches in Short Spotts, 4 in Branditch Shot, 15 in Short Furlong and Finch-way, the 2 single acres in the same shott, and upon 4 acres of the 12 between Michell's Path and Melbourne Way, in all upon 15 acres of gleabe land, besides a single stetch of my owne, which is about 3 bushells an acre and some small matter above.

Note: That these 2 dressings, making in all 5 quarters and 7 bushells, were all that lay in the malthouse.

Of the 3rd dressing, December 7, there was in all 5 quarters and half a bushell, whereof 4 quarters and half were sold to Cambridge at 1s. 10d. per bushell, for 3l. 6s. 0d., and the remaining 4 bushells and half were sold at home at the same rate, for 8s. 3d.

Of the 4rth dressing, December 20, there was in all 4 quarters 3 bushells and half, whereof half a bushell was tayle, put amongst the of corne, and one bushell and half were given to ye poore, and 3 bushells were spent by our selvs, and the remaining 3 quarters 6 bushells and half were sold at home at 1s. 10d. per bushell, for 2l. 15s. 11d.

Of the 5th dressing of rye, January 23rd, there was in all 5 quarters 2 [p.24] bushells and half, whereof one peck was tayle, put amongst the off-corne, and 2 bushells and 3 pecks were given to ye poor, viz to Widdows Carr., Dun, M. Dewc, J: Wolfe, Tho. Rain., Thom: Whitby, Rbrt Wood. 2 quarters and 3 pecks were sold at home at 1s. 10d. per bushell, for 1l. 10s. 8d. ob., and sold at 2s. per bushell 2 quarters 5 bushells and 3 pecks, for 2l. 3s. 6d., and the remaining bushell spent by our selvs.

Of the 6th dressing of rye, February 19, by William Brock, there was in all 7 quarters and 3 bushells, whereof above a peck tayle; and eleven bushells and half were sold at home at 2s. per bushell, for 1l. 3s. 0d., and 2 bushells and half were given to the poore; and 4 quarters and 5 bushells were sold at 2s. 2d. per bushell, for 4l. 0s. 2d., and sold againe at home at 2s. 6d. per bushell 4 bushells and half, for 11s. 3d., and 3 bushells and a peck (all but about half a peck taken out of tayle wheat to make measure) were spent by our selvs, the last of it sent to mill March 20. This is the last dressing out of the first mow on the right hand in the new barn.

Of the 7th dressing of rye, March 20, by Will: Brock, there was in all 6 quarters 7 bushells and half, of which 3 pecks were tayle, laide by it self for little piggs, and one bushell and a peck were given to the poor, viz Will: Whitby, Widdows Dovy, Carrington, and 3 bushells that were promised before were sold at 2s. 2d. per bushell, for 6s. 6d., and there were sold of it at 8 groats per bushell 2 quarters 4 bushells and a peck, for 2l. 14s. 0d., and sold at 3s. per bushel 2 quarters one bushell and half, for 2l. 12s. 6d., and sold at 3s. 6d. per bushell 4 bushels and a peck, for 14s. 10d. ob.; and sold againe at 10 groats 3 bushells and half, for 11s. 8d.; and spent by our selvs 4 bushells, which make but 6 quarters 6 bushells and half, so that there is a bushel lost nevertheless.

Of the 8th dressing of rye, May 13th, there was in all 3 bushells and 3 pecks, whereof one bushell was spent by our selvs; and 2 bushells and 3 pecks were sold to Goodman Woollard, Casbourne, Thrift, at 10 groats per bushell, for 9s. 2d.

[p.25] Of the 9th dressing of rye, June 23, there was in all 5 quarters and 5 bushells, whereof one bushell and a peck was tayle; and 3 pecks were given to Widdow Cornhill and Tho: Reinold at the town's ende, 2 bushells were spent by our selvs, and one quarter and a bushell were sold at home at 8 groats per bushell, for 1l. 4s. 0d.; and 2 bushells and 3 pecks more sold at 2s. 6d., for 6s. 10d. ob.; and one bushell and a peck were sold at 7 groats, for 2s. 11d. The remaining 3 quarters and 4 bushells were added to the next dressing and carried out at 7 groates a bushell.

Of the 10th dressing, June 30, there was in all 4 quarters and 6 bushells, whereof one bushell was tayle, it being much eaten; and 2 quarters and 4 bushells (being made up 6 quarters from the last dressing) were carryed to Royston at 7 groats and sold for 5l. 12s. 0d., and one quarter and a bushell were sold at home at the same price, for 1l. 1s., and 4 bushells were sold to Thomas Page of Holton (July 17, who fetched them at home also, sending his man for them that day 1684) at the same price, for 9s. 4d.; and the remaining 4 bushells were spent by our selvs, the last of which was sent to mill August 7.

Of the eleventh dressing, about the same time, there was in all 6 quarters and 3 bushells, whereof (it being very much eaten with mice) 3 bushells were tayle, put amongst of corne. 2 bushells that were promised long before were sold at 2s. 2d., for 4s. 4d. 2 quarters and 3 bushells besides were sold at 7 groats per bushell, for 2l. 4s. 4d.; and were sold afterwards at 8 groats per bushel 2 quarters one bushell and half, for 2l. 6s. 8d.; and 5 bushells were spent by our selvs, which makes in all but 5 quarters 6 bushells and half, so that there is 4 bushells and half lost or else it was not set downe. I suppose the bin might not be empty for 6 months together and it was threshed two or three or more times, so that there must needes be some loss, though I wonder at so much.

Notandum: Of the whole crop of rye this yeare there was in all 52 quarters one bushell and 3 pecks.

An account of the barly brought in and threshed that yeare [p.26]

Of the first dressing of barly, by W: Thrift, November 12, there was in all 6 quarters and one bushell, whereof the odd bushell was tayle and laide amongst the off corne; and the remaining 6 quarters were sold to Mr. Glenister at Royston at 16s. per quarter, for 4l. 16s. 0d. and carryed in on Thursday November 15.

Of the 2nd dressing, November 14, by J: Casbourn, there was in all 7 quarters 2 bushells and half, whereof one bushell was tayle laide amongst the off corne and half a bushell was sold to W: Thr: at home for 1s. 0d.; and 6 quarters and half were carryed in to Mr. Glenister of Royston at 16s. per quarter also, for 5l. 4s. 0d. 5 bushells remaine, added to the 3rd dressing and sold with it.

Of the 3rd dressing threshed by J: Casb:, November 22nd, there was in all 6 quarters and one bushell, whereof the odd bushell was tayle, and the 6 quarters, together with the 5 bushells of the former dressing, were sold to M: Thurgood of Royston at 16s. per quarter and 1s. over, for 5l. 7s. 0d.

Of the 4rth dressing, November 20, by J: Casbourne, there was in all 4 quarters, which with 2 quarters of the next dressing were sold to M: Thurgood at Royston at 16s. 4d. per quarter, for 4l. 18s. 0d.

Note: That these 3 last dressings came all out of the middlestead in the new barne, being in all 17 quarters 3 bushells and half. It was supposed 2 quarters came of ye rye.

Of the 5th dressing, November 23, by Wil: Thrift, there was in all 8 quarters and 7 bushells, whereof one bushell was tayle; 2 quarters sold with the last dressing; the remaining 6 quarters and 6 bushells sold to the same man at ye same price, for 5l. 10s. 3d.

Note: That this dressing and the first, making in all 15 quarters, came out of the middlestead in the old barne.

Of the 6th dressing, December 12, by W: Thrift, there was in all 6 quarters [p.27] and 3 bushells, whereof 6 quarters and 2 bushells were sold to Mr. Rumbald at Royston at 17s. per quarter, for 5l. 6s. 3d., and the remaining bushell was tayle and put amongst the off corne.

Of the 7th dressing, December 17, by W: Thrift, there was in all 6 quarters and a bushell, of which the bushell was tayle; and the 6 quarters were sold to M: Thurgood of Royston, att 17s. per quarter, for 5l. 2s. 0d.

Of the 8th dressing, December 22, by W: Thrift, there was in all 6 quarters 3 bushells and about half, whereof one bushell was tayle againe; and one bushell sold at home for 2s. 1d. and 6 quarters and one bushell were sold to M: Thurgood of Royston at 17s. per quarter, for 5l. 4s. 1d.; remains half a bushell, whereof one peck sold to J: Haycock for 6d. and the other served to make up the next dressing even 6 quarters [and] sold with it.

Of the 9th dressing, January 1rst, by J: Casbourne, there was in all 5 quarters 7 bushells and 3 pecks. There was some tayle, but inconsiderable. The rest was made up even 6 quarters by a peck of the former dressing and sold to M: Thurgood at Royston at 15s. 9d. per quarter, for 4l. 14s. 6d.

Of the 10th dressing, January 7, by J: Casbourne, there was in all 5 quarters 5 bushells and half, of which one bushell was tayle, and 5 quarters and 4 bushell was sold to Mr. Mackrer of Ashwell at 17s. per quarter, for 4l. 13s. 6d., and one peck sold at home for 6d. ob., and the other to J: Casbourne for 6d. ob. likewise.

Of the 11 dressing, January 22, by Thrift, there was in all 10 quarters and 2 bushells, which were all carryed in on January 23 to Goodman Kefford of Triplow to be maulted for my owne use.

[*p.28*]   Of the 12th dressing, by Thrift, January 23rd, there was in all 12 quarters and one bushell, whereof 3 bushells were tayle (there being none reckoned out of the dressing the day before, though it was of one heape) put into the garner, and the remaining eleven quarters and 6 bushells were carryed in on Friday January 25 to Goodman Kefford of Triplow also, to be maulted for my owne use.

Of the 13th dressing, also by W: Thrift, January 31, there was in all 9 quarters and 2 bushells, whereof one bushell was sent to mill for our owne use, one bushell sold at home, and 6 quarters and 2 bushells carryed to M: Thurgood at Royston at 15s. per quarter, for 4l. 15s. 7d. ob.; and half a bushell sold at home againe to W: Thrift for 11d. ob., and 2 quarters and half a bushell were sold afterwards with the next dressing to the same man at the same prise; and 5 bushells were lintells taken out of this dressing (all but a small matter of tayle), which were sowne upon an acre and half a rood of my owne and gleabe together in Waterden Shott — the upper end of it.

Of the 14th dressing, by Thrift, February 14, there was in all 9 quarters one bushell and half, of which 2 bushells were tayle laide behind the garner door with the rest of the off corne; and the remaining 8 quarters 7 bushells and half (being made up even eleven quarters by the addition of 2 quarters and half a bushel of the former dressing) were sold to M: Thurgood at Royston also at 15s. per quarter, for 8l. 5s. 0d.

Note: That these two last dressings were all tithe, and almost all the tithe, of the Northfeild, which was brockfeild this yeare, and laid upon a rick together at last, my owne brock land barly being brought in before the rest of the towne began in that feild, these two dressings making in all 18 quarters 3 bushells and half.

Note: Also that the eleventh, 12th and 15th dressings, making in all 34 quarters and 5 bushells, were of my owne brockland corne, brought in altogether before the raine.

Of the 15th dressing, by Thrift, February 23, there was in all 12 quarters [*p.29*] and 2 bushells, of which 4 bushells were very course and unfit to put amongst the rest, though I maulted it, and therefore was put amongst the tayle; and the remaining eleven quarters and 6 bushells were the same day carryed in to Goodman Kefford's of Triplow to be maulted for my owne use.

Of the 16th dressing, March 4, in W: Thrift's barne, there was in all 12 quarters 7 bushells, whereof one bushell was tayle put amongst the off corne behind the garner doore, and the 12 quarters and 6 bushells were carryed out at 2 loades, whereof the last went out March 13, to M: Thurgood at Royston at 16s. per quarter, for 10*l*. 4s. 0d., this being the first dressing that came of from the great ricke.

Of the 17th dressing, March 4, also by Thrift etc., there was in all eleven quarters and one bushell, whereof ye odd bushell was tayle laide amongst the off corne behind the garner door, and the eleven quarters were sold to M: Thurgood at Royston at 16s. 9d. per quarter, for 9*l*. 4s. 3d., being the 2nd dressing from the great rick.

Of the 18 dressing of barly, March 14, by Thrift, there was in all 13 quarters and 6 bushells, whereof one bushell was tayle laid behind the garner doore with the rest of the off corne; and 6 quarters and 2 bushells were carryed in to G: Kefford on Munday March 24 to be malted for my owne use, and on Saturday March 29 were carryed in to him 4 quarters and 4 bushells more for the same purpose, and 4 bushells were sold at home to Henry and William Fordhams, at 9s., and the remaineing 2 quarters and 3 bushells were added to the next dressing and sold with it.

Of the 19th dressing, March 22, by Thrift also, there was in all 3 quarters 5 bushells and a peck, whereof a peck was sold to J: Casbourne at 6d., and the rest (being made up even 6 quarters by the addition of 2 quarters and 3 bushells of the former dressing) were sold to one of Ashwell at 18s. 2d. per quarter, for 9*l*. 9s. It was delivered to an Ashwell cart at Royston, March 29. This is 4th dressing from the rick, as also all the barly that is behind comes of from the same ricke.

Of the 20 dressing of barly, threshed by W: Brock in the new barne, there [*p.30*] was in all 6 quarters 4 bushells and a peck; of which there was 5 pecks tayle laide among the off corne, and one bushell and a peck sold at home for 2s. 5d.; and 5 bushells and 3 pecks were sent to mill for the sowes and piggs; and the remaining 5 quarters and half (together with 10 quarters and half that were bought for 8*l*. 7s. 8d., in all 16 quarters) were sowne upon 21 acres and a rood in Southfeild, being the brockfeild this yeare, and upon 19 acres and a rood in ye Barrfeild, or tilthe — in all upon 40 acres and half (of which 2 acres and 3 roods are my owne land) which is a small matter above 3 bushells an acre, one with another, red and white land, great and small measure.

Of the 21 dressing, Aprill 9, by W: Thrift, there was in all 12 quarters and 2 bushells, whereof one bushell was tayle laid amongst the off corne, 2 bushells and 3 pecks were sold at home, to Good: Spilman, Thrift, Casbourne, for 5s. 11d. ob., and a bushell sent to mill for the sowes etc.; and there were sold to M: Thurgood of Royston, at 17s. 6d. per quarter, 11 quarters and 5 bushells, for 10l. 3s. 5d; and a peck remaines, sold to Thrift at 6d. ob.

Of the 22nd dressing, by Will: Thrift also, there was 10 quarters 4 bushells and a peck, of which 2 bushells were tayle fitt for nothing but of corne. 5 quarters were carryed to Triplow and sold to Goodman Charles this Aprill 16 at 17s. per quarter, and 5 quarters carryed to him also at 17s. per quarter Aprill 19th, for 8l. 10s. 0d. The 2 bushells and peck remaining were sold to Thrift and Casbourne (together with a bushell fanned out of the next dressing) for 7s. 1d. ob.

Of the 23rd dressing, April 24, by Will: Thrift, there was in all 6 quarters 2 bushells and half, whereof 3 pecks were tayle, and a bushell and 3 pecks sold at home to Will Thrift for 4s. 1d. (as is partly mentioned in the last dressing), and the remaining 6 quarters were sold to M: Thurgood of Royston at 19s. per quarter, for 5l. 14s. 0d.

Of the 24rth dressing, Aprill 28, by Will: Thrift, there was in all 6 quarters one bushell and half, whereof one bushell and half was tayle put amongst the off corne, and the 6 quarters remaining were sold to M: Thurgood at Royston at 19s. per quarter also, for 5l. 14s. 0d.

[p.31] Of the 25th dressing of barly, by Will: Thrift, May 6, there was in all eleven quarters and 7 bushells, whereof there was a bushell tayle reserved for off corne, and the remaining 11 quarters and 6 bushells were sold to J: Ostler at Ashwell at 19s. 6d. per quarter, for 11l. 9s. 1d.

Of the 26th dressing, May 12, by W: Thrift, there was in all 9 quarters and a bushell, whereof one bushell was tayle. A quarter and 3 pecks were sold at home to Good: Thrift, Casb:, Payne, Watson, Gurner, Brooks, Mr. Bourne, etc., and 7 quarters and 4 bushells were sold to J: Ostler of Ashwell at 19s. 6d. per quarter, for 7l. 6s. 3d; and one bushell sent to mill for our owne use also.

Note: That these last eleven dressings came all of from the great rick at the end of the barne, and make in all 104 quarters 1 bushell and 3 pecks of barly, besides oates that were at the one end of the ricke and 2 jaggs of lintells never threshed. Now there was of the oates 2 dressings, making in all 21 quarters and one bushell, so that there was of all that rick in barly and oates that yeare 125 quarters 2 bushells and half. I laid abroade besides above 18 quarters of brockland barly, so that of barly alone there was above 6 score quarters laide abroad that yeare.

And there was of the whole crop of barly that yeare 220 quarters 2 bushells and half.

[p.32] **An account of the oates brought in and threshed that harvest**

Of the first dressing of oates, December 27th, there was in all eleven quarters and 6 bushells, which were part of the little ricke that came at last of from the Heath, and the old oates being just done, were all laide into the garner for our owne expence.

Of the 2nd dressing, January 2 (being the remaining part of that little rick from the Heath), there was in all seven quarters and one bushell, of which one quarter and half were put into the garner amongst the former, and 5 quarters and 5 bushells, being the heade of them, were laid by themselvs for seed, which were all sowne (upon 5 stetches in London Way Shott and 13 stetches in Waterden Shott, and upon the little rood and the single acre and 2 acres in Waterden Shott — in all upon 10 acres, besides the stony ends of 3 acres and a rood betwixt the broad balk and Harborow Hill) excepting about 4 bushells spent by our selvs.

Of the 3rd dressing, February 7th, there was in all 11 quarters and one bushell, whereof 3 quarters and 4 bushells were sold to Good: Nash and one of Foxton that fetched them, at 1s. 3d. per bushell, for 1l. 15s. 0d., and 6 quarters more were sold to Thom. Godfrey and Richard Wallis of Triplow at divers times, and to Mr. Maulden, Mr. Reinold, George Spilman, and other neighbours at 1s. 4d. per bushell, for 3l. 4s. 0d. The remaining one quarter and 5 bushells are spent by our owne horses.

Of the 4rth dressing of oates, May 21, there was in all 8 quarters and 3 bushells, whereof 6 quarters were sold to Mr. Reinold at 1s. 6d. per bushell and 6d. over, for 3l. 12s. 6d., and one quarter and 4 bushells afterwards to Thom: Godfrey at 19d. per bushell, for 19s., and the remaining 7 bushells also were sold at home by parcels at 1s. 8d. per bushell, for 11s. 8d.

Note: That this dressing was part of the great rick at the end of the barne, as also the 7th dressing, which make together 21 quarters and a bushell.

Of the 5th dressing of oates, June 4, there was in all 13 quarters, whereof [*p.33*] 3 bushells were sold at home at 1s. 8d. per bushell, for 5s., and the rest laide into the granary for our owne use. This was part of the middle rick that stood against the study window.

Of the 6th dressing, June 12, there was in all 6 quarters laide into the malthouse chamber for our owne use, which were the remaineing part of that middle rick, making both together 19 quarters.

Of the 7th and last dressing of oates there was in all 12 quarters and 6 bushells, all likewise laide into the malthouse chamber for our owne use, being part of the great rick; sold out since 4 bushells and half to Mr. Reinolds at 20d. per bushell, for 7s. 6d. The rest spent.

Note: That of the whole cropp of oates this yeare there was 70 quarters and one bushell, whereof there were sowne 5 quarters and a bushell and sold 16 quarters 3 bushells and half, in all 21 quarters 4 bushells 2 pecks. Remaines for expence 48 quarters 3 bushells and half.

An account of the pease brought in and threshed this yeare [*p.34*]

Of the first dressing, November the 2nd (which was cutt out of the side of the ricke and threshed in the malthouse), there was in all just one quarter, which was threshed for the fatting hoggs.

Of the 2nd dressing, December 31, there was in all but 3 bushells, which were the tithe of Dod's Close, and when they were threshed were but course, lying at the bottome of the mow in the old barne — reserved for fatting hoggs.

Of the 3rd dressing, January 8th, there was in all 9 quarters and one bushell, whereof about 2 pecks were given to Widdows Dovy, Bayes, Will Whitby and John French. 2 quarters 4 bushells and half were sowne upon 5 acres one rood and half in Southfeild (which is about 4 bushells an acre) and 6 quarters and 3 bushells were sold at divers times to neighbours and such as fetched them at home at 2s. 7d. per bushell, for 6l. 11s. 9d. The odd bushell was otherwise spent by our selvs.

Of the 4rth dressing, February 2nd, which was threshed in the new barne, there was againe 9 quarters and 3 bushells in all, which were for the present all laid into the garner; and out of them the hoggs were fatted that were killed in March; and 3 bushells and an half were sold at 2s. 7d., for 9s. 0d., the last peck of which was sold on June 29, 1684. The rest remaine still for our owne use.

Of the 5th dressing, June 6, there was in all 2 quarters and 2 bushells, which was part of the middle ricke against the study window which had nothing else of it but oates besides these pease. They remaine for our owne use in the garner.

Note: That all the other pease excepting the 2nd and this last dressing came all of from a pease rick at the orchard gate.

Notandum: Of the whole cropp of pease this yeare there is 22 quarters one bushell.

[p.36] **Lintells threshed that yeare**

Of the first dressing, December 11, there was but 3 bushells and half, threshed because they lay on top of the barly, whereof J. Casb: had 2 bushells for 3s. and W: Thrift had ye remaining bushel and half for 2s. 3d. They were but course.

Finis hujus anni

[p.37] An account of the quarter of corne of all sorts, sold or spent, that there was of my ninth cropp gathered in in harvest 1684, when harvest began generally on Munday July the twenty first (excepting that the lintells were most mowne and brought in the week before) and yet was compleatly ended on Saturday August the ninth (there being not one whole letting day wherein they could doe nothing all the harvest), lasting just three weekes. Notandum: That the Barr Feild this yeare was tilth and the Southfeild the broake cropp. The whole crop (excepting barly, which was generally good) was very thinn and went very neare together, I, for my owne part, not having one mow of wheat and rye together, when the yeare before I had full 3 mowes and half.

HARVEST ACCOUNTS 1684

An account of the cropp brought in in that yeare, and first of wheat, white or Kentish [*p.38*]

Of the first dressing, which was red wheate, August 26, there was in all 2 quarters, whereof 2 pecks being course were put amongst the rye, and 3 bushells were sold to Mr. Reynolds at 4s. 6d. per bushel, for 13s. 6d., and 9 bushells and half were sowne upon the 5 stetches in Brook Shot and ye 2 cross stetches, and 2 in More Shot, and one almost at the mill — in all, ten stetches, id est 3 acres and 3 roods in North Feild. That is somewhat above 3 bushells an acre; and the remaining 3 bushells were spent by our selvs.

Of the 2nd dressing, which was all white wheate, October 19, there was in all 14 bushells, whereof half a bushell sold at home at 2s. and a peck more at 1s. 3d.; 4 bushells and a peck were spent by our selvs; the remaining 9 bushells were mingled with the rye to make it good miscellaine and sowne upon the 2 acres in Brookshott.

Note: These 2 dressings, being in all 3 quarters and 6 bushells, were all that came out of the mault house.

Of the 3rd dressing, February 26, of red wheate, there was but 4 bushells and 3 pecks, whereof the 3 pecks were sold to M: P:, J: C: and W: T: for 3s. 6d., and the remaining 4 bushells were spent by our selvs.

Of the 4rth dressing of red wheate, March 17, there was in all 3 bushells and 3 pecks, whereof one peck was course, sold to J: Casbourne for 6d. 2 pecks more were sold to Math: P. and W: Thr. for 2s. 4d. The remaining 3 bushells were spent by our selves.

Of the 5th dressing of red wheate, Aprill 10th, there was in all 6 quarters 3 bushells and half (and almost a peck to supply the want that would be in the measure at the taking out), [*p.39*] whereof 2 pecks were tayle and a bushell and half besides was very course, yet sold at home for 4s. 4d.; one bushell and half more were sold at 5s. per bushell, for 7s. 6d., and one quarter 4 bushells and a peck were sold, 4s. 8d. per bushell, for 2l. 17s. 2d., and 2 quarters 2 bushells and half a peck sold at 4s. 4d. a bushell, for 3l. 18s. 6d.; and 2 quarters and one bushell were spent by our selves — in all 6 quarters 3, wanting half a peck, so that there [is] lost about 3 pecks. [*p.39*]

Of the 6th dressing, Aprill 28, there was in all 4 quarters and 7 bushells, which was all white wheate (laide in Mr. Maulden's chamber), of which one bushell and a peck tayle was sold nevertheless at 3s. 4d. and one bushell sold at home for 5s., and one bushell more sold att 4s. 8d., and 3 bushells and 3 pecks sold at 10 groats a bushell, for 12s. 6d. 3 quarters 7 bushells and half spent and half a bushell and a small matter more lost in lying half a year in the chamber and better.

Notandum: That the whole cropp of wheat this yeare was but 16 quarters and one bushell.

An account of the rye or miscellaine brought in and threshed in that yeare [*p.41*]

Of the first dressing of rye, August 30th, there was in all (with about 2 bushells of that which was shaled of the wheat and rye together with unloading in harvest) three quarters and six bushells, which was all sowne in the Northfeild, being tilthe crop to come in another yeare.

Of the 2nd dressing, September 4rth, there was in all one quarter and five bushells, where of 3 bushells and half were sold at home for seed to Good: Law and Carrington at 2s. 10d. per bushell, for 9s. 11d.; and the remaining 9 bushells and half were all sowne. There is sowne then of both these dressings 5 quarters wanting half a bushell, which were sowne upon 4 stetches in the Lower Shott, upon 8 more in the Middle Shot, and upon 9 in Northmoor Shot and the rood next the mill hedge, and upon six in the More Shott, all gleabe land, and 2 stetches of my owne in the same shott, and upon the 4 acre peice in North-more Shott — in all upon 15 acres and something more in Northfeild, which is something above 2 bushells and half an acre one with another.

Of the 3rd dressing of rye, October 18th, there was in all (with 2 pecks of the first dressing of wheat that was course put to it) three quarters and 2 bushells, whereof one bushel was sold at home for 2s. 10d. and 2 pecks more for 1s. 6d., and 2 quarters and 2 bushells sold at 3s. 6d. per bushell, for 3l. 3s. 0d. The remaining 6 bushells and half were spent by our selves, the last of which went to mill January 26.

Notandum: That these 3 dressings, making in all full 8 quarters 4 bushells and half, came all out of the maulthouse.

Of the 4 dressing, February 6, there was in all 2 quarters one bushell and half, whereof 6 bushells and 3 pecks given to the poor, 7 bushells and 3 pecks sold at 3s. 8d. a bushell, for 1l. 8s. 5d. The remaining 3 bushells spent.

[p.42]   Of the 5th dressing of rye, March 18, there was in all 7 bushells and a peck, whereof 5 bushells were sold at home at 3s. 8d. per bushell, for 18s. 4d., and 2 bushells spent by our selves, and about a peck devoured by little piggs.

Of the 6th dressing, April 11, there was in all 4 quarters 2 bushells and half, whereof one quarter and a peck was sold at 4s. per bushell, for 1l. 13s. 0d., and one quarter 6 bushells and half more were sold at 11 groats per bushell, for 2l. 13s. 2d., and 7 bushells more sold at 10 groates per bushell, for 1l. 3s. 4d., and 4 bushells and half spent by our selvs. About a peck lost.

Of the 7th dressing, May 12, there was in all 5 quarters 7 bushells and 2 pecks, whereof 2 pecks were tayle for little piggs, and 5 quarters and 5 bushells were sold at Royston att 10 groats per bushell, for 7l. 10s. 0d. The remaining 2 bushells sold at home at the same prise, for 6s. 8d.

Of the 8th dressing, May 22, there was in all 8 quarters and half a bushell, whereof one bushell and half was tayle saved for little piggs; and 3 quarters and 3 bushells were sold at Royston at 3s. 3d. per bushell, for 4l. 7s. 9d., and 2 quarters and 2 bushells more were sold at the same market at 3s. 4d. per bushell, for even 3l. 0s. 0d., and 2 quarters one bushell more were sold at home at the same rate, for 2l. 16s. 8d., and the remaining bushell was spent by our selves.

Of the 9th dressing, June 3, there was in all 4 quarters 4 bushells and half, whereof half a bushell was tayle. 4 quarters and 2 bushells were sold at home by the peck and the bushell, at 10 groats a bushell, for 5*l.* 13*s.* 4*d.*, and the remaining 2 bushells were spent by our selvs, the last of which went to mill June 26.

Of the 10th dressing of rye, by John Casbourne, June 10th, there was in all 3 quarters 5 bushells and half laid into ye garner, whereof 4 bushells were sold at home at 10 groats per bushell, for 13*s.* 4*d.*, one bushell was sold for 2*s.* 8*d.*, half a bushell was given to Ralph Chil:, and one quarter was sold at 7 groates a bushell, for 18*s.* 8*d.*, and one quarter and 3 bushells was spent by our selvs — in all 3 quarters and half a bushell, so that there is 5 bushels lost or else I want an account of it. [*p.43*]

Notandum: that of the whole cropp of rye this yeare there was but 38 quarters one bushell and 3 pecks.

An account of the barly brought in and threshed in that yeare [*p.44*]

Of the first dressing of barly, November 12, by William Thrift, there was in all 8 quarters and a peck, whereof 3 bushells and a peck were sold at home to Thrift, Good: Hay:, Payn, etc., at 2*s.* 10*d.* per bushell, for 9*s.* 3*d.*, and 6 quarters and 2 bushells were sold to Mr. Glen: of Royston and [*sic, for* 'at'] 23*s.* 6*d.* per quarter, for 7*l.* 6*s.* 10*d.*, and a bushell more sold to W: Thrift at 2*s.* 10*d.* The remaining 10 bushells were added to the next dressing and sold with it.

Of the 2nd dressing, November 14, by J: Casbourne, there was in all even 7 quarters, whereof 5 quarters (having the ten bushells of the former dressing added to them, making againe 6 quarters and 2 bushells) were sold to Mr. Glenister of Royston at 1*l.* 3*s.* 6*d.* per quarter and 1*s.* over in the loade, for 7*l.* 7*s.* 11*d.* The 2 quarters remaine to be sold, of which W: Thrift had another bushell and Math: Payne 2 bushells and half at 2*s.* 10*d.* per bushell, for 9*s.* 11*d.* One bushell was boyled for our owne fatting hoggs and 3 bushells and half more steeped for them. The remaining quarter was added to the next dressing and sold with it.

Of the third dressing, November 21, by Will: Thrift, there was in all 7 quarters 2 bushells and one peck, of which the odd peck was sold to Widdow Cornhill at 9*d.* 5 quarters and 2 bushells (by addition of one quarter of the last dressing being made up 6 quarters and 2 bushells) were sold againe to Mr. Glenister of Royston at 1*l.* 3*s.* 5*d.* per quarter, for 7*l.* 6*s.* 4*d.*, and 2 bushells and half more were boyled for fatting hoggs; and one bushell was put to the next dressing to make up even measure, and 1 quarter 4 bushells and half remaine in the sacks, of which 5 bushells more were steeped for hoggs, 2 bushells and half sold at home to Math: Payn for 7*s.* 4*d.*, a bushell to Mr. Maulden at 3*s.* The remaining 4 bushells were sold with the 5th dressing.

Of the 4rth dressing, November 22, by J: Casbourne, there was in all 6 quarters and one bushell, which, with one bushell of the former dressing (being 6 quarters and 2 bushells) were sold againe to Mr. Glenister at 1*l.* 3*s.* 5*d.* per quarter, for 7*l.* 6*s.* 4*d.*

[p.45] Of the 5th dressing, by William Thrift, there was in all 6 quarters and 4 bushells, which were made up even 7 quarters (by the addition of 4 bushells of the third dressing) and sold to Mr. Glenister againe at 1*l.* 3*s.* 10*d.* per quarter, for 8*l.* 6*s.* 10*d.*, November 29.

Note: That this dressing and the first and the third, making in all 21 quarters 6 bushells and half, came all out of the middlestead in the old barne.

Of the 6 dressing, by J: Casbourne, December 20, there was in all seven quarters and one bushel, whereof 6 quarters and 2 bushells were sold to Mr. Glenister at 1*l.* 3*s.* 10*d.* per quarter, for 7*l.* 9*s.* 0*d.* 2 bushells sold at home for 6*s.* to Mr. Maulden and Math: Payne, and 3 pecks sold to W. Thrift, J: Watson, Widdow Cornhill, for 2*s.* 3*d.*, and 4 bushells and a peck steeped for our fatting hoggs.

Note: That the 2 and 4 dressings and about 3 quarters of this as was guessed (making in all 16 quarters) came out of the middlestead in the new barne.

Of the 7th dressing, December 9th, by William Thrift, there was in all 9 quarters and 7 bushells, which being made up 10 quarters by the addition of a bushell of the next dressing, were sold to John Ostlar of Ashwell at 24*s.* per quarter, for 12*l.* 0*s.* 0*d.*

Of the 8th dressing, December 16, by Will Thrift also, there was in all 11 quarters, whereof 10 quarters were sold also to the same man at the same prise, for 12*l.* 0*s.* 0*d.* Only one bushell was throwen into the store and one bushell made up the last dressing, and 2 bushells and half sold at home to Mr. Maulden, Jude, Amb: Bonnis, J: Haycock, J: Watson at 3*s.* per bushel, for 7*s.* 6*d.*, and half a bushel sent to mill for f[owles]; and 3 bushells were steeped for fatting hoggs (besides a bushell and od of another dressing).

Notandum: That these 2 last dressings, making in all 20 quarters and 7 bushells, came out of the first left hand mow in the old barne.

[p.46] Of the 9th dressing, December 22, by John Casbourne, there was in all 9 quarters and 4 bushells, whereof 3 pecks sold at home to W: Thrift, and 2 pecks to J: Watson, and 2 pecks to Amb: Bonnis, and 2 bushells to J: Law, and 2 pecks to M: Payne, at 3*s.* per bushell — in all 4 bushells and a peck — for 12*s.* 9*d.*, and another peck sold at home for 9*d*; and sent to mill for fowles 2 pecks, and one bushell steeped for fatting hoggs; and the remaining 8 quarters and 6 bushells (by the addition of ten bushells of the next dressing being [made] up even ten quarters) were sold to Mr. Glenister at Royston at 1*l.* 3*s.* 5*d.* per quarter, for 11*l.* 14*s.* 10*d.*

Of the 10th dressing, December 31, by John Casbourne also, there was in all 11 quarters and 6 bushells, of which a bushell sold at home to Thrift, G: Bonnis and Widdow Cornhill for 3*s.*; half a bushel was steeped for the hoggs with some of the former dressing, and half a bushel boyled for horses being blooded, and ten bushells were sold with the former dressing, and ten quarters were sold to one of Baldock at 1*l.* 3*s.* 3*d.* per quarter, for 11*l.* 12*s.* 6*d.*, and one peck sold againe to Law for 9*d.*, and the remaining bushell and 3 pecks were steeped againe for fatting hoggs.

Of the 11th dressing, January 6th, by John Casbourne also, there was in all 5 quarters 3 bushells and 3 pecks, whereof 5 quarters were sold to M: Thurgood of Royston at 1*l.* 3*s.* 8*d.* per quarter, for 5*l.* 18*s.* 4*d.*, and 2 bushells and half were sold at home to Athan: Carrington, Casbourne, Thrift, and a bushell and a peck remaine. The bushell was sold to Amb: Bonnis, J: Payne, Mr. Bourne, for 3*s.*, and the peck was boyled for hoggs.

Notandum: That these 3 last dressings and 4 quarters of the 6th dressing, making in all 30 quarters 5 bushells and 3 pecks, came out of the first right hand mow in the new barne.

Of the 12 dressing, January 16, by J: Casbourne also, there was in all 9 quarters and 6 bushells, whereof 9 quarters and half were sold to Mr. Glenister at 24*s.* per quarter, for 11*l.* 12*s.* 0*d.*, a bushell sold to J: Casbourne for 3*s.*, and 3 pecks were boyled for fatting hoggs, and the remaining peck sold to Math: Payne for 9*d.*

Of the 13th dressing of barley, January 20, by William Thrift, there was in all 10 quarters and a peck, whereof the 10 quarters were sold to J: Roberts of Sandon and delivered at Royston at 23*s.* 8*d.* per quarter, for 11*l.* 16*s.* 6*d.*, and the remaining peck sold to Thrift at 9*d.* [*p.47*]

Of the 14th dressing, January 27, by Will: Thrift also, there was in all 10 quarters and 6 bushells, whereof one bushell sold to Ambrose Bonnis at 3*s.*, and 3 pecks to Thrift himself at 2*s.* 2*d.*, and a peck tayle; and a bushell sent to mill for the sow; the remaining 10 quarters and 3 bushells were sold to Anthony Phage at Baldock and delivered at Royston at 1*l.* 3*s.* 0*d.* per quarter, for 11*l.* 18*s.* 7*d.* *ob.*

Notandum: That these 2 dressings, making in all 20 quarters 6 bushells and a peck, came all out of ye middle mow on the left hand in the old barne.

Of the 15th dressing, January 31, by John Casb., there was in all 9 quarters and 6 bushells, whereof one bushell and half were fanned up and sold to Ambr: Bonnis at 4*s.* 6*d.*, and a bushell and half more were sold to W: Thrift and J: Casbourne for 4*s.* 4*d.* The remaining 9 quarters and 3 bushells were sold to the same Anthony Phage at 23*s.* per quarter, for 10*l.* 15*s.* 7*d.* *ob.*

Of the 16 dressing, February 2, by J: Casbourne, there was in all 6 quarters one bushell and a peck, whereof half a peck tayle. A peck and half he had himself for 1*s.* 0*d.*, and 3 pecks more to Math: Payne and Will Thrift, at 2*s.* 1*d.* The remaining 6 quarters were sold to Mr. Glenister of Royston at 1*l.* 3*s.* 2*d.* per quarter, for 6*l.* 19*s.* 0*d.*

Of the 17 dressing, February 10th, by William Thrift, there was in all 8 quarters and 3 bushells, whereof a bushell was sent to mill for the sow, 3 pecks sold to Mr. Bourn, Mat: Payne and J: Casb:, for 2*s.* 1*d.* *ob.*, and 2 bush[*p.48*]ells sold to Mr. Maulden and Ambr: Bonnis at 5*s.* 8*d.*; and 7 quarters were carryed in to Mr. Glenister at 22*s.* 8*d.* per quarter, for 7*l.* 18*s.* 8*d.*, and a peck more sold to Will: Whitby at 8*d.* *ob.*, and half a bushell to Will: Thrift at 1*s.* 5*d.*; and a bushell and half more sent to mill for the sow. The remaining 5 bushells were added to the next dressing and sold with it. [*p.48*]

Of the 18th dressing, February 17, by Will: Thrift, there was in all 7 quarters and 2 bushells, which (being made up 7 quarters and 7 bushells by the addition of 5 bushells of the former dressing) were all sold, together with another dressing, to one Peirson of Royston at 22*s*. 8*d*. per quarter, for 8*l*. 18*s*. 6*d*.

Notandum: That these 2 last dressings, making in all 19 quarters and five bushells, came out of the end mow in the old barne next the rick, haveing both the ilands lintells.

Of the 19th dressing, February 18, by J: Casbourne, there was in all 9 quarters and one bushell, whereof one bushell was sold to Math: P: and Amb: Bon: for 2*s*. 10*d*.; and 4 quarters were saved for seed; and 4 quarters and 6 bushells were sold with the former dressing (it being better corne and helping to put it of) to the same man at the same price, for 5*l*. 7*s*. 8*d*. 2 pecks sold to Rob: Ward, 1*s*. 5*d*., the remaining bushell and half to Amb: Bonnis and Will: Thrift for 4*s*. 3*d*.

Of the 20th dressing, February 24, by J: Casbourne, there was in all 5 quarters and 3 bushells, whereof one peck was sold to Math: Payne for 8*d*. *ob*., and all ye rest laide by for seed and sowne.

Of the 21 dressing, March 4rth, by J: Casbourne, there was in all 6 quarters and 5 bushells, whereof a bushell sent to mill for sow etc. 4 bushells and a peck more sold at home at 2*s*. 10*d*. per bushel, for 12*s*., to Widdow Fordham, Thrift, Whitby, M. P:, J: Casb. etc. 15 bushells added to the next dressing carryed in to Mr. Reynolds to be malted. The remaining 4 quarters and 3 pecks (together with 4 quarters of the 19th and 5 quarters 2 bushells and 3 pecks of the 20 dressing) were all sowne upon 36 stetches and 3 roods, and 12 acres and half in peices — in all upon 26 acres and 3 [*p.49*] roods in Barr-feild — and besides 4 acres [*p.49*] upon the Heath; so that there is sowne in all 30 acres and 3 roods (whereof a stetch is my owne land). And of seed there is sowne in all 13 quarters 3 bushells and half, which is one with another 3 bushells and half to an acre, and 2 bushells and half for ye odd 3 roods, which is large measure.

Of the 22nd dressing, March 16, by J: Casb:, there was in all 7 quarters 7 bushells and an half, whereof 2 pecks were sold to J: Casb: and W: Whitby for 1*s*. 5*d*. The remaining 7 quarters and 7 bushells (being made up 9 quarters 6 bushells by the addition of 15 bushells of the former dressing) were carryed in to in this March 16th anno *1685*, to Mr. Reynolds, to be maulted for my owne use.

Notandum: That these 4 last dressings, making in all 29 quarters and half a bushell, came all out of the end mow in the new barn, next the shed.

Of the 23rd dressing, March 28, by J: Casbourn, there was in all 10 quarters and 4 bushells, whereof 2 bushells were sold to Good: Thrift, Carrington, Bonnis, Widdow Dovy, for 5*s*. 8*d*., and the remaineing 10 quarters and 2 bushells were carryed in this day to Mr. Reynolds to be maulted for my owne use.

Of the 24th dressing, Aprill 4, there was in all ten quarters 2 bushells and a peck, whereof one bushell tayle, 2 bushells and a peck sold at home to Thrift himself, J: Whit:, Fren., Math: P:, Harvy, Bourne, for 6*s*. 4*d*. 4

bushells were changed for mault. The remaining 9 quarters and 3 bushells were sold to Mr. Glenister at 1*l.* 2*s.* 3*d.* a quarter, for 10*l.* 8*s.* 3*d.*

Of the 25th dressing, April 23, by Thrift, and the last of my barly, there was in all 15 quarters and 3 bushells and 3 pecks, whereof one bushell tayle put amongst the off corne, another stony reserved for the sow, 2 bushells 3 pecks sold at home for 7*s.* 4*d.*, and 11 quarters 2 bushells sold to Haydon at 1*l.* 1*s.* 6*d.* per quarter, for 12*l.* 1*s.* 10*d.*, and the remaining 3 quarters and 5 bushells laide in the garner for expence or sale, of which also [*p.50*; 'of... also' *repeated in ms.*] 11 bushells and half a peck were sold out at home at 8 groats per bushell, for 1*l.* 9*s.* 8*d.*, and half a bushell sent to mill for fowles by my wife. [*p.50*]

Notandum: That of the whole crop of barly this yeare there was in all 217 quarters and one peck.

An account of the oates brought in and threshed that yeare [*p.51*]

Of the first dressing, January 5, by W: Th:, there was in all 15 quarters, whereof about 5 quarters and a bushel were sowne upon 21 stetches and half the 2 acres against Lamps Acre and the 4 acres in Fawdon Hill and More Shott — in all upon 12 acres and 3 roodes. The remaining 9 quarters and 7 bushells were laide in the malthouse chamber for our owne use.

Of the 2nd dressing, February 28, by the saide W: Th:, there was in all 10 quarters and 4 bushells, which were all laide up for our owne expence in the malthouse chamber likewise.

Of the 3rd dressing, March 7, by Will: Thrift, there was in all eleven quarters and 5 bushells, of which 5 bushells were laide into the malthouse chamber and 4, being course, were put directly into binn in the stable, and also 6 bushells more of the best — in all 2 quarters wanting a bushell; and there remaines of the best 9 quarters and 6 bushells, whereof Mr. Reynolds had 6 quarters, and 3 quarters and 6 bushells were also sold at home to Thom: Godfrey, Widdow Wallis, Richard Wallis, Will: Symonds, etc., all at 16*s.* per quarter, for 7*l.* 16*s.* 0*d.*

Of the 4rth dressing, March 24, by Will Thrift also, there was in all 8 quarters 3 bushells and half, whereof againe 1 quarter 3 bushells and half were againe laide into the malthouse chamber; the 7 quarters remaining (of the best) were laide into garner, whereof 3 quarters were also sold at home at 16*s.* per quarter to Good: Sell, Goodman Prime of Triplow, and to Henry Wallis, etc., for 2*l.* 8*s.* 0*d.*, and the remaining 4 quarters were sold, 4 bushells to Mr. Reynolds and the rest to Richard Wallis at the same prise, for 3*l.* 4*s.* 0*d.*

Of the 5th dressing of oates, by Will: Thrift also, on May 2, there was in all 8 quarters and five bushells, whereof a bushell and half were pease given to fatting hoggs, and 2 bushells and half more were course and carryed directly into stable; and one quarter sold for 16*s.*, and half a bushell more were sold for 11*d.* The remaining 7 quarters and half a bushell were laide into malthouse chamber for our owne use. [*p.52*]

Of the 6th dressing of oates, by Will: Thrift also, May 9, there was in all five quarters and 4 bushells, whereof 2 bushells and half tayle carryed into stable, and a bushell and half course pease reserved for hoggs etc., and the remaining 5 quarters spent by our owne horses.

Notandum: Of the whole crop of oates this yeare there was 59 quarters 5 bushells and half, whereof there was sowne 5 quarters 1 bushel and half and were sold 17 quarters 6 bushells and half — in all 23 quarters. The remaining 36 quarters were all spent by our selvs.

[*p.53*] **An account of the pease brought in and threshed that yeare**

Of the first dressing, January 8, by Will Thrift, there was in all but one quarter 4 bushells and half, whereof at least a bushell of them were tayle and full of oates (though about a peck of them sifted out made horse meat) and about 11 bushells were sowne upon my owne 2 little acres and about half of the 2 acres of gleabe lying in Finchway Shott, and about a bushell left in the garner.

Of the 2nd dressing, March 9, there was in all 3 quarters and 4 bushells, whereof 4 bushells, being very full of oates, were put amongst the oates in the malthouse chamber. 4 bushells more being of a middle sort, I fatted hoggs with, and 2 quarters and 4 bushells were laide in the garner amongst the old ones out of which one quarter and one bushell were sold at home at 4*s.* per bushell, for 1*l.* 16*s.* 0*d.* The rest, being but 11 bushells, were spent and remaine to be spent.

Memorandum: That of the whole crop of pease this yeare there was but 5 quarters and half a bushell.

THE HARVEST ACCOUNTS

II 1685–1689

[County Record Office, Cambridge, 691/A1]

An account of the quarters of corne of all sorts, sold or spent, that there was of my tenth crop gathered in in harvest anno 1685, when harvest generally on Wednesday, July 22, was begun (though all but myself and the Lordship began 2 dayes before), and was ended on Bartholomew Day, all but an half acre of lintells of the farme, so that those who were shortest in harvest were a full month and 5 dayes. The weather was very wett all harvest, even from Swithin. There was not a loade of corne brought in in any tolerable case betwixt Saturday night, August 15, and Friday night, August 21. The North Feild this yeare was tilth, and the Barrfield the broak crop. The crop was very short; my own barnes held all the corne and the middlestead in the new barne was empty besides (that is, it had nothing in it but what the mowes would have held very well), and yet I had 2 acres of pease and 3 stetches of rye of my owne proper land besides the gleabe; many oates especially were brought in very badly. [691/A1, p.1]

[*In the manuscript there follows a contents list to the remainder of the book with page references*]

An account of the crop brought in in harvest that yeare, and first of wheate, white or Kentish [p.2]

Of the first dressing, on October 8, which was all white wheate, there was in all but 7 bushells and a peck, whereof one peck was given to R: Woollard, being poor and sick. 2 pecks more were spent by our selves. 3 bushells and half were sold at 3s. a bushel, for 10s. 6d., and the remaining 3 bushells were sold at 8 groates per bushell, for 8s. This was all the white wheate that lay in the malthouse.

Of the 2nd dressing, which was all red wheat, on October 16, there was but 5 bushells and a peck, of which 3 bushells and a peck were sold at 8 groats a bushell, for 8s. 8d., and the remaining 2 bushells were spent by our selves. This was all the red wheat that lay in the malthouse.

Of the 3rd dressing, December 15, by J. Casbourne, that was all red wheate; there was 5 quarters 4 bushells and a peck, of which 3 pecks were tayle sold to J: Casb: for 1s. 3 quarters and 5 bushells and a peck more were all sold at home at 3s. per bushell, for 4l. 7s. 9d. The remaining one quarter and 6 bushells were spent by our selves, so yt there is about a peck loss, though some of it be left.

Of the 4th dressing of red wheate, March 27, there was in all 2 quarters wanting half a bushell, whereof 2 pecks were tayle; and 2 pecks were given to Widdow Reynolds; and 6 bushells were spent by our selves; and the remaining 8 bushells and 2 pecks were sold at home, at 3s. per bushell, for 25s. 6d.

Of the 5th dressing, Aprill 19, which was all white wheate, there was in all 5 quarters and 5 bushells, whereof half a bushell was tayle and another bushell was but course, sold to J: Casbourne for 1s. 4d.; and one peck sold for 9d., and 4 quarters 6 bushells and 3 pecks sold at home at 8 groates per bushell, for 5l. 3s. 4d.; and 2 pecks given to Math: Payne. The remaining 4 bushells were spent by our selves; the last was sent to mill June 7.

Of the 6th dressing, on Aprill 26, which was all white wheate, there was in all 3 quarters and half, of which 2 pecks were tayle sold with the tayle of the [p.3] next dressing for 8d.; and sold at home at 8 groats per bushell one quarter one bushell a peck and half, for 1l. 5s., and spent by our selves one quarter 7 bushel 2 pecks and half. So there is 2 bushells and half either that I have no account of or that is loste with mice or divers times skreening, or the like.

Of the 7th dressing, of red wheate, May 6, there was in all 3 quarters and 5 bushells, which was this day laide into a chamber at Edward Godfreyes, whereof one bushell was tayle, for off corne; and a bushell and 3 pecks more were of a better sort of tayle, sold to taskers at 2s., and one bushell and 3 pecks of it were sold at 8 groats per bushel, for 4s. 8d., and one quarter and 5 bushells and 3 pecks more were sold at 3s. per bushel at home, for 2l. 1s. 3d.; and were spent by our selves one quarter 2 bushells and 3 pecks.

Of the 8th dressing, May 10th there was even 6 quarters, whereof one quarter and half was sold at home at 3s. per bushel, for 1l. 16s. 0d.; and 4 bushells and half more were sold at home also, at 3s. 8d. per bushel, for 16s. 6d., and 3 quarters 3 bushells and a peck were spent by our selves; so that there is 4 bushells and something more lost out of this and the former dressing, they being both together and being often skreened at first and being about 11 months in spending. It was all red wheate.

Notandum: Of the whole crop of wheate this yeare there was 27 quarters 6 bushells and one peck, whereof 6 bushells was tayle or at least very course.

An account of the rye or miscellaine brought in in harvest that yeare [p.4]

Of the first dressing, which was fanned up at severall times between September 2 and the 7th, there was in all 3 quarters and 5 bushells, which were all sowne as they were dressed up in the Rowes and Waterden Shott.

Of the 2nd dressing, which was fanned up between September the 8th and 14th, there was in all (with about a bushell of shalings fanned up in harvest) 5 quarters and half a bushell, which with the former dressing was all sowne upon 16 stetches and 3 acres in Waterden Shott – in all 9 acres – and upon 5 stetches in London Way Shott and 5 stetches and a rood in the Rowes and the 5 acres and acre – in all upon 21 acres wanting a rood, which is made up by the stony lands' ends of the acre and stetches that are to be sowne with wheat in Waterden Shott (which is about 3 bushells and a peck an acre and something above, one with another there being at least 10 acres of large measure) of which 21 acres 6 stetches are my owne and not gleabe.

Of the 3rd dressing, September 18, there was in all two quarters and a bushell, whereof one quarter and 6 bushells were sold at home att 7 groats a bushell, for 1*l*. 12*s*. 8*d*. The remaining 3 bushells were spent by selves.

Notandum: That these 3 dressings (making in all ten quarters 6 bushells and 2 pecks) came all out of the maulthouse.

Of the 4rth dressing, on December 3rd, there was in all 4 quarters 3 bushells and 3 pecks, whereof 3 quarters 7 bushells and 3 pecks were sold at home at 2*s*. per bushell, for 3*l*. 3*s*. 6*d*. The remaining 4 bushells were spent by ourselves; the last sent to mill on January 5.

On the 5th dressing, on January 16, by J: Casbourne, there was in all 5 quarters and 7 bushells, whereof one peck very neare was tayle, and another peck better tayle was given to horses; sold at home at 2*s*. per bushell 4 quarters and half a bushell, for 3*l*. 6*s*. 0*d*.; and 7 bushells and 3 pecks were spent by our selves (whereof about a bushell upon piggs), and the remaining 6 bushells and one peck were given to the poore.

Of the 6th dressing, on March 11, there was in all one quarter 2 bushells [p.5]
and half, whereof one quarter and a peck were sold at home at 2*s*. a bushell, for 16*s*. 6*d*., and the remaining 2 bushells and a peck were spent by our selves.

Of the 7th dressing, on April 8, there was in all 7 quarters and one bushell, of which half a bushell was tayle; and 3 quarters and 3 bushells and a peck were sold at home at 2*s*. per bushell, for 2*l*. 14*s*. 6*d*., and 2 quarters 7 bushells and 3 pecks more were sold at home also, at 1*s*. 10*d*. per bushell, for 2*l*. 3*s*. 6*d*.; and half a bushell was given to Will: Whitby, and 5 bushells remaining were spent by our selves; about half a peck above measure laide in the garner wasted.

Of the 8th dressing, May 13, there was in all even ten quarters, whereof 3 pecks were tayle; and sold at home at 1*s*. 10*d*. per bushell 2 quarters and 6 bushells, for 2*l*. 0*s*. 4*d*.; and 4 bushells more were sold at Royston at 19*d*. *ob*. per bushell, for 6*s*. 6*d*.; and 4 quarters 2 bushells and half more were sold at home at 1s. 8d. per bushell, for 2l. 17s. 6d.; and 3 bushells more sold at 2s. per bushell, for 6s.; and given to poor – Ralph Chilterne, Rob: Woollard, J: Watson – 2 bushells and half; and spent by our selves one quarter and 4 bushells. The remaining 5 pecks wasted by often skreening and ground downe by mice into the chaff house underneath.

Of the 9th dressing, May 22, there was in all 9 quarters and 4 bushells, whereof one peck was tayle; and one bushell and half was changed for new rye for seede, and one quarter and one bushell were spent by our selves; and 8 quarters and one bushell were all sold at home at 2s. per bushel, for 6l. 10s. 0d., and the remaining peck lost with skreening and mice.

Of the 10th dressing, June 3, there was in all ten quarters 4 bushells and half, whereof the half bushell was tayle; 5 quarters and 4 bushells were sold at home and to J: Thorne of Barly at 1s. 9d. per bushell, for 3l. 17s. 0d., and 3 quarters and 7 bushells were sold at home afterwards out of the garner at 2s. per bushell, for 3l. 2s. 0d.; and 2 bushells spent by ourselves; and 7 bushells remaining were lost by often skreening, it being also ground downe very much into the chaff house underneath by mice.

[p.6] Of the 11 dressing of rye, on June 11th, there was in all 5 quarters and three bushells, whereof there was taken out of it for tayle 3 bushells and a peck; and 4 bushells were sold at Royston at 1s. 9d. per bushel, for 7s., and 2 quarters and 3 bushells and 3 pecks were sold out of it at home, at 2s. per bushell, for 1l. 19s. 6d.; and 2 bushells more were spent by ourselves; and 5 bushells and half were sowne, because we wanted seed of the new rye, and 5 bushells more were put to the 12 dressing to make it up even 8 quarters, and so laide into a chamber att W: Thrift's, so that there is lost out of this dressing still 3 bushells and half; and the 5 bushells that were put to the next dressing were also all lost, for it lying in his chamber a year round there was no more taken out than there was laid in without that 5 bushells, so that there is lost out of this dressing in effect one quarter and half a bushell.

Of the 12 dressing, on June 22, there was in all 7 quarters and 5 bushells, whereof half a bushell was tayle; and 2 quarters and 3 bushells were spent by our selves; and 5 quarters and one bushell and half more were sold at home at 7 groates (that is 2 quarters and 7 bushells), for 2l. 13s. 8d., and the remaining 2 quarters 2 bushells and half at 2s. per quarter, for 1l. 17s. 0d. So that there is not so much lost out of the last dressing as was supposed by half a bushell.

Notandum: Of the whole cropp of rye this yeare there was in all 72 quarters 5 bushells and a peck, of which 8 quarters 5 bushells and half were sowne and 16 or 20 quarters of it being sold at 20d. and 21d. and 1s. 10d. per bushell, did not arise to much; and above 2 quarters of it were clearly lost by mice and often skreening and such like.

[p.7] **An account of the barley brought in in harvest that yeare**

Of the first dressing of barly, by J: Casbourne on November 3rd, there was in all 5 quarters and 4 bushells, whereof one bushell was tayle; and the five quarters and 3 bushells, being very course, were sold to Mr. Glenister of Royston at 16s. 4d. per quarter, for 4l. 7s. 9d.

Of the 2nd dressing, by J: Casbourne on November 17, there was in all 6 quarters, of which one bushell was tayle. 5 quarters and 4 bushells were sold to Mr. Rowly at Chiswick Hall at 17s. 6d. per quarter, for 4l. 16s. 3d.; and the 3 bushells remaining were put to the next dressing (making that 6 quarters and 4 bushells of the best, besides a bushell tayle).

Of the 3rd dressing, November 24, by J: Casb: also, there was in all 6 quarters and 2 bushells (to which were added 3 bushells of the former dressing, making in all 6 quarters and 5 bushells), of which one bushell was tayle. 6 quarters and 2 bushells were sold to a widdow at Lidlington at 18s. 2d. a quarter, for 5l. 13s. 6d.; one bushell sold at home at 2s. 3d. The remaining bushell was put to the next dressing.

Of the 4th dressing, by William Thrift on December 2nd, there was in all just five quarters, whereof one bushell was tayle; the other (being made up even 5 quarters againe by the addition of an odd bushell of the former dressing) were sold to the same person at Lidlington at 18s. 6d. per quarter, for 4l. 12s. 6d.

Of the 5th dressing, by William Thrift on December 9th, there was in all 6 quarters and one bushell, whereof one bushell was tayle, a peck sold at home for 7d., 2 pecks sent to mill and 5 quarters sold to M: Thurgood at Royston at 18s. 8d. per quarter, for 4l. 13s. 4d., 7 bushells and a peck remaining. The 7 bushells were sold with 4 quarters and a bushell of the next dressing and the peck was sold at home for 7d.

Of the 6th dressing, by Will: Thrift also, on December 17, there was in all [p.8] 5 quarters and 2 bushells, whereof one bushell was tayle, and 4 quarters one bushell (being made up even 5 quarters by the addition of 7 bushells of the former dressing) were sold at 19s. per quarter to one Goodman Ilet of Barkway, for 4l. 15s. 0d., and the remaining quarter was added to the next dressing.

Of the 7th dressing, on December 23, by Will: Thrift also, there was in all 9 quarters and 3 bushells, whereof one bushell was tayle, and the 9 quarters and 2 bushells (together with a quarter of the former dressing, making up ten quarters and 2 bushells) were all sent in to Good: Kefford's on January 2 to be malted for my own use.

Notandum: That the 4, 5, 6 and this 7 dressings, making in all 25 quarters 6 bushells, came all out of the first left hand mow in the old barne

Of the 8th dressing, on December 31, there was in all 5 quarters and 3 bushells, whereof one bushell was tayle; the 5 quarters and 2 bushells sold to M: Thurgood at 19s. per quarter, for 4l 19s. 8d. It was threshed by J: Casbourne.

Of the 9th dressing, by J: Casbourne also, January 7, there was in all 8 quarters and 2 bushells, of which one bushell was tayle; and the remaining 8 quarters and one bushell was sent in to Good: Kefford's to be malted on January 8 by Mat: P: and Brook.

Of the 10th dressing, on January 12 by Will: Thrift, there was in all 8 quarters and a peck, of which a bushell was tayle, a peck sold to Tho:

Whitby for 7*d*.; 7 quarters and 7 bushells remaining were sent in on January 14 to Good: Kefford to be maulted for our own use.

Of the 11th dressing, on January 20 by Will: Thrift, there was in all 6 quarters and 2 bushells, whereof one bushell was tayle, one bushell sold at home for 2*s*. 4*d*., 5 quarters sold to M: Thurgood at 19*s*. 2*d*. per quarter, for 4*l*. 15*s*. 10*d*. The remaining quarter sent in with the next dressing to be malted at Good: Kefford's.

[*p.9*] Of the 12th dressing, on January 22, by J. Casbourne, there was in all 4 quarters and 2 bushells, whereof one bushell was tayle, and the remaining 4 quarters and one bushell (together with a quarter of the former dressing making in all 5 quarters and a bushell) were carryed in on Saturday January 23 to Good: Kefford to be maulted.

Of the 13 dressing, on January 29, by Will Thrift, there was in all 6 quarters and five bushells, of which a bushell and above half was tayle, and half a bushell sold at home to J: Casbourne, at 1*s*. 1*d*. *ob*. 5 quarters sold to M. Thurgood at 18*s*. 9*d*. per quarter, for 4*l* 13*s*. 9*d*. The remaining quarter and 3 bushells were sold to the same person, for 1*l*. 2*s*. 6*d*.

Notandum: That the 10, 11, and this 13 dressings, making in all 20 quarters 7 bushells and a peck, came out of the 2nd mow on the left hand in the old barne.

Of the 14th dressing, on February 2, by J. Casbourne, there was in all 5 quarters and 6 bushells, whereof a bushell and half tayle, half a bushell fanned up for himself, 1*s*. 2*d*.; and 3 quarters and 6 bushells were sold to M: Thurgood at 18*s*. per quarter, for 3*l*. 7*s*. 6*d*.; one quarter and 6 bushells remaining were added to the next dressing and sold with it.

Of the 15 dressing, on February 12, by J. Casbourne also, there were in all 5 quarters 5 bushells and 3 pecks, whereof 3 pecks were sold to himself and Mat: Payne, for 1*s*. 8*d*., and a bushell was tayle; and a bushell sent to mill for piggs and fowles etc.; 3 quarters and 4 bushells (together with one quarter and 2 bushells of the former dressing, making in all 5 quarters and 6 bushells) were sold to M: Thurgood at 17*s*. 9*d*. per quarter, for 4*l*. 13*s*. 2*d*.; and the remaining one quarter 7 bushells, being made up 5 quarters out of the next dressing, were changed for seed with Mr. Reynolds.

[*p.10*] Of the 16th dressing, on February 18, by J. Casbourne, there was in all 4 quarters and half, [*p.10*] of which 3 quarters and one bushell were changed for seed, as before, and 3 pecks were tayle; and 2 pecks sold at home for 1*s*. 1*d*.; and the remaining 9 bushells and 3 pecks were sowne.

Of the 17 dressing, on March 2, by John Casbourne, there was in all 7 quarters and half a bushell, whereof one bushell and half was tayle; and 6 quarters and 2 bushells were sold to one of Ashwell at 17*s*. 10*d*. per quarter, for 5*l*. 11*s*. 5*d*., and 5 bushells remaining, whereof 5 pecks sold at home at 2*s*. 3*d*. per bushell, for 2*s*. 9*d*., and 3 pecks spent upon piggs, and 3 bushells added to the next dressing.

Of the 18 dressing, March 9, by J: Casbourne also, there was in all 4 quarters and 2 bushells, whereof one bushell was tayle, and the remaining 4 quarters one bushell (being made up 4 quarters 4 bushells by the addition of 3 bushells of the former dressing) were carry'd in on March 10th to

Good. Kefford to be maulted by Henry Law and Wil: Brook, being Wednesday.

Of the 19 dressing, on March 17, by John Casbourne, there was in all 3 quarters one bushell and half, whereof 2 pecks were sent to mill, and 2 quarters were sowne (together with 5 quarters changed with Mr. Reynolds and 5 quarters bought at Cambridge, in all 12 quarters, which were sowne upon 10 acres in Southfeild, and all North Feild generally, and 2 stetches of my owne, in all on 30 acres and something above half, which is something above 3 bushells an acre one with another) and one bushell remainder sent to mill afterwards for the sow.

Of the 20th dressing, by J: Casbourne, on March 24th, there was in all 5 quarters and five bushells, whereof one bushell was tayle. 3 pecks were sold at home for 1s. 9d., 5 quarters and 3 bushells were carryed in on Thursday March 25 to Good: Kefford to be maulted, the remaining peck made a mash for the horses, being blooded.

Of the 21 dressing, April 6, by William Thrift, there was in all 6 quarters and 2 bushells, of [p.11] which half a bushell was tayle laid into garner with the off corne, 2 bushells were course, that were saved for the sow, and three bushells more were spent for fatting hoggs; half a bushell sold at home for 1s. 1d., and 5 quarters and half were sold to Nut[hamp]st[ea]d at 16s. 8d., for 4l. 11s. 8d. [p.11]

Of the 22nd dressing, Aprill 13, there was in all 5 quarters and 7 bushells, whereof one bushell was tayle fit for nothing but of corne, and one other bushell was sold at home, at 2s. 2d., and ye remaining 5 quarters and 5 bushells sold to one of Ashwell at 16s. per quarter, for 4l. 10s. 0d. This was dressed by Will: Thrift also.

Of the 23 dressing, Aprill 22, by William Thrift also, there was in all 8 quarters and 4 bushells, of which there was one bushell tayle, one bushell sent to mill for sow and piggs; 3 bushells were sold at home for 5s. 10d. to Good. Payne, Harvy, Thrift, and 5 quarters and 5 bushells were sold to J: Thak of Nut[hamp]stead at 15s. 6d. per quarter, for 4l. 7s. 2d., and the remaining 2 quarters and 2 bushells were laid into garner.

Notandum: That these 3 last dressings, making in all 20 quarters and 5 bushells, came all out of the 2nd mow in the old barne next the orchard.

Notandum: There was of the whole crop of barly this yeare even 138 quarters.

An account of the pease that were brought in that harvest [p.12]

Of the first dressing of pease, threshed by Mathew Payne on afternoon etc., and dressed up on October 22nd, there was but 4 bushells and half and something better, which were all Mr. Reynolds his tithe and were much in the wett and were very course; and comeing last in, were laid in the middle steade of the new barne where there was nothing else that yeare. They were used for fatting hoggs.

Of the 2nd dressing, November 20th, by W: Th., there was in all 2 quarters of the best, laide into garner for our owne use; and a bushell and half tayle, most of them oates which the horse keeper carryed directly into stable. These lay in the middle stead in the old barne besides 19 quarters of oates of the first 3 dressings.

Notandum: Of the whole crop of pease this yeare there was but 2 quarters and 6 bushells and yet I had 3 acres sowne of my owne.

[p.13] **An account of the oates that were brought in that yeare**

Of the first dressing of oates, by William Thrift on October 23, there was in all 4 quarters and 3 bushells (which I was forced to thresh sooner than ordinary for my horses); of these, 5 bushells were carried directly into stable, the rest laide into garner for our owne expence.

Of the 2nd dressing, by William Thrift on November 2nd, there was in all 6 quarters and 6 bushells and about a half a bushell tayle, all laid into garner for our owne use; but 2 bushells and half were sold at 1s. 4d. per bushell, for 3s. 4d., and an other bushel for 1s. 4d.

Of the 3rd dressing, November 24, by Will: Th:, there was in all 7 quarters and 7 bushells, of which the 7 bushells, being courser than the rest, were carried directly into stable. The 7 quarters were laide into garner, and sold out of them a peck for 4d. and another peck for 4d. and a bushell for 1s. 4d. and another bushell and half for 2s.

Of the 4th dressing, February 3, by Will: Thrift, there was in all 5 quarters 2 bushells and half, of which there was a coomb very course carryed directly into stable. The remaining 4 quarters 6 bushells and half were laid into garner (this is ye first dressing that came off of the mow next the pond).

Of the 5th dressing, by Will: Thrift on February 19, there was in all 8 quarters and 2 bushells, of which 4 bushells were tayle carried directly into stable. 7 quarters and 6 bushells were laide into garner, but out of them were sold to Mr. Reynolds 4 bushells, for 5s. 4d., and 2 pecks more to Good: Payne and Thrift, and the rest remaine for our owne use.

[p.14] Of the 6th dressing, by Will: Thrift also, on March 4, there was in all 9 quarters and 6 bushells, whereof the 6 bushells were tayle, yet eaten by horses, and 6 bushells more were carryed directly into stable. The remayning 8 quarters and 2 bushells were laid into garner amongst the rest.

Of the 7th dressing there was in all 9 quarters and 6 bushells, March 19 by Wil: Thrift, whereof 7 bushells were course yet eaten by horses, and 8 quarters and 7 bushells were laid into malthouse chamber for our owne use.

Of the 8th dressing, by William Thrift also, on March 29, there was in all 8 quarters and 3 bushells, whereof there was one quarter very course that were carryed directly into stable, and 7 quarters and 3 bushells laid into malthouse chamber to the last dressing.

Memorandum: That of the whole crop of oates this yeare there was in all 60 quarters and 4 bushells.

An account of all the corne that there was of my eleventh cropp that [*p.15*] was gathered in in harvest anno 1686, which harvest began generally on Munday July 19th (though I began my self but on July 20th and yet sent away most of my men (tithe men and those that work all the yeare excepted) 3 dayes before others made an ende generally) and was ended wholly on Saturday, August 14th (all but a little rye carryed to Foxton by Good. Eversden). So that this harvest lasted a compleate month. The weather was good, all but 2 dayes in the third week and about 3 more in the last weeke, or else all would have finished three dayes sooner. The North Feild this yeare was brock and the South or Heath Feild was the tilth, and was not so good for the part as the North Feild was. Rye especially was full of weedes and had a short dwindling eare. It was a cold Spring and a dripping Summer, and the crop was sowne very early in the Spring, I suppose somewhat at the soonest for that Heath Field

An account of the wheate, white or red, that was brought in in that harvest, viz 1686 [*p.16*]

Of the first dressing, September 2, which was all white wheate, there was very neare ten bushells, of which about 2 bushells were mingled with the seed rye, and the other 8 bushells were sowne all upon the 2 acres on Fawdon Hill, and 4 little stetches next it on this side and downe to the Moore, which is scarcely 3 bushells to an acre.

Of the 2nd dressing, September 20th, which was all red wheate, there was in all 2 quarters 1 bushell and half, whereof 10 bushells were sowne upon 3 acres and half downe to the Moore and upon Fawdon Hill, and mingled with rye, and 4 bushells and half more were sold at 3*s.* per bushell, for 13*s.* 6*d.*, and the remaining 3 bushells were spent by our selves. These were all in the malthouse.

Of the 3rd dressing, on Aprill 5, there was in all 5 bushells of red wheate, of which there was sold at home 3 bushels and half at 3*s.* 6*d.* per bushel, for 12*s.* 3*d.*, and the remaining bushell and half was spent by our selves.

Of the 4rth dressing, on May 6, which was al redd wheate also, there was in all 2 quarters and half a bushel, of which 2 pecks were tayle yet sold to J: Casbourn for 1*s.* 0*d.*, and one quarter 2 bushells and a peck more were sold at home at 3*s.* 6*d.* per bushel, for 1*l.* 15*s.* 10½*d.*; and the remaineing 5 bushells and 3 pecks were spent by our selves.

Of the 5th dressing, of white wheate, on Saturday May 14th, there was in all 4 quarters 7 bushells and a peck, of which 5 pecks were tayle, sold nevertheless for 2*s.*; and one bushel and one peck were sold at 3*s.* 6*d.* per bushell, for 4*s.* 4*d. ob.*, and 7 bushells and 3 pecks more were all sold at home at 11 groats per bushell, for 1*l.* 8*s.* 5*d.*, and one quarter and one bushel more were all sold at home at 10 groats per bushel, for 1*l.* 10*s.* 0*d.*, and the remaining 2 quarters and 4 bushells were all spent by our selvs.

Of the 6th and last dressing, which was all redd wheate, on Tuesday May [*p.17*] 24, there was in all but 4 quarters 6 bushells and half, of which 2 bushels were tayle, sold nevertheless for 4*s.*; and one bushell was sold at 3*s.* 8*d.*, and six bushells and a peck more were sold at 3*s.* 6*d.* per bushell, for 1*l.* 1*s.* 10*d. ob.*; sold at 10 groates 3 pecks more, for but 2*s.* 6*d.*, sold at 3*s.* one bushell and a peck more, for 3*s.* 9*d.*, and sold at 8 groates at home, as all the rest were, 6 bushells and 3 pecks more, for 18*s.* 0*d.*; and 2 quarters and 4 bushells spent by our selvs; and the remaining half bushel lost by skreening, mice, etc.

Notandum: Of the whole cropp of wheate this yeare there was but 15 quarters 6 bushells and 3 pecks, of which 4 bushells wanting a peck were tayle, and 2 quarters 4 bushells were sowne, so that there was 12 quarters and 7 bushels sold and spent.

[*p.18*] **An account of the rye or miscellaine brought in in that harvest, viz 1686**

Of the first dressing of rye, September 9th (with a bushell of the shalings that were gathered up in harvest), there was in all 3 quarters and a bushell, which were all sowne upon 3 stetches in Branditch Shott and upon 8 stetches in Short Shotte and the 3 half acres and 3 stetches on the brow of Fawdon Hill, on the further side of the 12 acres – in all upon 6 acres and 3 roods – and upon some other parcells which are to be mentioned in the next dressing.

Of the 2nd dressing, September 14, there was in all 2 quarters 7 bushells and a peck, which was all sowne upon the 3 2-furlong stetches and 2 3-furlong stetches at the towne side and the 3 roods and a single stetch in Finchway, and the 2 White Acres and Lamps Stetch and Acre, and 3 stetches betwixt that and the 12 acres, in all (with the 6 acres before) upon 17 acres wanting half a rood of the gleabe and 2 acres of my owne – in the whole 19 acres wanting half a rood, so that besides all my seed in the malthouse there was 6 bushells and half sowne of old rye. Only a bushell and half was changed for new, that is above 3 bushells per acre one with another (reckoning the wheat put to it).

Notandum: That these 2 dressings, making in all but 6 quarters and a peck, were all that I had out of the malthouse.

Of the 3rd dressing, November 20, by J. Casbourne, there was in all 4 quarters and 5 bushells, whereof 6 bushells and 3 pecks were given to the poore and 3 quarters wanting a peck were sold all at home at 2*s.* per bushell, for 2*l.* 7*s.* 6*d.*; and 6 bushells were spent by ourselves. The remaining half bushell was lost by skreening and mice. This lay in the middlestead and was much of it mowne.

Of the 4rth dressing, March 11, by J. Casbourne, there was in all 6 quarters 3 bushells and half, whereof one peck tayle laide behind the garner door and 5 quarters sold to Caucutt at 2*s.* a bushel, for 4*l.* 1*s.* 0*d.*; and one quarter one bushell and a peck more sold at home at 2*s.* a bushell, for 18*s.* 6*d.*, and the remaining 2 bushells were spent by our selves.

[*p.19*] Of the 5th dressing of rye, on Thursday March 31, there was in all 5 quarters and half a bushell, of which 2 pecks were tayle; and a bushel was given to Widdow Bayes and Widdow Woollard, and 4 quarters and one bushel were sold at home at 2s. a bushel, for 3l. 6s. 0d., and the remaining 6 bushells were spent by our selvs.

Of the 6 dressing, on April 19th, Tuesday, there was in all 6 quarters 5 bushells and half, of which a peck was tayle; and 4 bushells were sold at Royston for 8s. 6d., and 5 quarters more were sold to the miller of Shepereth at 2s. 1d. per bushel, for 4l. 3s. 4d., and 4 bushells sold at Royston againe for 8s.; and 5 bushells sold next day at Royston againe for 10s., and the remaining peck sold at home for 6d.

Of the 7th dressing, on Tuesday May 3, there was in all 6 quarters and 2 bushells, whereof one peck was tayle; and one quarter and five bushells were sold at Royston againe, at 2s. per bushel, for 1l. 6s. 0d., and 4 quarters and half were sold to one of Bourne at the same price, for 3l. 12s. 0d., and the remaining 3 pecks were sold at home, to Mr. Bourne and Good: Brock for 1s. 6d.

Of the 8th dressing, on Tuesday May 10th, there was in all 5 quarters 6 bushells and 3 pecks, of which one peck was tayle; and 3 quarters and 2 bushells were sold at Royston at 2s. per bushel, for 2l. 12s. 0d., and 2 bushels spent by our selvs; and the remaining 2 quarters 2 bushells and half were sold at home at 2s. per bushell, for 1l. 17s. 0d.

Of the 9th dressing, on Thursday June 2, there was in all but 4 quarters and 7 bushells, of which (it being the last and much eaten) there was a bushell tayle; and the remaining 4 quarters and 6 bushells were sold to one of Ashwell at 2s. 2d. ob. per bushell, for 4l. 3s. 11d.

Notandum: Of the whole crop of rye this yeare there was but 45 quarters 6 bushells and an half, whereof 6 quarters and 7 bushells were sowne.

An account of the barley brought in in harvest 1686 and threshed [p.21]

Of the first dressing of barly, by William Thrift, October 26, there was in all 8 quarters, which were carryed in partly on October 28 to William Thorowgood, and sold to him at 18s. 8d. per quarter – that is 5 quarters of it.

Of the 2nd dressing, October 29, by Will: Thrift also, there was but one quarter and 4 bushells, which was added to 3 quarters of the former dressing and sold to the same person at the same price. Both the dressings together make 9 quarters and half, which were sold for 8l. 17s. 4d.

Of the 3rd, November 4th, by Will: Thrift, there was in all 5 quarters and 7 bushells, which were sold to Good: Bowes at 18s. 8d. per quarter, for 5l. 9s. 8d.

Notandum: That these 3 dressings (making in all 15 quarters and 3 bushells) came out of the middle stead in the old barne.

Of the 4rth dressing, November 6, by J. Casburn, there was but one quarter and a bushell, being nothing but the cutting of the mow side in that barne, and it was sold with the former dressing at the same price, for 1l. 1s. 0d.

Of the 5th dressing, November 16, by William Thrift, there was in all 7 quarters and 2 bushells, whereof one bushel was tayle, and one bushell sold at home, at 2s. 4d., and 5 quarters and 5 bushells sold to Mr. Glenister at 19s. per quarter, for 5l. 6s. 10d.; and the remaining quarter and 3 bushells were sold to Mr. Fliton of Lidlington at 18s. 10d. per quarter, for 1l. 5s. 10d.

Of the 6th dressing, November 27, by Will: Thrift, there was in all 6 quarters and 7 bushells, whereof 2 pecks were tayle; and 3 quarters were sold to Mr. Fliton of Lidlington at 18s. 10d. per quarter, for 2l. 16s. 6d. The remaining 3 quarters 6 bushells and half were sold to J: N: at Shaftno End at 19s. per quarter, for 3l. 13s. 7d.

[p.22] Of the 7th dressing of barly, on November 30th, by John Casbourne, there was in all 4 quarters and six bushells, whereof half a bushel was tayle, and the remaining 4 quarters five bushells and half (because they made up the three quarters six bushells and half of the former dressing even measure, that is just 8 quarters and half) were sold together with that to John Norris at Shafnoe End at 19s. per quarter, for 4l. 7s. 11d.

Of the 8th dressing, by J: Casbourne, on December 9th, there was in all 5 quarters and 6 bushells, of which there was one bushell tayle laide into garner, and the remaining 5 quarters and 5 bushells were sold to Urias Bowes at 18s. 8d. per quarter, for 5l. 5s. 0d.

Of the 9th dressing, on December 20, by J. Casbourn, there was in all 5 quarters 5 bushells and 3 pecks, whereof the 3 pecks were tayle laide into garner; and one bushell sent to mill for the boare; and the remaining 5 quarters and half were sold to Steven Hawker of Chishill at 19s. per quarter, for 5l. 4s. 6d.

Of the 10th dressing, on December 30th, by John Casbourne also, there was in all 6 quarters, whereof a bushell was tayle laide into garner, 5 quarters and 6 bushells were sold to M: Thurgood at Royston at 19s. 2d. per quarter, for 5l. 12s. 3d., and the remaining bushell was sold at home for 2s. 4d.

Of the 11 dressing, on January 7, by John Casbourne also, there was in all 4 quarters and 7 bushells, of which one bushell was tayle laide into garner. The remaineing 4 quarters and 6 bushells were sold to M: Thurgood at 19s. per quarter, for 4l. 10s. 3d.

Notandum: That the 4th and these 5 last dressings, making in all 28 quarters 1 bushell and 3 pecks, came all out of the first mow on the right hand in the new barne, excepting about a quarter or there abouts of from the end mow.

[p.23] Of the 12 dressing of barly, January 14, by J: Casbourne, there was in all 5 quarters and 6 bushells, whereof a bushell was tayle laide into garner. The remaining 5 quarters and 5 bushells were carryed into M: Thurgood at Royston at 19s. per quarter, for 5l. 6s. 10d.

Of the 13 dressing, January 21, by J: Casbourne also, there was in all 5 quarters and three bushells, of which one bushell was tayle; and 5 quarters and one bushell was sold to M: Thurgood at Royston at 19s. 4d. per quarter, for 4l. 19s. 1d., and the remaining bushell was sold at home at 2s. 4d.

Of the 14 dressing, by J: Casbourne also, February 1, there was in all 7 quarters 6 bushells and half, of which 2 bushells and above half were tayle;

and 7 quarters 4 bushells wanting a quarter were reserved for seede in the garner, and all sowne with a peck more.

Of the 15th dressing, by J. Casbourne also, on February 7, there was in all 6 quarters and one bushell and 3 pecks, of which a bushell and half was tayle, and the remaining 6 quarters and a peck were all laide into garner likewise and reserved for seed, and all sowne, together with 4 quarters 4 bushells bought besides.

Notandum: That these 4 last dressings, making in all 25 quarters and a bushell and one peck (together with about a quarter before of the 11th dressing as was guessed) were all that came out of the end mow in the new barne next the shedd, and at that rate there was but 26 quarters one bushell and a peck in all that mow.

Of the 16 dressing, by Will: Thrift, on February 8th, there was in all 7 quarters and 2 bushells, whereof one bushell was tayle. 5 quarters and 5 bushells were sold to one Godfry at Ashwell, at 2s. 6d. per bushel, for 5l. 12s. 6d., and the remai[ning] 12 bushells were sold at 19s. 4d. to J: Brand of Royston with the next dressing, for 1l. 9s. 0d.

Of the 17th dressing of barly, February 12, by Will: Thrift, there was in all [p.24] 4 quarters and 6 bushells, whereof one bushell againe was tayle, laide behind the garner door; and the remaining 4 quarters and 5 bushells were sold with part of the former dressing to J: Brand of Royston at 19s. 4d. per quarter, for 4l. 9s. 5d.

Of the 18th dressing, February 19, by Will: Thrift also, there was 5 quarters and one bushell, whereof the odd bushell was tayle laide in the garner. The 5 quarters remaining were sold to my neighbour Bowes at 19s. per quarter abateing one penny, for 4l. 14s. 7d.

Notandum: That these 3 last dressings, making in all 17 quarters and a bushell, came out of the first mow on the left hand in the old barne.

Of the 19 dressing, February 26, by William Thrift, there was in all 5 quarters 4 bushells and half, of which the half bushel was tayle; and the 5 quarters and 4 bushells were carryed in on Monday March 7 to Good. Sell's at Triplow to be malted for our own use.

Of the 20 dressing, March 7, by William Thrift also, there was in all 6 quarters 2 bushells and 3 pecks, whereof one bushell was tayle; and 5 quarters and 4 bushells more were carryed in on Saturday March 12 to Richard Sell's of Triplow to be malted for my owne use; and 5 bushells and 3 pecks remaine, which were added to the next dressing and sold with it.

Of the 21st dressing, March 14, by Will: Thrift, there was in all 7 quarters one bushell and a peck, of which one bushell was tayle. 4 quarters 4 bushells and a peck (together with 5 bushells and 3 pecks of the former dressing, making up together 5 quarters 2 bushells) were sold to Mr. Glenister at 1l. 0s. 2d. per quarter, for 5l. 5s. 10d. The 2 quarters 4 bushells remaining were sold to Esqr. Cook at 19s. 6d. a quarter, for 2l. 8s. 9d.

Of the 22nd dressing, on March 22, by W: Thr., there was in all 6 quarters and 2 bushells, whereof one peck made an end of sowing, 5 pecks tayle laid into garner, 6 quarters were sold to J: Norris, for 5l. 17s. 6d., and the half [p.25] bushell remaining laide into garner for our owne use.

Of the 23rd dressing, March 28, 1687, by Will: Thrift also, there was in all 5 quarters and 3 bushells, whereof a bushell tayle was laide behind the garner door; 4 quarters and half more were sold to Esqr. Cook at Upper Chishill at 19*s*. 6*d*. a quarter (together with 20 bushells of the 21 dressing, making in all even 7 quarters), for 4*l*. 7*s*. 9*d*.,one bushell and a peck more sold at home at 3*s*. 1*d. ob*. to Mr. Killenb: and J: Gurner, and the remaining 4 bushells and 3 pecks were laide into garner. So that with the half bushel of the former dressing there is saved in the garner 5 bushels and a peck in all; this peck was sold afterwards for 7*d*.

Notandum: That of the whole cropp of barly this yeare there was but 130 quarters 5 bushels and an half, whereof there was sowne (besides 4 quarters and half bought at Cambridge) 13 quarters 4 bushells and an half; and malted for our owne use 11 quarters, and spent one bushell, and saved for expence 5 bushels. There was tayle 2 quarters 2 bushells and half besides, all which together make 27 quarters and 5 bushells, to which, if the 4 quarters and half that I bought for seed be added, the whole ariseth to 32 quarters and one bushel; so that I sold clearly that yeare but 98 quarters 4 bushells and half.

[p.26] **An account of the pease that I had of my crop in that yeare, being all tithe, for I had none sowne of my owne**

Of the first dressing of pease, November 2, by John Casbourne (because they lay in the middle steade of the new barne) there was in all 2 quarters and 3 bushells, whereof a peck was tayle, and given unto hoggs at the first shutting up; and 7 bushells and a peck at divers times were sold to J: Casbourne and Math: Payne at 2*s*. 2*d*. per bushell, for 15*s*. 8*d. ob*, and the remaining 11 bushells and half were laide into garner and spent by our selves.

Of the 2nd dressing, November 30, by Wil: Thrift, there was in all but one quarter, whereof one bushell was course, oates and pease together, carryed in to stable; 3 bushells sold to Wil: Thrift at 2*s*. 2*d*. per bushell, for 6*s*. 6*d*.: remaine 4 bushells in the garner, whereof one more was sold to him againe and another to Anthony Parish, for 4*s*. 4*d*., and the rest spent.

Of the 3rd dressing, December 6, by Wil: Thrift, there was but 6 bushells that lay by themselves amongst the oates in the barne and therefore a bushell of them were reckoned tayle, being partly oates; and 2 bushells he had himself at 4*s*. 4*d*. and the other 3 bushells were spent by our selves.

Of the 4rth dressing, January 13, by W: Thrift, there was in all 5 quarters and about 3 pecks, whereof 1 quarter and 5 bushells and 3 pecks were sowne upon the 3 half acres and a stetch beyond it and the 2 acres in Horsted Shot – in all upon 4 acres wanting half a rood – and 4 bushells tayle mingled with the oates. The remaining 2 quarters and 7 bushells were laide in the garner for our owne use.

Of the 5th dressing, by Will: Thrift, on Thursday April 28, there was in all 5 quarters and half, of which one quarter was tayle, that is as many oates as pease, and the remaining 4 quarters and half were all laide into garner for our use.

Notandum: That of the whole cropp of pease this yeare there was 14 quarters 5 bushells and 3 pecks, and yet I had none sowne – they were all tithe. [p.27]

An account of all the oates I had of my cropp in that yeare [p.28]

Of the first dressing of oates, December 7, by W: Thrift, there was in all 2 quarters and 7 bushells, whereof one bushell was given to Edward Godfrey (together with a sack of chaff) for killing and selling my bull etc., and the remaining 2 quarters 6 bushells were laide into malthouse chamber and mingled with the old ones, and spent.

Of the 2nd dressing, December 14, by Wil: Thrift, which were all white oates, there was but one quarter and 7 bushells, whereof a peck was sold at 4*d.*, 2 bushells that were most tayle carryed into stable. The rest, though carryed into malthouse chamber, yet being course were not mingled with the rest but lye to be spent by themselves.

Of the 3rd dressing, January 4, by Wil: Thrift, there was in all 7 quarters and 6 bushells, whereof 2 bushells and half were tayle carryed directly into the stable. Half a bushell more were most pease, but, having many stones amongst them, were given to fatting hoggs at theire first shutting up, and 2 pecks were sold for 8*d.* The remaining 7 quarters 2 bushells and half were laide into malthouse chamber for expence.

Of the 4rth dressing, January 24, by Will Thrift, there was in all 9 quarters, whereof 2 pecks were sold to taskers for 8*d.*, and 2 quarters 5 bushells and half were sowne upon about 7 acres in the South Field, upon all the stetches that lye outward in Royston Roade and some of the stony lands end, and some of them, and the remaining 6 quarters and 2 bushells were laide into malthouse chamber for expence.

Of the 5th dressing, on Aprill 4rth, there was 7 quarters and one bushell, of which 5 pecks sold to taskers and others for 1*s.* 8*d.* 7 quarters wanting a peck laide into malthouse for our owne use.

Of the sixth dressing of oates, there was in all 7 quarters 7 bushells and an half, which were dressed on Saturday April 16. Of these, 4 bushells were tayle and carryed into stable, and ye remaining 7 quarters 3 bushells and half were laide into malthouse chamber for our owne use. [p.29]

Of the 7th dressing, on Saturday April 23rd, there was in all 6 quarters 4 bushells and 2 pecks, whereof, they being the bottome of the ricke, there was one quarter tayle: and the remaining 5 quarters 4 bushells and half laide into the malthouse chamber for our owne use. Only five sold and 5 pecks more, at 1*s.* 8*d.* Nevertheless, there is about 19 or 20 quarters now in the malthouse chamber, April 26.

Notandum: That of the whole cropp of oates this yeare there was in all 43 quarters and one bushel, of which there hath been sold 4 bushells and 3 pecks. There was sowne of them 2 quarters 5 bushells and half; and there was of tayle 2 quarters and a bushell, which yet were spent, which make in all 5 quarters 3 bushells and a peck. The rest are spent and remaine for expence.

[*p.30*] An account of the corne that there was of my twelfth crop, gathered in in harvest anno 1687, which harvest (though some began on Munday July 18) began generally on Thursday July 21. There were divers very wett days, but yet such as did not much hinder work, though they prevented carrying of corne. The harvest was compleatly finished on Friday August 12th, and had been ended the day before only for the wett; so that we were generally 3 weekes and 2 dayes in harvest. The Barr Field was the tilth this yeare and the South Field was the broak cropp, which was thought to exceede the tilth for its part, and indeed the barly was very good there; when in the tilth, although there was some very good, yet there was some also that was very bad. Rye and wheate was generally good, though some of my owne wheate was very ordinary, being sowne upon oate land and not dunged. Sheafe corne was wel inned, excepting some towards the end, and other corne also was inned tolerably well. The corne that was good had generally a very large eare, both rye and barly, but barly especially. I had 7 stetches of my owne barly this yeare, and 2 acres rye.

[*p.31*] **An account of the wheate, white or red, brought in in that harvest**

Of the first dressing, which was white wheate, on September 6th, there was in all but one quarter 3 bushells and half, whereof 5 bushells were sowne upon 6 stetches of glebe next the mill in Northfield; and 4 bushells and half were sold at home at 8 groates per bushell, for 12*s.*, and one bushell was given to Thrift, and the remaining bushell was spent by our selves.

Of the 2nd dressing, on October 18th, there was in all 3 quarters 6 bushells and 3 pecks, of which one bushell was tayle, sold nevertheless to Wil: Thrift for 1*s.*, and one quarter and 2 bushells was sowne upon the 4 acres in North Field; and one quarter one bushell and an half sold at home at 8 groats a bushel likewise, for 1*l.* 5*s.* 4*d.*; and a bushel was given to J: Casbourne; and all the rest, viz one quarter one bushell and a peck, spent by our selves. This dressing was all red wheate.

Notandum: That these 2 dressings, making in all 5 quarters 2 bushels and a peck, was all laide in the malthouse.

Of the 3rd dressing, which was al red wheat, on January 20, there was in al 4 quarters and 2 bushels, of which 2 quarters and 4 bushels were sold at home at 8 groats per bushel, for 2*l.* 13*s.* 4*d.*, and one quarter and 6 bushels of it were spent by our selves, although a bushel of it was tayle.

Of the 4rth dressing, of red wheate, on Saturday April 14, 3 quarters 2 bushells and half, whereof 2 pecks and half were tayle, sold nevertheless to J: Casbourne for 1*s.*; and 2 quarters 6 bushells and 3 pecks more were sold at home at 2s. 8d. per bushel, for 3l. 0s. 8d., and 3 bushells were spent by our selves. The remaining half peck was lost with mice or skreening.

Of the 5th dressing, of red wheate also, there was on Thursday May 10 7 quarters 4 bushels and half, whereof one bushel and half was tayle, sold to the taskers nevertheless at 1s. 6d. the bushel, for 2s. 3d.; and one quarter 6 bushels and half more were sold at 8 groates per bushel, for 1l. 18s. 8d., and 6 bushels and half more were sold at 7 groats a bushel, for 15s. 2d., and 4 quarters and 4 bushels were spent by our selves; so that still there is 2 bushels loss by skreening etc.

Of the 6th dressing of wheate (which was all white wheate), on Saturday May 19th, there was in all ten quarters and 4 bushels, of which 2 bushels were tayle, but sold nevertheless for 2s. 6d., and the ten quarters and 2 bushels were laide into Goodman Bunning's chamber, where they lay till Michaelmas following, and then 5 quarters were sold to Puckeridge at 2s. 3d. per bushel, for 4l. 10s. 0d.; and 2 quarters more were sold at home afterwards at 7 groats per bushel, for 1l. 17s. 4d., and one quarter and 7 bushells more sold at 8 groates per bushel, for 2l. 0s. 0d., and one quarter one bushel and half were spent by our selves; so there is one bushel and half lost by skreening and measuring and such like at least. [p.32]

Notandum: Of the whole crop of wheate this yeare, yt is 30 quarters 7 bushels and a peck, whereof 12 quarters wanting half a bushel were white wheate and all the rest redd.

An account of the rye or miscellaine brought in in that yeare [p.33]

Of the first dressing of rye or miscellaine, the last whereof was dressed up on September 16, there was in all about 5 quarters and 3 bushells, of which there was about a peck tayle taken out of the 5 bushells of shalings that were dressed first, and 5 bushells were sold at Royston at 2s. per bushel, for 10s., and all the rest were sowne upon 14 acres and about half; that is upon all the land I have in the North Field excepting the 4 acres and 6 stetches betwixt that and the mill, whereof that under the mill hedge is reckoned one. That is about 2 bushells and 3 pecks an acre, one with another, there being very little large measure. Of this, 2 stetches are my owne.

Of the 2nd dressing, on September 22, there was in all 4 quarters 3 bushells and 3 pecks, of which 2 pecks were tayle, and the 4 quarters 3 bushells and a peck were sold all at 2s. a bushel, for 3l. 10s. 6d., at home.

Notandum: That these 2 dressings, makeing in all 9 quarters 6 bushels and 3 pecks, came all out of ye malthouse.

Of the 3rd dressing, December 26, there was in all 12 bushells, whereof one quarter and half a bushell was sold at home at 16s. per quarter, for 17s.; and 3 bushells more were spent by our selvs, half a peck tayle and rest boyled for the horses when blooded.

Of the 4rth dressing, Friday February 3, there was in all 5 quarters 2 bushells and half, whereof a peck was tayle; and 2 bushells were given to Widdow Carrington, Widdow Chilterne, Widdow Reynolds, Widdow

Woolward, and 4 quarters one bushel and a peck were sold at home at 2*s*. per bushel, for 3*l*. 6*s*. 6*d*., and the remaining 7 bushels was spent by our selvs.

Of the 5th dressing, on Tuesday April 10th, there was in all 5 quarters and 6 bushells, of which 6 pecks were tayle, and yet sold for 2*s*., and 4 bushels and half were sold at home at 2*s*. per bushel, for 9*s*., and the remaining 5 quarters were sold to Shepreth Mill at 1*s*. 9*d*. per bushell, for 3*l*. 10*s*. 0*d*.

[*p.34*] Of the 6th dressing, on April 24th, there was in al 6 quarters 3 bushells and half, of which 7 bushels were sold at 2*s*. per bushel, for 14*s*.; and 3 bushels more were sold at [*p.34*] 1*s*. 11*d*. per bushell, for 5*s*. 9*d*., and 4 quarters 2 bushells and half more were sold at 1*s*. 10*d*. per bushel, for 3*l*. 3*s*. 3*d*.; and spent by our selvs 6 bushels; so that still there is one bushell loss with skreening, fanning, vermin.

Of the 7th dressing of rye, on Friday May 4, there was in all 10 quarters, of which 2 bushels and half were tayle, sold to taskers for 1*s*. 9*d*., and 2 bushels were skreened out and spent by the horses; and 7 bushells were given out of it to the poor; and 2 quarters and 7 bushels were sold at 1*s*. 10*d*. per bushel, for 2*l*. 2*s*. 2*d*., and 4 quarters one bushel and half more were sold at 1*s*. 8*d*. per bushel, for 2*l*. 15*s*. 10*d*., and the remaining 1 quarter 4 bushel were spent by our selvs in housekeeping.

Of the 8th dressing, May 23, there was in all 8 quarters, whereof there was 5 bushels a sort of tayle out of which the horses had 5 pecks; and 3 bushels and 3 pecks were sold for 3*s*. 10*d*., and 4 bushels of the best were sold at Royston at 1*s*. 8*d*., for 6*s*. 8*d*.; and 4 bushels more againe for 6*s*. 6*d*., and five quarters and 4 bushels more were sold at 19*d*. per bushel, for 3*l*. 9*s*. 8*d*.; and 6 bushels and a peck were sold to Henry Law at the same price, being it lay in his chamber, for 9*s*. 6*d*.; so that there is 3 pecks loss by skreening or otherwise.

Of the 9th dressing of rye, on Wednesday May 30, there was in all 13 quarters and one bushel, wherof 4 bushels being tayle, or skreened out afterwards, were given to the horses. 4 quarters and one peck were sold at 1*s*. 6*d*. per bushel, for 2*l*. 8*s*. 4*d*. *ob*., and one quarter one bushel and a peck more were sold also at home at 20*d*. per bushel, for 15*s*. 5*d*. Five bushels were spent by our selvs and 4 bushels were sold at Royston at 1*s*. 10*d*. per bushel, for 7*s*. 4*d*.; and 2 quarters and 3 bushels were sold to Foxton at 20*d*. per bushel, for 1*l*. 11*s*. 8*d*. A bushel and half fetched out for little piggs. So that at this rate there is still 3 quarters and 6 bushels lost and squandered. There was a good deale ground downe at time into the chaff house underneath, but there must be some other way to wast it besides that, or there could not be so much loss.

[*p.35*] Of the tenth dressing of rye, on Munday June 4, there was in all six quarters and one bushel, whereof one bushel was tayle yet sold to J: Casbourne for 1*s*. 4 bushels were sold at Royston and 5 quarters and half carryed to Aswell, all at 20 per bushel, for 4*l*. 0*s*. 0*d*.

Of the 11 dressing, on June 9, there was againe 6 quarters one bushel and half, whereof there were sold at home at 20*d*. per bushel 1 quarter five bushels and half, for 1*l*. 2*s*. 6*d*.; and 2 quarters 4 bushels and half more sold at home at 1*s*. 10*d*. per bushel, for 1*l*. 17*s*. 7*d*., and 6 bushels sold at Hare Street at 2*s*. per bushel, for 12*s*.; and one quarter was given to the poore the next yeare, before the new was threshed; a peck spent with little piggs, about a peck more lost and squandered, and the remaining bushel spent by our selvs.

Of the 12th dressing, on June 23, there was in all 10 quarters laide into a chamber at Anthony Parishes, and there lay in all above 16 monthes. Out of it there was given to poore one bushel, and there was sold out of it at 1*s*. 10*d*. per bushel 3 quarters 3 bushels and one peck, for 2*l*. 9*s*. 11*d*. *ob*.; and sold againe out of it at 20*d*. per bushel 4 quarters and 4 bushels and 3 pecks, for 3*l*. 1*s*. 3*d*., and one quarter and 2 bushels spent out of it by our selvs; so that there is five bushels squandered or misreckoned or the like.

Notandum: That the whole crop of rye this yeare amounts to the quantity of 82 quarters 2 bushels and a peck – a very great crop.

An account of the barly that was brought in in that harvest [p.36]

Of the first dressing, by J: Casbourne on Friday November 4th, there was in all six quarters and five bushells, of which 6 quarters and four bushells were sold to Mr. Rumbald of Royston at 16*s*. 8*d*. per quarter, for 5*l*. 8*s*. 4*d*., and the remaining bushell was sav'd and sold with the next dressing out of the other barne.

Of the 2nd dressing, by Wil: Thrift on November 9, there was in all 6 quarters and 2 bushells, of which 3 bushells were tayle, some given to the piggs, and almost 3 bushells laide behind the garner door; the five quarters 7 bushells remaining (being nevertheless made up even 6 quarters by one bushel of ye former dressing) were sold to M: Thurgood of Royston at 17*s*. per quarter, for 5*l*. 2*s*. 0*d*.

Of the 3rd dressing, by Will: Thrift also, on November 12th, there was in all five quarters and 4 bushells, of which 3 bushells againe were tayle laide into garner, and 5 quarters were carryed to Lydlington at 17*s*. per quarter, for 4*l*. 5*s*. 0*d*., and a bushell remains, sold afterwards with the fifth dressing.

Of the 4th dressing, by John Casbourne on November 16th, there was in all 6 quarters and 5 bushells, of which 6 quarters and half were sold to M: Thurgood at Royston, for 5*l*. 13*s*. 9*d*., at 17*s*. 6*d*. per quarter, and a bushel remaineth, sold since to Edward Godfrey for 2*s*. 2*d*.

Of the 5th dressing, by Will Thrift on November 23, there was in all 8 quarters, whereof 2 bushells were tayle. The remaining 7 quarters and 6 bushells (together with one bushel remaining of the 3rd dressing) were all sold to Mr. Glenister at 17*s*. 6*d*. per quarter, for 6*l*. 19*s*. 9*d*.

Of the 6th dressing, by J: Casbourne on Saturday November 26, there was in all 7 quarters and five bushells, whereof one bushell was tayle put into garner, and the 7 quarters and half remaining were sold to Mr. Glenister at 17*s*. 6*d*. likewise, for 6*l*. 11*s*. 3*d*.

Note: That about 17 quarters of the first and 4rth and of this dressing came out of the middle stead in the new barne, and about 24 quarters out of the middlestead in the old barne.

[p.37] Of the 7th dressing of barley, by Will: Thrift on Wednesday November 30, there was in all 5 quarters and 7 bushells, of which 4 bushells were tayle; and one bushell was sold at home, at 2s. 2d., and the remaining 5 quarters and 2 bushells were sold to one Smith of Upper Chishill at 17s. 9d. per quarter, for 4l. 13s. 2d.

Of the 8th dressing, by Will: Thrift on December 9, there was in all 7 quarters and one bushell, whereof one bushell was tayle; and 5 quarters and 6 bushells were sold to Mr. Glenister at 17s. 6d. per quarter, for 5l. 0s. 6d. The remaining one quarter and 2 bushells were afterwards sold with the next dressing to Will: Sell of Ashwell at 18s. per quarter, for 1l. 2s. 6d.

Of the 9th dressing of barly, on December 13th, by W: Thrift also, there was in all 6 quarters and 7 bushells, of which 3 bushells were tayle, almost all seedes; and one bushel was sold at home for 2s. 2d., and the remaining 6 quarters and 3 bushells were sold to the saide W: Sell of Ashwel at 18s. per quarter, for 5l. 14s. 9d.

Of the 10th dressing, by J: Casbourne on December 16th, there was in all 9 quarters and five bushells, of which there was a peck tayle; and 3 pecks were sold at home to Widdow Bayes and J: Watson, for 1s. 7d. ob., and 7 quarters and 5 bushels more were sold to the same man at the same price, for 6l. 17s. 3d.; and one quarter and 7 bushels remaining were sold with the next dressing to the same man at 18s. 4d. per quarter, for 1l. 14s. 4d. ob.

Of the 11th dressing, by J: Casbourne also, on December 22, there was in all 6 quarters and five bushells, whereof one bushell was tayle, and the remaining 6 quarters and 4 bushells were sold to the same Will: Sel at Ashwel at 18s. 4d. per quarter, for 5l. 19s. 2d.

Of the 12th dressing, by Wil: Thrift on December 23, there was in all 6 quarters and 4 bushells, of which 2 bushells were tayle, and the remaining 6 quarters and 2 bushells were sold to the same Will: Sel at 18s. 4d. per quarter, for 5l. 14s. 7d.

Of the 13th dressing, by J: Casbourne on January 7, there was in all 9 quarters and five bushells, of which one bushel was tayle; and 3 bushells more made up the last dressings, to Will: Sel 15 quarters at 18s. 4d. per quarter, being sold for 6s. 10d. ob.; and 5 quarters more were sold to James [p.38] Pitty at 17s. 5d. per quarter, for 4l. 7s. 0d. And the [p.38] four quarters and one bushell remaining were sold to one Good: Course of Lidlington at 17s. per quarter, for 3l. 10s. 1d., together with 11 more of the 14 dressing.

Of the 14th dressing, by J: Casbourne, January 13th, there was in all 6 quarters and 7 bushells, of which one bushel was tayle; one bushel sold at home for 2s. 1d. ob., and one quarter and 3 bushels more were sold with the last dressing at 17s. per quarter, for 1l. 3s. 4d. ob.; and the remaineing five quarters and 2 bushells were sold to one Goodman Godfrey of Ashwel at 17s. 4d. per quarter, for 4l. 11s. 0d.

Of the 15th dressing, by J: Casb: also, January 26, there was in all 5 quarters one bushell and half, whereof the one bushel was tayle; and half a bushel was sent to mill. The remaining five quarters were carryed in the same day to Richard Sel of Triplow to be malted.

Of the 16th dressing, by Will: Thrift on January 27, there was in all ten quarters and two bushells, whereof the 2 bushells were tayle; and all the ten quarters were carryed in to Richard Sell of Triplow to be malted on Saturday January 28.

Of the 17th dressing, on Friday February 3rd, by W: Th: also, there was in all 6 quarters and 3 bushells, of which one bushell was tayle; and the remaining 6 quarters and 2 bushells were sold to James Pitty of Upper Chishill at 16s. 7d. per quarter, for 5l. 3s. 6d.

Of the 18th dressing, Friday February 10, by Wil: Thrift also, there was againe 6 quarters and 3 bushells, whereof one bushel was tayle, and the 6 quarters and 2 bushells were sold at 16s. 8d. to M: Thurgood at Royston, for 5l. 4s. 2d.

Of the 19th dressing, February 14, by J: Casbourne, there was in all 9 quarters and 4 bushells, of which one bushell and half was tayle; and half a bushel sold to J: Casbourn himself at 1s.; and 5 quarters were sold to one Morley of Ashwel at 16s. 10d. per quarter, for 4l. 4s. 2d., and the remaining 4 quarters and 2 bushels were carryed in on Wednesday February 15 to Rich: Sell's at Triplow to be malted.

Of the 20th dressing, by Wil: Thrift on Tuesday February 21, there was in all 9 quarters and one bushell, of which 2 bushells againe were tayle, most seedes; half a bushel sold at home, at 1s. 1d.; half a bushel more was sent to mill for our owne use. The remaining 8 quarters and six bushells were carryed in to Rich: Sell to be malted on Friday and Saturday February 24 and 25.

Of the 21st dressing of barly, by John Casbourne on Friday February 24, [p.39] there was in all 8 quarters and 2 bushells, of which two bushels and 3 pecks were tayle, almost all seedes; and one peck was sold to J: Casbourne for 6d., and the 7 quarters and 7 bushels remaining were carryed in to Rich: Sel's to be malted on Saturday February 25 and on Tuesday February 28.

Of the 22 dressing, by Wil: Thrift on Wednesday February 29, there was in all 6 quarters and five bushells, of which one bushell was tayle laide into garner, and the 6 quarters and 4 bushels remaining were sold to one Laurence of Barkeway at 16s. 9d. per quarter, for 5l. 8s. 10d.

Of the 23rd dressing, by J: Casbourne on Tuesday March 6, there was in all 8 quarters and two bushells, whereof one bushell was tayle; and the remaining 8 quarters and one bushell was sold to Mr. Jackson the glazier at 16s. 4d. per quarter and carryed in all at once, for 6l. 12s. 8d.

Of the 24th dressing, by J: Casbourne on March 15, there was in all 8 quarters one bushel and half, of which one peck was tayle and one peck sold at home for 6d.; and the 8 quarters and one bushell was carryed in on Friday March 16 to Rich: Sell's to be malted for my owne use.

Of the 25th dressing, on Friday March 23 by J: Casbourn, there was in all 6 quarters and 2 bushells, which was all sold to Mr. William Chapman of Ashwel at 16s. 6d. per quarter, for 5l. 3s. 0d.

Of the 26th dressing, on Thursday March 29 by J: Casbourne also, there was in all 5 quarters and one bushel, whereof half a bushel was tayle and a peck sold at home for 6*d*.; and 5 quarters were sold to Mr. Whitham at Royston at 16*s*. per quarter, for 4*l*. 0*s*. 0*d*., and the remaining peck was spent.

Of the 27th dressing, on Saturday April 7 by W: Thrift, there was in all 9 quarters and 7 bushels, of which 5 pecks were tayle; 2 bushels and a peck sold at home for 4*s*. 6*d*., and 9 quarters and 3 bushells were on April 10th carryed to Richard Sel to be malted; and the half bushel remaining was sold to himself at 1*s*.

[*p.40*] Of the 28th dressing of barly, April 18, by Will: Thrift, there was in all 8 quarters and 3 bushels; of which neare 3 bushels were tayle; and the 8 quarters were al sold and carryed at once to Mr. Keitley of Royston at 15*s*. 10*d*. per quarter, for 6*l*. 6*s*. 6*d*.

Of the 29 dressing, on Friday April 28 by Will: Thrift, there was in all 8 quarters and 2 bushels, whereof one bushel was tayle; and 6 quarters were sold to Steven Haggar of Upper Chishil at 16*s*. 4*d*. per quarter, for 4*l*. 18*s*. 0*d*., and the remaining 2 quarters and one bushel were laide into garner for our own expence, being the last.

Notandum: That the whole cropp of barly this yeare amounteth to the quantity of 215 quarters and 6 bushels, whereof 2 quarters and half was spent, being of the best or thereabouts; about 5 quarters tayle spent also.

[*p.41*] **An account of the oates that were brought in in that harvest**

Of the first dressing of oates, on October 28, there was in all 2 quarters and 4 bushels, which were the tithe oates of the Grange and, being brought in last, were brought into the hay house and threshed in the malthouse and all spent by our selvs, haveing few old ones left.

Of the second dressing, January 5, there was in all 9 bushells (which were most of them shaled, being brought of from the ricke into barne by a cart) and were almost as many pease as oates, and were laide into malthouse chamber for our owne expence, the chamber haveing been quite empty a month before.

Of the 3rd dressing, on January 17th, there was in all 15 quarters and half a bushel, of which 2 bushels were fanned up before hand; and 5 bushells were carryed directly into stable, being tayle; and one quarter was paide where I borrowed it, and 13 quarters and 2 bushels and half remaining were laide into malthouse chamber and spent by our selvs.

Of the 4rth dressing, on Friday March 16, there was in all 9 quarters and 6 bushells, of which ten bushells were carryed into stable at divers times, and 8 quarters and half were laide into malthouse chamber for our owne use.

Of the 5th dressing, on Friday March 23, there was in all 9 quarters and two bushels, whereof 6 bushells courser than the rest were carryed into stable, and the remaining 8 quarters and 4 bushells were laide into malthouse chamber for our owne use.

Of the 6th dressing, on March 28, there was in al five quarters and 6 bushels, of which 4 bushels were paide to Mr. Grey for 3 of barly I borrowed of him for seed; and 3 quarters and 4 bushels were sold at home to Henry Wallis, Henry Welsh and Ed: Peck the butcher, and 4 bushels more to Tho: Godfrey. The rest spent.

Notandum: The whole crop of oates this yeare is 43 quarters 3 bushels and half.

An account of all the pease that were brought in in that harvest [p.43]

Of the first dressing of pease, by Wil: Thrift on Thursday January 5th, there was in all 8 quarters and six bushels; whereof there were 2 quarters sold at home to men in the yard and Mr. Reyn: and J: Watson, Edw: Godf:, Widdow Brooks, at 2s. 3d. per bushel, for 1l. 16s. 0d., and 2 quarters and 6 bushells more were sowne upon 4 acres and 3 stetches in Barr-field, in all upon 5 acres and half a roode, whereof one stetch was my owne and not gleabe (it is about 4 bushells and half per acre) and five bushells more sold afterwards to J: Watson and Wil: Thrift againe, at the same price, for 11s. 3d., and the rest spent by our selvs.

Of the second dressing, on Friday March 9 by Will: Thrift, there was in all seven quarters and six bushels, whereof 3 bushels and 3 pecks were a sort of tayle, oates and pease and other stuff together, and so carryed directly into stable; and the remaining 7 quarters and 2 bushells and about a peck of the best laide all that day into the garner, out of them have been sold since 4 bushels and a peck, the last of which were to Rich: Haycock on April 17, 1688, and 2 bushels and half more sold since to Henry Wallis and Mr. Killenbeck, April ultimo the other spent.

Notandum: The whole cropp of pease this yeare is 16 quarters and an half.

An account of all the corne that there was of my thirteenth cropp, [p.44] gathered in in harvest 1688, when harvest began generally on Thursday July 26 (some began to reape on the Munday before, but no corne was carryed till that day because they reaped some corne greener than they commonly doe in this towne, and so left it the longer in the field) and it was completely ended on Wednesday August 22nd, about 4 a clock in the afternoon. The day before was wet or else it had been finished then, so that harvest lasted a compleate month. It was generally a very good harvest, and all sorts of corne were carryed in very well, though there came some rayne upon the wheate as it lay in the gavel; yet there came good weather upon it, so that it was very well inned. The North Field this yeare was the tilth and the Barr Field was the brock. There was very good barly upon the Heath – that half of it that lyeth next Norwich Road. I had of my owne land besides gleabe 2 stetches of rye, one of pease and 2 single acres of barly in both fields.

[p.45] **An account of the wheate, white or redd, that was brought in in that harvest**

Of the first dressing, of white wheate, about September 8th, there was in all but 6 bushells and an half, which was all sowne upon the 2 acres and stetch in Horsted Shott and upon the stetch and part of the half acre cross London Way at the first knapp.

Of the 2nd dressing, which was all red wheat, on October 16, there was in all 2 quarters 4 bushels and an half, whereof 2 quarters 2 bushels and something above half were sowne upon the remaining part of the half acre, the little stetch and 3 half acres, and 3 other stetches beyond it, and the five acres – in all about 8 acres – and the remaining 2 bushels almost were spent by our selvs.

These 2 dressings, making in all 3 quarters and 3 bushels, were all threshed out of the malthouse.

Of the 3rd dressing, of red wheate, on Friday January 18, there was in all 4 quarters and six bushels, whereof one bushel was tayle yet spent by ourselvs; 3 bushels more were sold at 2s. 6d. per bushel, for 7s. 6d., and 2 quarters and six bushels more were sold al at home at 8 groates per bushel, for 2l. 18s. 8d., and the remaining one quarter and 4 bushels were spent by our selves.

Of the 4rth dressing, of red wheate, on Friday April 19, there was in all five quarters and 2 bushels, whereof one bushel and half was tayle, sold nevertheless to taskers for 1s. 3d.; and one bushel sold of the best, for 1s. 10d., and one quarter and six bushels more were sold at home at 8 groates per bushel, for 1l. 17s. 4d.; and 2 bushels more sold at 2s. 6d. per bushel, for 5s., and one quarter and 4 bushels more sold at home for 7 groats a bushel, for 1l. 7s. 6d., and one quarter and 2 bushels spent by our selvs; and the remaining bushel and half lost or not accounted.

[p.46] Of the 5th dressing, of red wheate, on May 7th, there was in all 9 quarters and 3 bushels, whereof 2 bushels were tayle, sold to taskers at 1s. 3d. per bushel, for 2s. 6d.; and five quarters more were sold to Baldock at 2s. 6d. per bushel, for 5l.; and 3 quarters more were sold at home at 7 groats per bushel, for 2l. 15s. 0d., and six bushels more were spent by ourselvs, so that there is still lost by mice or rats and fanning etc. 3 bushels, or not accounted when delivered out.

Of the 6th dressing, on May 17th, which was all white wheate, there was in all 6 quarters five bushels and 3 pecks, whereof 5 pecks were such tayle as was laide behind the garner doore; and 3 bushels and 3 pecks were a better sort of tayle and sold to the taskers, for 4s. 6d., and, of the best, 4 bushels were sold to Good: Gladwel at 2s. 6d. per bushel, for 10s., and 3 bushels more were sold at 8 groats, for 8s.; and one quarter and 3 bushels more were sold at 3s. per bushel, for 1l. 13s. 0d., and the remaining 3 quarters 6 bushels and 3 pecks were spent by our selvs, the last of it sent to mil April 9, 1690. It lay in a chamber at John Aspenal's, whose house had stood empty a yeare, and though the wheate lay almost a yeare there was very little lost at last.

Of the whole crop of wheate this yeare there was 7 quarters 4 bushels and a peck of the white wheate and 21 quarters 7 bushels and 2 pecks of the red wheate; so that there is in all of both of them 29 quarters 3 bushels and 3 pecks.

An account of the rye or miscellaine brought in in that harvest [p.47]

Of the first dressing of rye, on September 12, there was 2 quarters one bushel and half, which were all sowne upon 10 stetches or selions in Shortfurlong, and Duck Acre Shott in Southfield, and upon other parcels of land mentioned in the next dressing.

Of the 2nd dressing, on September 18th and some a day or 2 before, there was in all 3 quarters, whereof 3 bushels and a peck were sold at 1s. 6d. per bushel, for 4s. 10d. ob., and 6 bushells and half more were sold at 1s. 8d. per bushel, for 10s. 10d.; one peck spent by our selvs, and the remaining one quarter and 6 bushels were sowne upon 8 stetches in the Rowze and the single acre and the 4 acres, besides a 3 roods of my owne – in all 12 acres and half with those stetches mentioned before, which is about 2 bushels and half to an acre, there being sowne in al 4 quarters wanting 2 pecks.

Note: That these 2 dressings, making in all 5 quarters one bushel and half, were threshed out of the malthouse.

Of the 3rd dressing, on Tuesday March 26, there was in all 6 quarters and 7 bushels, of which 2 bushels had many pease in them (the pease being laid upon that mow and shaled amongst it) and more laide amongst the off corne. 2 bushels more were a sort of tayle reserved for little piggs. 4 bushels more were sold at Royston at 1s. 6d. per bushel, for 6s.; 5 quarters and half more sold to Buntingford at 1s. 8d., for 3l. 13s. 4d., and 2 bushels and half were sold at home at 1s. 8d., for 4s. 2d.; and the remaining 2 pecks were given to poor.

Of the 4rth dressing, on Thursday April 11th, there was in all 5 quarters and three bushels, whereof a bushel and 3 pecks were tayle, laide into garner for little piggs; one peck sold to Robert Coleman at 5d., and the remaining 5 quarters and one bushel were sold to George Miller of Buntingford at 1s. 9d. per bushel, for 3l. 11s. 9d.

Of the 5th dressing of rye, on Wednesday May 22nd, there was in all 7 [p.48] quarters 4 bushels and one peck, whereof the 4 bushels were tayle, and yet 7 pecks of it were sold for 1s. 5d. ob.; and 2 pecks the horse keeper had, and the rest spent by weaned piggs. Of the best, half a bushel was given to Good: Ingrey; and 6 quarters and one bushel were sold at home at 1s. 6d. per bushel, for 3l. 13s. 6d.; and 6 bushels spent by our selvs; and the remaining 3 pecks lost by skreening and fanning and vermin.

Of the 6th dressing, on Thursday June 6th, there was in all 7 quarters and 6 bushels, of which 2 bushels and half were tayle, spent by weaned pigs etc.; and 6 quarters sold to Baldock at 1s. 6d. ob. per bushel, for 3l. 14s. 0d., and 2 bushels more were spent by our selvs; and a quarter sold at home at 1s. 6d. per bushel, for 12s. The remaining bushel and half lost by skreening, fanning, vermin, etc.

Of the 7th dressing, on Thursday June 13, there was in all 9 quarters and 3 bushels, of which one bushel and half was tayle; and 4 bushels were sold to Good. Davyes and carryed to Hare Street for 6s. 3d.; and 5 quarters more were sold to Mr. C: at Cambridge at 1s. 6d. per bushel, for 3l. 0s. 0d. One quarter more sold to Will Ingrey at once, for 12s., and 2 quarters 4 bushels and a peck sold out of the garner by the peck at 1s. 6d. per bushel also, at 1l. 10s. 5d. ob., and one bushel spent by our selvs; and the remaining peck lost in measuring or the like.

Of the 8th dressing of rye, on June 22, there was in all ten quarters and one bushel, whereof one bushel tayle; and out of the best there was sold at home at 20d. per bushel five bushels and half, for 9s. 2d., and at 22d. per bushel 7 bushels and 3 pecks more, for 14s. 2d., and there were sold at 2s. per bushel 4 quarters 6 bushels and 3 pecks, for 3l. 17s. 6d., and at 2s. 4d. per bushel there was sold 2 quarters and 4 bushels more, for 2l. 6s. 8d.; and 6 bushels were spent by our selvs, some spent by weaned piggs; about a bushel and half lost by skreening, vermin, etc.

[p.49] Notandum: Of the whole crop of rye this yeare there was in all fifty two quarters one bushel and three pecks.

[p.50] **An account of the barley brought in in that harvest**

Of the first dressing of barly, by William Thrift on Thursday November 8th, there was in all even six quarters, which were sold all to my neighbour Urias Bowes at 14s. 2d. per quarter, for 4l. 5s. 0d.

Of the 2nd dressing, on Friday November 9th, by John Casbourne, there was but 4 quarters and one bushel, which were likewise sold to the same chapman and at the same prise, for 2l. 18s. 2d.

Of the 3rd dressing, by J: Casbourne on November 21, there was in all five quarters and five bushels, which were all sold to M: Thurgood at Royston at 14s. 2d. per quarter, for 3l. 19s. 8d.

Of the 4rth dressing, on Saturday November 24 by John Casbourne, there was in al 5 quarters and 6 bushels, whereof one bushell was sent to mil for hoggs, etc., and the remaining 5 quarters and five bushels were sold to one Rumbald of Royston at 14s. 2d. per quarter, for 3l. 19s. 8d.

Of the 5th dressing, by Will: Thrift on November 28, there was in all 7 quarters and 6 bushels, which were al sold, partly to M: Thurgood and partly to one Rumbald, at 14s. 2d. per quarter, for 5l. 9s. 9d.

Of the 6th dressing, on Thursday November 29 by Will: Thrift also, there was in all 5 quarters and 3 bushells, sold at 14s. 2d. per quarter to one Rumbald of Royston also, for 3l. 16s. 1d.

Of the 7th dressing, also by Wil: Thrift, on Munday December 10th, there was in all 8 quarters and 2 bushels, which were all sold to Mr. Witham at Royston at 14s. 4d. per quarter, for 5l. 18s. 3d.

[p.51] Of the 8th dressing of barly, by John Casbourne on Tuesday the 11th of December, there was in all 7 quarters and six bushels, which were all sold to Mr. Witham also, at 14s. 4d. per quarter, for the summe of 5l. 11s. 1d.

Of the 9th dressing of barly, on Wednesday December 19th by Will:

Thrift, there was in all 4 quarters 3 bushels and one peck, whereof one peck was sold to Widdow Dovy for 5*d. ob.*, and the remaining 4 quarters and 3 bushels were carryed into Goodman Kefford at Triplow on Friday December 21 to be malted for me.

Of the 10th dressing, on Saturday December 22nd by J: Casbourne, there was in all 6 quarters and 4 bushels, of which 3 bushels were taken out as tayle, the corne being very course, laide behind the garner door. One bushel sold to Ed: Godfrey for 1*s*. 9*d*., and 2 pecks more to Widdow Dovy for 11*d*., and 5 quarters and 7 bushels were carryed in the same day to Goodman Kefford also to be malted for me; and the remaining 2 pecks were sold to J: Casbourne for 10*d. ob.*

Of the 11th dressing, on Munday January 7 by John Casbourne, there was in all ten quarters and six bushels, whereof one bushel was tayle; and one bushel more was sold at home to Edw: Godfrey for 1*s*. 9*d. ob.*, and one bushel and a peck more were sold to Mat: Payne, J: Casb: Will: Brock, Widdow Dovy, at 2*s*. 2*d. ob.*; and ten quarters and 2 bushels were carryed on January 8 to Goodman Kefford of Triplow to be malted for me; and half a bushel was fanned up for a mash for horses being blooded; and a peck remaines, that was sold afterwards to J: Casbourne for 5*d. ob.*

Of the 12th dressing, on Saturday January 12, by Wil: Thrift, there was in all 15 quarters and 2 bushels, which were all carryed in on Tuesday January 15 to Richard Sell's of Triplow to be malted for me. This was all the middle mow in the old barne, and something more, viz a flooring from the end mow next the ricke.

Of the 13 dressing of barly, on Thursday February 7th by John Casbourne, there was in all 12 quarters five bushels and a peck, of which the peck was fanned up and sold to Richard Haycock for 5*d. ob.*; and 3 bushels were fanned up and a bushel on the dressing day (in all 4 bushels) were sent up to mill for the 2 sows and fowles, etc., and the remaining 12 quarters and one bushel were sent into Richard Sell at Triplow to be malted for me at two several times, viz five quarters and five bushels on Thursday February 7 and 6 quarters and 4 bushels on Friday February 8, when they brought back to carry to Ware 9 quarters of the old malt. [*p.52*]

Of the 14th dressing of barly, by Will: Thrift on February 12 (which was all in the end mow next the rick), there was in all 8 quarters 7 bushels and half, whereof one bushel was fanned up and sent to mill. Half a bushel more was sold to Good: Harvy for 11*d*., and 8 quarters and five bushells were carryed in on Saturday February 16 to Rich: Sell's to be malted, when they brought back all that was left of the old malt, viz 8 quarters 3 bushels and a peck, that were sent to Ware also; and 2 pecks remaining sent to mill for our selvs afterwards with 2 pecks of the next dressing.

Of the 15th dressing, on Tuesday February 19, there was in all 10 quarters and five bushels and a peck, of which 2 pecks were sent to mill with the former dressing and one peck sold to J: Casbourne for 5*d. ob.*; and ten quarters wanting one peck were saved for seede, that were all sowne on Northfield and Southfield; and 2 bushells and 3 pecks more were sent to mill for hogs and fowl, and the remaining 2 bushels were sent in with 8 quarters of the next dressing to be malted for me.

Of the 16 dressing, on Thursday February 28 by Will: Th., there was in all twelve quarters, whereof (2 bushels being added to them of the former dressing) 8 quarters and 2 bushels were carryed in on Friday March 1 to Rich: Sell's of Triplow to be malted for me; and the remaining 4 quarters were sowne, so that of this and the 15 and 17th dressings there are sowne in all 16 quarters 5 bushels and half upon 20 acres and half in Northfield, and 14 and half in Southfield – in all upon 35 acres, most of it in Waterdeane Shot and 4 stetches there of my owne.

[p.53] Of the 17 dressing of barly, by Will: Thrift on Saturday March 9, there was in all ten quarters wanting one bushel, whereof 3 bushels were spent upon sowes and piggs and fowles, etc.; and 6 quarters and 6 bushels were carryed in on Thursday March 14 to Rich: Sell's of Triplow to be malted for me. The remaining 2 quarters and 6 bushels were sowne all but a peck, which was sold to Richard Fairechild for 5$d.$ $ob.$

Of the 18th dressing, on Thursday March 21 by Wil: Thrift, there was in all ten quarters wanting a peck, whereof one bushel and 3 pecks were sold to Rich: Fairechild and Math: Payne for 3$s.$ 3$d.$, one bushel more was sent to mill for sowes, etc., and five quarters and 4 bushels were sold to Lidlington at 14$s.$ 10$d.$ per quarter, for 4$l.$ 1$s.$ 6$d.$; and the remaining 4 quarters and one bushel were carryed in on Saturday March 23 to Richard Sell's to be malted for me, by Math: Payne, at which time he brought 2 bushels of my new malt for himself and John Casbourne.

Of the 19th dressing of barly, by Wil: Thrift on Thursday March 28, there was in all 9 quarters 7 bushels and half, whereof 2 pecks were sold at home for 11$d.$, one bushel sent to mill for sowes, etc., 5 quarters and 6 bushels sold to Mr. Chapman of Ashwel at 15$s.$ 3$d.$ per quarter, for 4$l.$ 7$s.$ 9$d.$, and 4 quarters were carryed in to Rich: Sel's of Triplow on Saturday March 30 to be malted for me.

Of the 20 dressing, by Wil: Thrift, of Friday April 5, there was in all 10 quarters 3 bushels and a peck, whereof a peck was sold at home for 6$d.$, and 2 bushels were tayle laid into garner, and 10 quarters and one bushel were carryed in on Saturday April 6 to Good: Kefford, to be malted for me, by Will: Thrift and Will: Dovy, when Will: Thrift brought back from Richard Sel's a bushel and half of malt for himself and by my appointment.

[p.54] Of the 21 dressing, on Friday April 12, by Will Thrift, there was in all 7 quarters and 2 bushels, [p.54] whereof one bushel and half was tayle, laide into garner amongst the off corne; a peck more sold to Widdow Whitby for 6$d.$ and a peck more spent by ourselvs; and one quarter was carryed the same day to Rich: Sel's of Triplow to be malted for me (he wanting about so much to make up a steeping), and the remaineing 6 quarters were carryed to Ashwel and sold at 15$s.$ 7$d.$ per quarter, for 4$l.$ 13$s.$ 6$d.$

Of the 22nd dressing, on April 16, Tuesday, by W: Thrift, there was in all five quarters and 3 bushels, whereof one bushel was tayle laide behind the garner doore amongst the off corne; and 2 bushels and half were sold at home at 2$s.$ per bushel for 5$s.$, and 4 quarters and half were carryed in the day after, viz April 17, by Math: Payne, when I was at London, to R: Sel's at

Triplow to be malted for me (at which time also they brought ten quarters of new malt from him and laid it in the garner at home). The remaineing 3 bushels and half, being the last of this yeare, were saved for our owne expense.

Notandum: Of the whole crop of barly this yeare there was in all 184 quarters 4 bushels and 3 pecks, whereof one quarter and half a bushel were tayle; and 2 quarters and 3 bushels more were spent upon fowles, hoggs, etc.; and 16 quarters five bushels and half were sowne. So that there is spent after this manner 20 quarters and a bushel, and all the rest, viz 164 quarters 3 bushels and 3 pecks, were malted and sold.

An account of the oates that were brought in in that harvest [*p.55*]

Of the first dressing there was but two bushells and about a peck and half, not all out, half a bushel of oates and pease and all. They were tithe of the end of the 22 acres in the Grange Field next Walden Way, which at that time Mr. Miles pretended a right to falsly. They were brought in the last of all my corne, and the barnes being ful and this ricke topt up without it, they were laid by themselvs in the hayhouse and threshed out presently after my seed rye and wheat in the malthouse, by Good. Thrift and Casbourne, and dressed up on Thursday October 18 by them.

Of the 2nd dressing of oates, December 21, there was 3 bushels and half, which were fanned up, being the shalings of part of the oate ricke when it was innd. The half bushel was most pease, but however the horses had that and the rest too, before any more were drest.

Of the 3rd dressing, on Thursday December 27th, there was in all 7 quarters 3 bushels and half, of which 6 bushels were carryed directly into stable, and 3 quarters were laid into malthouse chamber for our owne expence; and 3 quarters and three bushels wanting a peck were laide into garner for seed, and all sowne upon 12 acres of the Heath, together with 9 bushels of the 5th dressing.

Of the 4rth dressing, on Wednesday January 2, there was in all 9 quarters and seven bushells, of which 2 bushels were paide where they were borrowed, and the rest laide all into malthouse chamber for our owne use, that is 9 quarters and 5 bushells.

Of the dressing on Wednesday January 16, the 5th dressing, there was in all but one quarter and 7 bushels, whereof 6 bushels were carryed into stable and the remaining 9 bushels were laide into garner for seed, and sowne upon the Heath, together with 3 quarters and 3 bushels of the 3rd dressing.

Of the 6th dressing of oates, by John Casbourne in his barne, that came [*p.56*] from under the barly and pease upon the ricke, February 22, there was in all 3 quarters and one bushel, whereof 7 bushels went directly into stable; and 2 quarters 5 bushels and half were sold (or spent by Mr. Hony, horse, which I was paide for at 14*d. ob.* per bushel) for 1*l.* 6*s.* 0*d.*; so that there is 3 bushels and half sold out of the next dressing.

Of the 7th dressing, on Friday April 26 by W: Thrift, there was in all 7 quarters and 7 bushels, whereof 3 bushels, being tayle, were carried directly into stable; and 3 bushels and half (or is saide) were sold, or spent by Mr. Hony, horse; so that there was 7 quarters and half a bushel laide into malthouse chamber for our use.

Of the 8th dressing of oates, by Wil: Thrift on Thursday May 9, there was in al 5 quarters and 3 bushels, whereof 3 bushels, a sort of tayle, carried with 2 bushels more into stable. The rest laide into malthouse chamber for our owne use.

Memorandum: Of the whole cropp of oates this yeare there was in all 36 quarters 2 bushels and half, whereof 4 quarters and half were sowne and 2 quarters and 6 bushels were sold – in all 7 quarters and 2 bushels. Spent then by our selvs 29 quarters and half a bushel besides pease.

[p.57] **An account of the crop of pease that were gathered in in that harvest**

Of the first dressing there was in all but 12 bushels and something above half, dressed up on Friday December 21, being only a little corner of the ricke, out of which 1 bushel was sold for 2s.; and the rest laide into garner for fatting hoggs, etc., and 3 bushells and a peck more sold afterwards, for 6s. 6d., and the rest all spent by hoggs.

Of the 2nd dressing of pease, on Wednesday January 30th, there was in all 17 quarters and five bushels, whereof 2 bushels being tayle, the horsekeeper had them into stable directly. 3 bushels were fanned up and spent by our owne hoggs before hand, and 3 quarters and six bushels were laide into malthouse chamber and spent; and 6 quarters of them were sold at home at 2s. per bushel at divers times, for 4l. 16s. 0d. and 7 quarters and 2 bushels laide into garner which remaine for our owne use also.

Of the 3rd dressing, on Saturday February 16, there was in all 3 quarters 5 bushels and an half, whereof one peck was tayle. The rest all laide into garner for our owne expense.

Of the 4rth dressing, on Thursday April 18, there was in all but 7 bushels, all laide into the malthouse chamber and spent with the oates there.

Of the whole cropp of pease this yeare there was in all 23 quarters and 6 bushels.

[p.59] An account of the quarters of corne of all sorts, sold or spent, that I had of my fourteenth cropp, which was gathered in in harvest 1689. Harvest that yeare began generally on St. James his day, being Thursday July 25th (some began the Munday before to reape, but, corne being green, there was none carried till that day. They left it in gavels, or in the feild though bound up, the weather being very good), and it was compleatly ended betimes on Thursday morning August the fifteenth – all in our towne ended on Wednesday night, excepting about an half acre of Math: Payne's that was left alone in the feild till Thursday morning; so that harvest was begun and ended in three weekes just. The South Field was the tilthe feild this yeare and the North Feild brock. I had of my owne in South Feild 6 stetches, and 2 in the North Feild – in all 3 acres, besides gleabe. I had no tithe from the Grange till the Tuesday and Wednesday after we had made end, and had but 2 little jaggs, which was almost all oates. The weather was very good this harvest and corne brought in very dry. There was one wet night about the beginning of barly harvest which hindered carting, 2 or 3 hours in the evening and most part of the next day, but being faire weather afterwards, proved no hinderance nor damage.

An account of the wheate both white and red brought in in that harvest [*p.60*]

Of the first dressing, which was all red wheate, there was in all one quarter 2 bushels and a peck, which was all sowne upon 3 roods in Fincholn Shot, and 6 stetches and the 3 half acres in Short Spotts in Barr-feild – in all 4 acres and half; that is about 2 bushels and a peck per acre.

Of the 2nd dressing, on September 17th, which was also all red wheate, there was in all 3 quarters 2 bushels and half, of which 2 pecks given to Mat: Law, being new marryed; one bushel more spent by ourselvs, and the remaining 3 quarters and one bushel were sold at home at 3*s*. per bushel, for 3*l*. 15*s*. 0*d*., saveing that there is about a peck spent and given away and lost by vermin.

Note: That these 2 dressings, making in all 4 quarters and 4 bushels and 3 pecks, were all threshed out of the malt-house, besides 6 quarters of rye – vide page 62.

Of the 3rd dressing, of red wheate, on Thursday January 2nd, there was in all but one quarter six bushels and a peck, whereof one bushel was tayle, most eaten by the little piggs; the rest put amongst the rye; and 2 bushels and 3 pecks of it were sold at 3*s*. per bushel, for 8*s*. 3*d*.; and 1 quarter and half a bushel more were sold at 10 groats per bushel, for 1*l*. 5*s*. 6*d*., and the remaining 2 bushels were spent by our selvs.

Of the 4th dressing, of red wheate, on Tuesday February 11, there was in al 7 quarters 4 bushels and 3 pecks, whereof 4 bushels and half were tayle, yet sold to taskers at 2*s*. per bushel, for 9*s*.; and 1 quarter 3 bushels and half sold at home at 10 groats a bushel, for 1*l*. 17*s*. 6*d*., and 5 quarters and a bushel more sold to Buntingford at 3*s*. 6*d*., for 7*l*. 0*s*. 0*d*.; and 3 bushels spent by ourselvs, so that there are 3 pecks missing which were misreckoned or lost by mice and rats.

Of the 5th dressing, April 12, which was all red wheate, there was in all 4 [*p.61*] quarters and 6 bushels, whereof one bushel was tayle, yet sold to taskers for 1*s*. 8*d*. 4 quarters and half more were sold to Royston at 3*s*. 1*d*., for 5*l*. 11*s*. 0*d*., and the remaining bushel sent to mill for our selvs.

Of the 6th dressing, which was all white wheate, on Saturday April 26, there was in all 8 quarters and 2 bushels, whereof 2 pecks were course tayle reserved for hoggs; and 3 bushels more were sold (though a sort of tayle) at

22*d.* a bushel, for 5*s.* 6*d.*; and one quarter and 7 bushels were sold at home at 3*s.* per bushel, for 2*l.* 5*s.* 0*d.*; and 3 quarters 5 bushels and half more were sold at home at 3*s.* 4*d.* per bushel also, for 4*l.* 18*s.* 4*d.*; and 2 quarters and 2 bushels wanting a peck were spent; so that there is about one peck squandered and spent.

Of the 7th dressing, on May 26, which was al red wheate, there was in all 8 quarters 2 bushels and a peck, whereof 2 bushels were tayle, sold to taskers at 1*s.* 8*d.* per bushel, for 3*s.* 4*d.*; and of the best, there was one quarter 2 bushels and half sold at 3*s.* per bushel, for 1*l.* 11*s.* 6*d.*, and 1 quarter 2 bushels and 3 pecks more were sold at 8 groats per bushel, for 1*l.* 8*s.* 8*d.*; and 2 quarters and one bushel more were sold at 2*s.* 6*d.* per bushel, for 2*l.* 2*s.* 6*d.*; and the remaining 3 quarters and 2 bushels were spent by our selvs.

Of the whole crop of wheate this yeare there was 8 quarters and 2 bushels of white wheate and 27 quarters of the red wheate. In all there is 35 quarters and 2 bushels.

[*p.62*] **An account of the rye or miscellaine brought in in that harvest**

Of the first dressing of rye, or miscellaine (there being some white wheate threshed among it), on Thursday September 12th, there was in all three quarters and 6 bushells, to reckon the shalings and all, that were shaled with unloading in harvest time, which was all sowne upon 4 stetches in Branditch-shott and Shortfurlong and Finchholn Shott in Barfield, or Fawdon Feild, as may be mentioned in the next dressing.

Of the 2nd dressing, September the 16, there was in all 2 quarters 2 bushels and half, of which 4 bushels sold at home at 18*d.* per bushel, and an odd peck at the same price, for 6*s.* 4*d. ob.*; the remaining 14 bushels and a peck all sowne upon 13 stetches in the 2 shots before mentioned, and ye 2 furthest stetches in Short Spotts, and upon 4 stetches on this side the 12 acres, and upon 2 single acres in the shots before mentioned and 4 acres of the 12 next the Southfield, and Lamps Acre – in all 16 acres wanting a rood and half of gleabe – and upon my owne 2 acres of coppy land, in all 18 acres wanting a rood and half, which is 2 bushels an[d] half per acre. It is most of it smal measure.

Note: That these 2 dressings, making in all 6 quarters and half a bushel, were all threshed out of the malt house, with a few white wheate sheaves amongst it.

Of the 3rd dressing, February 14, there was in all 12 bushels and half, whereof one peck was tayle; and one quarter 2 bushels and 2 pecks were sold at home at 7 groats per bushel, for 1*l.* 4*s.* 6*d.*; a bushel sent to mill for our selvs; a peck left, that was spent by little piggs. The remaining 2 pecks were forgotten to be accounted or miscounted.

Of the 4rth dressing, March 8, there was in al 13 bushels and half, whereof a peck tayle; and 2 pecks given to Thom: Ward; 2 bushels sent to mill, and the remaining 10 bushels and 3 pecks were sold at 7 groats also, for 1*l.* 5*s.* 1*d.*

[*p.63*] Of the 5th dressing of rye, March 20th, there was in all 6 quarters and one bushel, whereof a bushel and half tayle was reserved for little piggs; and half a bushel was given to Wil: Rand, and 6 bushels were sold at home at 7 groats per bushel, for 14s., and five quarters more were sold to Simon Beaman of Shingey at the same rate, for 4l. 13s. 4d., and the remaining bushel was spent by ourselves.

Of the 6th dressing of rye, April 2nd, there was in all six quarters and a peck, whereof the peck was tayle; and 7 bushels and half were given to the poor; and 2 bushels more spent by ourselvs; and 4 quarters and 5 bushels more were sold at home at 7 groats a bushel, for 4l. 4s. 0d. The remaining bushel and half spent by little piggs or forgotten to be set downe.

Of the 7th dressing there was in al 8 quarters 3 bushels and half, May 20, whereof one peck was given to Rob: Macer; and 4 quarters 5 bushels and half were sold at home at 7 grots a bushel, for 4l. 9s. 2d., and 3 quarters and a peck more were sold at 2s. 2d. per bushel, for 2l. 12s. 6d. ob., and five bushels were spent by our selvs; the odd half bushel squandred and lost.

Of the 8th dressing of rye, on Saturday May 31, there was in all 6 quarters 2 bushels and half, whereof one bushel and half was tayle; and 3 quarters and 6 bushels were sold at home at 2s. 2d. per bushel, for 3l. 5s. 0d., and 6 bushels of it were spent; the rest lost. So that there is one quarter and 5 bushels clearly lost out of this and the former dressings at least, which could not happen by loss in skreening or any such way, not without fraud of the horsekeeper or some other. Though the bin was not quite empty for 6 months together, yet so much loss could not be fairely.

Of the ninth dressing of rye, on Saturday June 14, there was in all five [p.64] quarters and 6 bushels, whereof 7 pecks were tayle laide by for piggs; and 2 quarters 3 bushels and a peck were sold at 2s. 2d. per bushel, for 2l. 1s. 8d.; and 2 quarters 5 bushels and a peck more were sold at 2s. per bushel, for 2l. 2s. 6d., and 3 bushels and 3 pecks and about half were spent by our selvs; and the remaining peck and half sold besides. But here it is to be noted that there was half a bushel of wheate put to it afterwards, of the yeare following, and yet there is still 4 bushels more than I looked for of the whole dressing besides. In the last dressing there was 19 bushels wanting, yet here is 4 more than I expected, so that as to this there is (I suppose) some mistake, and that there was 4 bushels more carryed into Good: Aspenal's chamber than I was told of.

The whole crop of rye this yeare is 41 quarters and half wanting a peck.

An account of the barly that was brought in in that harvest [p.65]

Of the first dressing of barly, on Wednesday October 30 by John Casbourne, there was in all but 4 quarters 5 bushels and 3 pecks, whereof one peck was sold to Math: Payne at 9d., and 2 pecks my wife had for fowles; and 4 quarters and half were sold to Mr. Chapman of Ashwel at 13s. 7d. per quarter, for 3l. 1s. 0d., and a bushel that was skreened afterwards sold with the next.

Of the 2nd dressing, by J: Casbourn also, on Munday November 11th, there was in all five quarters and 5 bushels, sold to Mr. Cook at Chisil at 13s. 5d. per quarter, for 3l. 17s. 1d., the bushel of the former dressing being included.

Of the 3rd dressing, on Tuesday November 12th, there was in all 6 quarters and 3 bushels, by Will: Thrift, which was all sold to Mr. Cook of Chisil also, at 13s. 5d. per quarter, for 4l. 5s. 6d.

Of the 4rth dressing, by Will: Thrift also, on Friday November 22, there was in all 9 quarters and five bushels, which with 5 bushels out of J: Casbourne's barne, being made up even 10 quarters and 2 bushels, were carryed in to Goodman Kefford's on Thursday November 28 by Math: Payne and Wil: Dovy to be malted for me; so Good: Kefford had then in all 10 quarters 2 bushels.

Note: That these 2 last dressings, being in all but 16 quarters, were all in the middlestead of the old barn.

Of the 5th dressing of barly, on Wednesday November 27, by John Casbourne, there was in all 7 quarters and 4 bushels, of which one quarter was sold with some of the former dressings to Mr. Cook at Chishill, for 13s. 7d. And five bushels as above said, and the remaining 5 quarters and 7 bushels, were carryed in on November 28, 1689, by Math: Payne and Will: Dovy to Richard Sell's of Triplow to be malted for me.

[p.66] Note: That this and the 1st and the 2nd dressings of barly, making in all 17 quarters 6 bushels and 3 pecks, came all out of the middlestead in the new barne.

Of the 6th dressing of barly, on Friday December 6th by John Casbourne, there was in all 8 quarters and 7 bushels, of which 8 quarters and 4 bushels were carryed in this day to Rich: Sel's of Triplow to be malted for me; and the remaining 3 bushels were sent to mill for the sowes and little piggs.

Of the 7th dressing of barly, on Saturday December 14, by John Casbourne also, there was in all 6 quarters and five bushels, which was all that very day carryed to Rich: Sel's by Math: Payne and Wil: Dovy to be malted for me.

Of the 8th dressing, on Munday December 16th by Wil: Thrift, there was in all 6 quarters and 7 bushels, whereof 5 quarters and 5 bushels were sold to Mr. Keytly of Royston at 13s. 2d. per quarter, for 3l. 13s. 10d., and the remaining ten bushels were carryed in on Tuesday December 17 to Richard Sel's of Triplow to be malted for me with the rest that he hath already.

Of the 9th dressing, by Wil: Thrift also, on Munday December 23, there was in all ten quarters and two bushels, whereof 3 bushels were sent to mill for the sowes; and 5 quarters and 4 bushels sold to Gadwel End at 13s. 5d. per quarter, for 3l. 13s. 9d.; and the remaining 4 quarters and 3 bushels were carryed in on Munday December 30th, to Richard Sel's to be malted for me, by Math: Payn and Wil: Dovy.

Of the 10th dressing, on Friday December 20 by John Casbourne, there was in all just 8 quarters, which was all carryed in by the same persons on that December 30th to Rich: Sell's to be malted for me; so that he had in all carryed in to him on that day full 12 quarters and 3 bushels.

Of the eleventh dressing of barly, by Will: Thrift on Wednesday January [p.67] 1, there was in all 11 quarters and three bushels and half, whereof one bushel was sent to mill for sowes, half a bushel boyled for the horses, and the remaining 11 quarters and 2 bushels were carryed in by Math: Payne and Will: Dovy on Friday January 3rd to Rich: Sell's to be malted for me.

Of the 12 dressing, on January 10, Friday, by Will: Thrift, there was in all 6 quarters and 3 bushels, which was all carryed in to Richard Sel's to be malted for me by Math: Payne and Will: Dovy on Tuesday and Wednesday January 14 and 15 – 5 quarters and 1 bushel one day, and 10 bushels the next.

Of the 13th dressing, on Tuesday January 14th by John Casbourne, there was in all 9 quarters 7 bushels and 3 pecks, whereof 3 pecks were sold to Mat: Payne at 1s. 3d., and one bushel sent to mill for sowes; and 4 quarters and 3 bushels were carryed in (when some of the former dressing was carryed also) to Rich: Sel's to be malted for me on Wednesday January 15 by Math: Payne and Wil: Dovy; and five quarters more of it were sold to M: Thurgood of Royston at 13s. 3d. per quarter, for 3l. 6s. 3d.; and 3 bushels more remain which were spent by sowes and piggs.

Of the 14th dressing, by John Casbourne also, on Friday January 24, there was in all nine quarters, which were all carryed in on Saturday January 25 by Math: Payne and Wil: Dovy to Rich: Sel's to be malted for me with the rest he hath already.

Of the 15th dressing, by John Casbourne also, on Friday January 31, there was in all five quarters and 3 bushels carryed all in to Richard Sel's to be malted for me on Saturday February 1 by Mat: Payn and W: Dovy.

Of the 16 dressing, on Munday February 17, there was in all 7 quarters [p.68] and 3 bushels, of which 3 pecks were sold to Good: Payne and Thrift, for 1s. 3d., and six quarters 7 bushels and one peck were sowne; and the remaining 3 bushels sold with the 18th dressing.

Of the 17th dressing, on Tuesday February 25 by Wil: Thrift also, as the former was, there was in all 5 quarters and 6 bushels, which were all sowne also, partly in Southfield, partly in Barr Feild or Fawdon Feild.

Of the 18 dressing, on Wednesday February 26 by J: Casbourne, there was in all ten quarters and 3 bushels, whereof 5 quarters and 7 bushels (having the 3 bushels of the 16th dressing added to them, and so being made up 6 quarters and 2 bushels) were sold to Urias Bowes at 12s. 10d. per quarter, for 4l. 0s. 2d.; and half a quarter sold to Will: Fordham for seed at the same rate, and a peck sold to W. Thrift at 5d., and the remaining 3 quarters 7 bushels and 3 pecks are reserved for seed, and all sowne together with some other of the former dressing and of the following.

Of the 19th dressing, by John Casbourne also, on Thursday March 6, there was in all 7 quarters 6 bushels and 3 pecks. One peck tayle, and 2 pecks, sold to Antony Parish for 10d., and one bushel sent to mill for piggs; 6 quarters and half sold to Urias Bowes at 12s. 10d. per quarter, for 4l. 3s. 5d., and the remaining 9 bushels also were sowne or changed for seed – that is tantamount.

Of the 20 dressing of barly, by Will Thrift on Munday March 10, there was in all 10 quarters and five bushels, whereof 10 quarters were carryed to Good: Kefford's of Triplow to be malted for me on Tuesday March 11 by Math: Payne and Wil: Dovy. 2 pecks were sold to Will: Thrift for 10*d.* and a bushel and half spent by our selves for fowles, hoggs, etc.; and the remaining 3 bushels were sowne.

[*p.69*] Of the 21 dressing, by Will. Thrift also, on Friday March 14, there was in all 6 quarters and one bushel, whereof 5 quarters and 4 bushels were sold to Royston at 13*s.* per quarter, for 3*l.* 11*s.* 6*d.*; and 3 bushels were sowne (together with 18 quarters and a bushel of the 16 and 17, 18, 19, 20, dressings sowne in all there 18 quarters and half, upon about 43 acres in both feilds (whereof 2 acres and a rood my owne land in the Southfeild), which is neare 3 bushels and half per acre), and the remaining 2 bushels added to the next dressing.

Of the 22nd dressing, by Will: Thrift on Thursday March 20, there was in all ten quarters and two bushels, which (being by the addition of the 2 bushels of the former dressing made up ten quarters and half) were all carryed to Good: Kefford's to be malted by Payne and Dovy on Tuesday 25 of March 1690.

Note: That these dressings are all out of the barnes, the following dressings all of from the rick

Of the 23rd dressing, there was in all, March 28, 6 quarters and 4 bushels, whereof one bushel was sent to mill and the remaining 6 quarters and 3 bushels were sold to Mr. Cooke of Chishil at 13*s.* per quarter, for 4*l.* 2*s.* 10*d.*

Of the 24rth dressing, April 10, there was in al ten quarters and 2 bushels, whereof 3 pecks and one bushel were sold at home, for 2*s.* 7*d.* *ob.*; and a peck spent by our selvs; and the remaining ten quarters were al sold to Urias Bowes at 11*s.* 10*d.* per quarter, for 5*l.* 18*s.* 4*d.*

Of the 25th dressing, on Friday April 18, there was in al 9 quarters five bushels and half, whereof 6 quarters and 6 bushels were sold to Barkway at 11*s.* 10*d.* per quarter, for 3*l.* 19*s.* 9*d.* 2 bushels and half were sold at home, for 3*s.* 9*d.*, and the remaining 2 quarters and five bushels saved for our owne expence, having 2 sowes to pigg.

Of the whole crop of barly this yeare there was in all 201 quarters 3 bushels and a peck.

[*p.70*] **An account of the oates brought in in that harvest**

Of the first dressing of oates, October 14, there was in al three quarters and half a bushel, which were about half the tithe oates that I had this yeare at the Grange. The ricke being full, these were laide into hay-house and threshed in the malthouse and laide into malthouse chamber for our owne use, my old oates being all done.

Of the 2nd dressing, by W: Thrift on Thursday January 23, there was in all 10 quarters and 7 bushels, whereof 7 bushels were partly fanned up before hand and carryed into stable now; and ten quarters remaining were

HARVEST ACCOUNTS 1689-1690

carryed into malthouse chamber for our expence and use.

Of the 3rd dressing, by Will: Thrift alone also, there was in all 8 quarters and half a bushell, whereof 4 bushels and half were carryed directly into stable; and the remaining 7 quarters and half were carryed this January 31, the day of theire dressing, into malthouse chamber for our owne expence.

Of the 4rth dressing, by W: Thrift and J: Fere together on Saturday February 8, there was in all 17 quarters 6 bushels and half, whereof one quarter and 2 bushels were carryed directly into stable, being a sort of tayle and partly some that were left of the seede; and 12 quarters one bushel and half were laide into malthouse chamber; and 4 quarters and 3 bushels were sowne in the South Feild upon 8 or ten acres. So that there is this 26 of February in the malthouse chamber for expence at least 29 quarters, some being taken out already, viz about 5 bushels.

Of the 5th dressing, May 12, being a few that were laid on the rick amongst the pease, there was in all but one quarter and 2 bushels, all spent by our selvs.

For the whole crop of oates 44 quarters and 2 pecks.

An account of the pease brought in in that harvest – I had none of my owne land sowne; they were all tythe [p.72]

Of the first dressing of pease, by Will: Thrift on Saturday December 7, there was in all 6 quarters and 3 bushels and neare a peck besides, whereof 2 quarters and 5 bushels were sold at home at 20d. per bushel, for 1l. 15s. 0d., and the rest all spent by our selvs, some being paide where they were borrowed before, and some fanned up before hand, viz about 9 bushels.

Of the 2nd dressing, by Will: Thrift on Friday May 9, there was in all six quarters and six bushels, whereof six bushels and half have been sold at 1s. 8d. per bushel, for 10s. 10d. The rest remaine in the granary for our owne use.

The whole crop of pease this yeare, being all tithe, I having none else, is 13 quarters one bushel and a peck or neare upon, whereof there is sold in all 3 quarters 3 bushels and half.

An account of the quarters of corne of all sorts that I had of my [p.73] fifteenth crop, which was gathered in in harvest 1690. Harvest began that yeare generally on Wednesday July 23 (although divers were reaping the Saturday before; Mr. Grey had 14 acre men at it, which he increased before I began to 18 and 20, though there was very little carried untill Tuesday in the afternoon) and was compleatly ended on Friday night, August 22; so that harvest lasted just 31 dayes, that is a month and three dayes. Sheafe corne was brought in most of it very well; but afterwards we had much wett, and great raines, which

REV. JOHN CRAKANTHORP

occasioned much barly to be growne. The top of my rick (just finished before the raine) is much growne. The continued raines occasioned loose corne to be brought in very badly, and was at least a weeke's hinderance in the whole harvest. The Barr Feild this yeare was tilth and the South Feild was the brock or each crop. I had this yeare 6 stetches in Southfeild and two acres and a stetch in Barrfeild of my owne temporal estate, that is 4 acres and two roods and half. On tuesday this August 12 there was a dreadful tempest of thunder and lightning, which began a little before supper time and burnt all William Botterel's houses of Chrissol with 9 horses in his stable and one cottage adjoyning, so that they had nothing left, the fire being sudden. Laus Deo, me et alios, sua omnipotentia protegenti. Amen.

[p. 74] **An account of the wheat, both red and white, brought in that harvest**

Of the first dressing, of white wheate, threshed out of the malthouse September 16, there was in all 7 bushels, whereof 2 bushels and half were sowne upon one of the 3-furlong stetches in Northfeild. The other 4 bushels and a peck were mingled with the rye to make it good miscellaine; and the remaining peck was sold for 9d.

Of the 2nd dressing, on October 9 and 13, which was all red wheate, there was in all 3 quarters 2 bushels and 2 pecks, whereof 12 bushels and a peck were sowne upon 16 stetches in Northfeild. 2 pecks more were put to the rye and the remaining 1 quarter 5 bushels and 3 pecks were sold at 2s. 6d. per bushel, and 6 bushels of it at 2s. 5d. - in all together for 1l. 13s. 10d.

Notandum: That these 2 dressings, making in all 4 quarters one bushel and half, were all that were in the malthouse.

Of the 3rd dressing, on Saturday February 7, there was in all 2 quarters and 7 bushels, whereof 7 bushels were spent by ourselvs; and one quarter 7 bushels more were sold at home at 3s. per bushel, for 2l. 5s. 0d., and the remaining bushel was tayle. This was all white wheat.

Of the 4th dressing, on March 10, there was in all five quarters and 6 bushels, all sold and delivered at Royston at 2s. 6d. per bushel, for 5l. 15s. 0d., being al red wheat.

Of the 5th dressing, March 14, which was all white wheat (though it was very black, that it could not be used without washing), there was in al 5 quarters and 4 bushels, whereof 5 quarters were sold to one about Hertford and delivered at Royston at 2s. 2d. per bushel, for 4l. 5s. 0d., and 2 bushels and half sold at the same rate at home, for 5s. 5d., and the remaining bushel and half at 2s. per bushel, for 3s.

Of the 6th, on Tuesday March 17, which was of the same black wheate, there was in all 4 quarters 4 bushels and a peck, whereof six bushels, being washed, were spent by our selvs. The remaining 3 quarters and 6 bushels and a peck were all sold at home at 2s. per bushel, for 3l. 0s. 6d.

[p. 75] Of the 7th dressing, Aprill 27, 1691, which was red wheate, there was in all 8 quarters and 5 bushells, whereof 2 bushels, being tayle, were put to the rye; and five quarters and 4 bushels were sold to Mr. Oliver or Valentine Lee at 2s. 6d. per bushel, for 5l. 10s. 0d., and one quarter 3 bushels and half more sold at home at 2s. 6d. per bushel also, for 1l. 8s. 9d.; and the remaining one quarter 3 bushels and half were spent by our selvs.

Of the 8th dressing, on June 9th, there was in all 8 quarters and 3 bushels, which was all red wheate also, whereof 2 bushels were tayle and put into 5 quarters and 5 bushels of rye; and 6 quarters of it were sold to one of Digswel and delivered at Royston at 2s. 5d. per bushel and 1s. over in the whole, for 5l. 13s. 0d.; and 6 bushels more sold at 7 groates per bushel at home, for 14s. 0d.; and 4 bushels and 3 pecks more sold at home at 8 groates per bushel, for 12s. 8d.; and the remaineing 6 bushels and a peck were spent by our selvs.

Of the 9th and last dressing, June 20, which was all red wheate also, there was in all 5 quarters and 7 bushels, whereof five pecks were tayle, sold nevertheless at 2s.; and 4 quarters and 3 bushels more were sold at market at 2s. 5d. per bushel also and 1s. in the whole, for 4l. 5s. 6d.; and 7 bushels and a peck sold at home at 3s. per bushel, for 1l. 1s. 9d.; and the remaining 3 bushels and half were spent by our selvs.

Note: That of the whole crop of wheate this yeare there is in all 45 quarters five bushels and 3 pecks.

An account of the rye or miscellaine brought in in that harvest [p. 76]

Of the first dressing of rye, which was dressed up at divers times betwixt September 20 and October 1, there was in al 4 quarters 3 bushels and half, whereof one bushel was sold at 2s. All the rest was sowne upon 20 stetches and 4 acres and 2 acres – in all upon 13 acres and half – in Northfeild, whereof 2 stetches are my owne; so there is sowne in all (with the white wheate that was put to it) 4 quarters and 7 bushels wanting a peck, which is 3 bushels an acre, one with another, wanting about half a peck per acre.

Of the 2nd dressing, on Tuesday October 7, there was in all but 1 quarter 3 bushels and half, whereof ten bushels and half were sold at home at 2s. 2d. per bushel, for 1l. 2s. 9d. The remaining bushel was spent by our selvs.

Note: That this five quarters and 7 bushels is all that was threshed in the malthouse this yeare.

Of the 3rd dressing, on Saturday December 20, there was in all 6 quarters and 4 bushels, whereof five quarters were sold to Good: Warner of Baldock at 15s. per quarter, for 3l. 15s. 0d., and the one quarter and half remaining were sold with the next dressing to one at Ashwel at the same rate, for 1l. 2s. 6d.

Of the 4rth dressing of rye, on Friday January 2, there was in all 4 quarters and 7 bushels, whereof 3 quarters and half (being made up 5 quarters with the 12 bushels of the former dressing) was sold to one Everet of Ashwel at 15s. per quarter, for 2l. 12s. 6d., at [sic, for 'and'] 2 bushels sold out of the barne at 1s. 10d. per bushel, for 3s. 8d., and a bushel spent by our selvs. The remaining quarter was sold with the next dressing at 1s. 11d. per bushel, for 15s. 4d.

Of the 5th dressing, January 10, there was in al 4 quarters, sold (with the former quarters) to the same Everet at Ashwel at 1s. 11d. per bushel, for 3l. 1s. 4d.

[p.77] Of the 6th dressing of rye, on January 29, there was in all one quarter and five bushels, by John Casbourne, which was al sold at home at 1s. 10d. per bushel, and some at Royston by the bushel at the same rate, for 1l. 3s. 10d.

Of the 7th dressing, February 3, there was in all five quarters and 4 bushels, whereof 2 quarters and 6 bushels were sold at 22d. per bushel at home, for 1l. 19s. 4d.; and 2 quarters more at 2s. per bushel for 1l. 12s. 0d. 2 bushels and half were given to the poor, viz Widdows Dovy, Rand, Thrift, Casbourne, and Tho: Ward. The remaining 3 bushels and half spent by our selvs.

Of the 8th dressing, February 28, there was in al 2 quarters 2 bushels and half, whereof above a peck was tayle reserved for little piggs; and one bushel was spent by our selvs and the remaining 2 quarters one bushel and a peck were sold at home at 2s. per bushel, for 1l. 14s. 6d.

Of the 9th dressing, March 20, there was in all 3 quarters 5 bushels and half, whereof the odd half bushel was tayle reserved for little piggs; and 2 bushels were spent by our selvs, and the remaining 3 quarters 3 bushels were sold at home at 2s. per bushel, for 2l. 14s. 0d.

Of the 10th dressing, Thursday May 7, there was in all 7 quarters five bushels and half. The odd half bushel was tayle reserved for little piggs. Out of the 7 quarters and five bushels were sold at 2s. per bushel inprimis 6 bushels and an half at home, for 13s. 0d.; and one quarter and five bushels were sold at 22d. per bushel, for 1l. 3s. 10d.; and 4 quarters 6 bushels and half more were sold at 20d. per bushel, for 3l. 4s. 2d.; and 3 bushels remaining were spent by our selvs.

Of the 11th dressing, on Tuesday May 19th, there was in all 9 quarters just, whereof half a bushel againe was tayle, reserved for little piggs; and 6 quarters were sold to Widdow Whitehead at Royston at 1s. 10d. per bushel, for 4l. 8s. 0d.; and 3 bushels more were sold at 1s. 9d. per bushel, for 5s. 3d.; and 2 quarters and 3 bushels more were sold at home at 1s. 8d. per bushel, for 1l. 11s. 8d.; and the remaining bushel and half were given to the poor, viz to Widdow Watson, Widdow Woollard, Henry Law, being sick.

[p.78] Of the 12th dressing of rye, on Saturday May 30, there was in all 8 quarters and 2 bushels, whereof five bushels were spent by our selvs; and five quarters were sold to Symonds the miller at Wittlesford at 1s. 10d.; for 3l. 13s. 4d., yet so that he fetched it at home; and the remaining 2 quarters and five bushels were sold at home also, at the same price, for 1l. 18s. 6d.

Of the whole crop of rye this yeare there was 59 quarters 2 bushels and 2 pecks.

[p.79] **An account of the barly also brought in in that harvest**

Of the first dressing of barly, by John Casborn on Munday November 3rd, there was in all five quarters and four bushels, whereof one bushel is sent to mill for the sow, a bushel sold at home to M: Payn for 1s. 4d. ob., and the remaining 5 quarters and 2 bushels sold to Mr. Bowes at 11s. per quarter, for 2l. 17s. 8d.

Of the 2nd dressing, by J: Casborne also, there was in all six quarters and one bushel, whereof one bushel was spent by hoggs; the remaining 6 quarters sold to Mr. Urias Bowes at 11s. per quarter and carryed in this November 13, for 3l. 6s. 0d.

Of the 3rd dressing, on Friday November 21 by J: Casb. also, there was in all just 6 quarters, which was sold to my neighbour Urias Bowes at 10s. 10d. per quarter, for 3l. 5s. 0d.

Of the 4rth dressing, by J: Casbourne also, on Saturday November 29, there was in all six quarters and 2 bushels, whereof 6 quarters and one bushel were sold to Mr. Pigg of Chrissol at 10s. 10d. per quarter, for 3l. 6s. 4d., and half a bushel sold at home to Math: Payn for 8d.; the remaining half bushel spent by hogs.

Note: That about 19 quarters and half, or neare 20 quarters, came out of the middlesteade of the new barne.

Of the 5th dressing, by John Casbourne also, there was in all 7 quarters five bushels and half, whereof the 7 quarters and 5 bushels were sold to Mr. Glenister at Royston at 10s. 6d. per quarter, for 3l. 16s. 6d., and a peck sold at home at 4d.; the remaining peck laide into garner for hoggs, December 3, the day of its dressing.

Of the 6th dressing, on December 3 also, by Will: Th., there was in all 8 quarters and 7 bushels, whereof 6 quarters and 2 bushels sold to Shafnoe End at 11s. per quarter, for 3l. 8s. 9d., and the remaining 2 quarters were sold to Mr. Glenister at 10s. 6d. per quarter, for 1l. 7s. 6d., because the dressing it was sold withal was wett and course.

Of the 7th dressing of barly, by William Thrift on Tuesday December 9, there was in all but 4 quarters and 3 bushels, whereof one bushel sold at home to Math: Payne for 1s. 4d., and the remaining 4 quarters and 2 bushels were sold also to Mr. Glenister at 10s. 6d. per quarter, for 2l. 4s. 6d. [p.80]

Of the 8th dressing of barly, by Willi: Thrift also, there was in all 14 quarters, whereof one bushel was sold at home to Mat: Payne, for 1s. 3d., and six quarters were sold to M: Thurgood at Royston at 10s. 9d. per quarter, for 3l. 4s. 6d., and the remaining 7 quarters and 7 bushels were carryed in to Rich: Sell's at Triplow, to be malted for me, by Math: Payne and Thom Barber on Saturday December 20.

Of the 9th dressing, on Wednesday December 24 by Wil Thrift also, there was in all 14 quarters and 6 bushels and a peck, whereof the 14 quarters and six bushels were carryed in by Math. Payne and Thom: Barber on Munday December 29 to Rich: Sel's of Triplow to be malted for me; the remaining peck spent by sow haveing piggs.

Of the 10th dressing, by Wil: Thrift also, there was in al 16 quarters, whereof one bushel was sold to Antony Parish at 1s. 4d., and 15 quarters and 4 bushels were carryed in on Wednesday January 7 to Rich. Sell's of Triplow, to be malted, by Math: Payne and Thom Barber; and the three remaining bushels were sent to mill for sow and piggs.

Of the eleventh dressing, by W: Thrift also, on January 12, there was in all 7 quarters 4 bushels and half, whereof the 2 pecks were sold at home for 8d.; and one bushel sent to mill; and the 7 quarters and 3 bushels remaining were carryed in by M: Payne and Tho: Barber on Munday January 19 to Rich: Sell's of Triplow to be malted by him for me.

Of the 12th dressing, by Will Thrift also, there was in all six quarters, all sold to Mr. Uri: Bowes at 10s. 3d. per quarter, for 3l. 1s. 6d.

[p.81] Of the 13 dressing of barly, by Will: Thrift also, on January 20, there was in all six quarters 7 bushels and half, whereof the six quarters and 7 bushels were sold also to Mr. Urias Bowes at 10s. 3d. per quarter, and carryed in with the last dressing on January 22, for 3l. 10s. 6d.; and the remaining 2 pecks were put behind the garner door for off corne.

Of the 14th dressing, by Will: Thrift also, February 3rd, there was in all 7 quarters and 3 bushels, of which 2 bushels were spent by sow and piggs; and the remaining 7 quarters and one bushel were carryed in by Tho: Barber and Math: Payne on Wednesday February 10 to Rich: Sel's of Triplow to be malted for me.

Of the 15th dressing, on February 9 by Will Thri. also, there was in all 6 quarters five bushels and an half, whereof 3 bushels were sold at home, at 3s. 8d., and 6 quarters and 2 bushels were sold to one Turpentine of Buntingford and delivered at Barly to his trustee att 11s. per quarter and 6d. over in the loade, for 3l. 9s. 3d. The remaining half bushell was put into garner to off corne.

Of the 16th dressing, by J: Casbourne on Tuesday February 17, there was in all 7 quarters and 7 bushels, whereof five quarters were sold to the saide Mr. Turpentine of Buntingford and delivered as before at 11s. 4d. per quarter, for 2l. 16s. 8d. 2 pecks of the same were sold at home at 8d., one bushel sent to mill for piggs, and 2 quarters and 5 bushels more sold at Royston at 11s. 6[d.] per quarter, for 1l. 10s. 2d.; remaining half bushel was laide into garner behind the door.

Of the 17th dressing, by J: Casbourne also, on Tuesday February 24th, there was in all 6 quarters and half, whereof one bushel was sold at home to Antony Parish, for 1s. 5d., and 2 quarters and 3 bushels sold at Royston at 11s. 6d. per quarter, for 1l. 7s. 3d.; and the remaining 4 quarters saved for seed and sowne afterwards.

[p.82] Of the 18th dressing, on Friday March 6th, there was in all eleven quarters and five bushels, [p.82] threshed by Will: Thrift, whereof 3 pecks were sold to Anto: Parishes wife and J: Casbourne, at 1s. 1d. ob.; and six quarters and 3 bushels were carryed in by Math: Payn and Tho: B: on Munday March 9 to Rich Sel's of Triplow to be malted for me; a bushel sold to Rob: Macer at 1s. 6d., a peck laide into garner for hogs, and the remaining five quarters sold to Mr. Turpentine of Buntingford and delivered at Barly at 12s. 1d. per quarter, etc., for 3l. 0s. 5d.

Of the 19th dressing, by Will: Thrift also, on Thursday March 12, there was in all 7 quarters and 2 bushels, whereof a bushel sold to Anton: Parish, and a bushel more sent to mill for our selvs; and 6 quarters and 2 bushels sold to Mr. Turpentine at 11s. 8d. per quarter, for 3l. 12s. 11d., and the 6 bushels remaining were sold, one to Antony Parish, and the other 5 were sold with the next dressing, at 11s. per quarter, for 8s. 3d.

Of the 20th dressing, on Saturday March 19 by Wil Thrift also, there was in all 8 quarters 3 bushels and half, whereof half a bushel sold to John Casbourn at 8d. ob.; and the remaining 8 quarters and 3 bushels were sold to Mr. Bowes at 11s. per quarter, for 4l. 12s. 1d.

Of the 21rst dressing, by John Casbourn on March 20, there was in all 3 quarters and 3 bushels, which were all sowne with 4 quarters of the 19 dressing and 5 that I bought, in all 12 quarters and 3 bushels (only 2 pecks sold out of it for 9d.) upon 31 acres or thereabouts – on the Heath 4 and all the rest in Barrfield, which comes to about 3 bushels and a peck to an acre.

Of the 22th, on Saturday March 28 by John Casbourne also, there was in all 4 quarters and five bushels, whereof 4 quarters and 4 bushels were the same day carryed in to Rich: Sell's to be malted for me, and the remaining bushel was sold afterwards to Will: Thrift for 1s. 4d. ob.

Of the 23rd and last dressing of barly, by John Casbourne on Friday April 3rd, there was in all but five quarters and one peck, whereof 3 quarters and a bushel were sold to Urias Bowes at 11s. per quarter, for 1l. 14s. 4d. ob., and 5 pecks were sold to John Casbourne at the same rate, for 1s. 8d. ob., and 2 bushels more to Math: Payne for 2s. 9d.; and the remaining one quarter and 4 bushels were laide into garner for our owne expense. [p.83]

Memorandum: That of the whole crop of barly this yeare there was in all 178 quarters and 7 bushels, whereof 12 quarters and 3 bushels were sowne and about 3 quarters more spent by hoggs and fowles etc.; the rest malted and sold.

An account of the oates that were brought in in that harvest [p.84]

Of the first dressing of oates, on Friday October 24, there was in all 7 quarters and 2 bushels. One quarter and five bushels were partly fanned up before hand and carryed directly into stable; the other five quarters and five bushels were laide into garner for our owne use and expence.

Of the 2nd dressing of oates, on Friday November 7, by Will: Thrift also, there was in all 7 quarters and 2 bushels, whereof the 2 bushels were carryed directly into stable, a peck sold at 5d., and the rest laid all into garner for our owne use.

Of the 3rd dressing, by W: Thrift on Friday November 14, there was in all 5 quarters 5 bushels; and neare a peck laide all into garner to the former.

Note: That all these oates lay in the middlesteade of the old barne.

Of the 4rth dressing, on Saturday January 31, there was in all 11 quarters, whereof 6 quarters laide into garner to the former; 5 quarters saved for seed, and all sowne but 6 bushels laide into garner afterwards.

Of the 5th dressing, on February 14, there was in al 16 quarters one bushel and half; all (but the bushel and half spent) laide into malthouse chamber.

Of the 6th dressing, Munday March 30, there was in all 8 quarters, whereof there were sold 4 bushels to Henry Wallis at 5s. 4d.; 5 bushels and half more - 3 to J: Rayner, 2 to G: Kefford - at 15d., for 6s. 3d., and 2 pecks at 8d.; the rest laide into malthouse chamber for our owne use.

Of the 7th dressing, there was in all (April 3) five quarters and 3 bushels, whereof the 3 bushels tayle were carryed into stable; the rest laide into chamber.

Note: That of the whole crop of oates this yeare there is in all 60 quarters 5 bushels 2 pecks.

[p.85] **An account of the pease brought in in that harvest 1690**

Of the first dressing of pease, by W: Thrift on Saturday November 15, there was in all but one quarter and six bushels which were pease, that lay in the middlestead of the old barne together with 20 quarters of oates; and one bushel of the 3 first dressings. They were laide into garner and expended by fatting hoggs.

Of the 2nd dressing, on Saturday November 22, there was in all 3 quarters and six bushels, whereof a bushel and neare half was tayle, and the 3 quarters and half and half a bushel laide into garner for our owne use. These came of from the top of the mow next the pond in old barne.

Of the 3rd dressing, on Munday February 2, there was in all 3 quarters and 4 bushels, whereof 3 pecks were sold to J: Casb: and Wil: Thrift, and 3 pecks to Good: Parish, one bushel to Henry Thompson, and 6 bushels more to Rich: Sell of Triplow - in all one quarter and half a bushel - at 1s. 8d. per bushel, for 0l. 14s. 2d.

Of the 4rth dressing, on Tuesday February 24, there was in all 6 quarters and five bushels, whereof 2 bushels and half were tayle, spent by the shoates at theire first shutting up; and the remaining 6 quarters 2 bushels and half laide into garner for our owne use.

Memorandum: That of the whole crop of pease this yeare there was in all fifteen quarters and five bushels.

[p.86] An account of all the corne that I had of my sixteenth cropp that was gathered in in harvest 1691. Harvest this yeare began generally on Thursday July 23 (although both the farme and the Lordship had divers acres men at it the Munday before but carryed none till Tuesday or Wednesday, and others also besides were reaping - wheate especially, which was mildewed this yeare, and so was reaped first) and (excepting 2 or 3 loade of oates of John Rayner's out of which I had my tithe) was compleatly ended on Saturday August 15, which was 3 weekes and three dayes. The harvest this yeare was tollerably good. Though we had 2 or 3 great raines, the weather was good after them. We had 3 great tempests this yeare, of thunder and lightning, once in harvest and 2 the week after, which, with the abundance of raine that fel, occasioned my rick to grow green before it could be thatched. Great harme was saide to be done in some places by those tempests, as in a towne near Stortford a barne ful of corne was burnt. The North Feild was tilth this yeare and the Barr Feild brock. I had 2 acres and a stetch of my owne land in the Barr Feild, and 2 stetches in North-feild, besides my glebe.

An account of my crop gathered in in that yeare, and first of wheate, white or red [*p.87*]

Of the first dressing of wheate, which was white wheate, on Munday September 7th, there was in all but 2 bushels and a peck, which were sowne upon 2 stetches at Hickses Hedges, and a part of a stetch in Horsted Shott, being the tithe of a piece of Good: Rayner's.

Of the 2nd dressing, which was all red wheate, on September 28, there was in all 2 quarters one bushel and 3 pecks, whereof five pecks of it were tayle, yet sold at 1*s.* 8*d.* per bushel, for 2*s.* 1*d.*; and the remaineing 2 quarters and 2 pecks were all sowne upon 6 stetches in Waterden, and one against Hickses Hedges and the stetch and half acre cross London Way at the first knap, and upon 4 acres in the Rowze – in al upon about 7 acres and an half, which is not 2 bushels and half per acre, it being all little measure, because stony ends in Waterden were sowne with rye before.

Of the 3rd dressing, which was all red wheate, on October 8, there was in all 3 quarters and one bushel, whereof one quarter 2 bushels and a peck were sold at home at 3*s.* per bushel, for 1*l.* 10*s.* 9*d.*; and 3 bushels and half more were sold at 10 groates per bushel, for 11*s.* 8*d.*; and the remaining one quarter 3 bushels and a peck were spent by our selvs, the last sent to mill December 7.

Notandum: All these 3 dressings, making in all five quarters and five bushels, were threshed out of the malthouse.

Of the 4rth dressing of red wheate, December 22, there was in all 4 quarters 5 bushels and a peck, whereof one bushel was tayle, yet sold for 2*s.* 6*d.*; 3 quarters one bushel and a peck were sold at home at 3*s.* 8*d.* per quarter, for 4*l.* 12*s.* 7*d.*; and one quarter and 2 bushels more spent by our selvs; and the remaining bushel either lost or misreckoned.

Of the 5th dressing, of red wheate, there was in all, April 27, 3 quarters 4 bushels and a peck, whereof 3 bushels were spent by our selvs; and 3 quarters and 3 pecks were sold at home at 3*s.* 8*d.* a bushel, only one quarter at 1*l.* [9*s.*] 10*d.*, for 4*l.* 11*s.* 5*d.* – so that half a bushel still remaineing is either lost or misreckoned.

Of the 6th dressing of wheate (which was all white wheate), on Tuesday [*p.88*] May 3, there was in all but one quarter and 3 bushels, whereof 3 pecks were sold to Will Symonds of Wittlesford at 11 groats per bushel, for 2*s.* 9*d.*; and about 9 bushels and 3 pecks were all spent by our selvs, so that there is about 2 pecks lost, it lying above 4 months; my self threw away at one time about a peck spoyled by wevils.

Of the 7th dressing, on Thursday June 2, which was all red wheate, there was in all 6 quarters 2 bushels and half, whereof 6 bushels and half were sold at home at 11 groates per bushel, for 1*l.* 3*s.* 10*d.*; and 2 bushels and half sold at home also at 4*s.* per bushel, for 10*s.*; and 5 quarters sold at Royston at 4*s.* 2*d.* per bushel, for 8*l.* 6*s.* 8*d.*; and the remaining bushel and half was tayle put amongst the rye.

Of the 8th dressing, which was all red wheat, on Tuesday June 14, there was in al 6 quarters and 2 bushels, whereof one bushel tayle, half of it being put behind the garner door, the other half amongst the rye; and 4 quarters sold to Lydlington at 4*s.* a bushel, for 6*l.* 8*s.* 0*d.*; and 7 bushels and half sold at home at 4*s.* a bushel, for 1*l.* 10*s.* 0*d.*; and the remaining one quarter one bushel and half was spent by ourselvs.

Note: That the whole crop of wheate this yeare is 27 quarters and six bushels, whereof one quarter and 5 bushels one peck was white wheate and all the rest redd.

[*p.89*] **An account of the rye or miscellaine gathered in in that yeare in al**

Of the first dressing of rye, at divers times, the last of them on Thursday September 17, there was in all six quarters, all sowne upon the five stetches cross London Way and upon 9 acres in Waterden Shott (that is the whole shott excepting the upper end of 6 stetches), and upon the acre and stetch towes in Waterden Shott – in all about 15 acres, that is 3 bushells an acre, and neare half a bushel over allowed for Waterden Shott, which is very great measure.

Of the 2nd dressing, September 21, there was in al but one quarter and 3 bushels, whereof 2 bushels sowne upon a stetch in the Rowze; and 5 bushels and half were picked and sowne upon the 2 acres in Horsted Shott for seed another yeare; and 2 bushels more were sold at 1*s.* 10*d.* per bushel, for 3*s.* 8*d.*, and the remaining bushel and half was spent by ourselves.

Note: That these 2 dressings, making in all 7 quarters and three bushels, were all threshed out of the malthouse.

Of the 3rd dressing, on November 25, there was in all one quarter 7 bushels and 3 pecks, whereof half a bushel was given to Athan: Carrington, and one bushel was spent by ourselves; and the remaining 14 bushels and a peck was sold at home and 7 groates per bushel, for 1*l.* 13*s.* 1*d.*

Of the 4rth dressing, December 1, there was in all 4 quarters and 6 bushels, whereof 2 pecks were given to Widdow George, and 4 quarters 2 bushels and half were sold at home at 2*s.* 4*d.* per bushel, for 4*l.* 0*s.* 6*d.*; and the remaining 3 bushels were spent by our selvs.

Of the 5th dressing of rye, on Tuesday February 9, there was in all eleven quarters and 3 bushells, whereof one peck tayle laide behind the garner door; and 2 quarters and 7 bushels sold at home at 7 groates, for 2*l.* 13*s.* 8*d.*, and five quarters sold at Royston, for 5*l.* 1*s.* 0*d.* Three bushels and half were given to Thom Ward and Widdows Ingrey, Numan, Rand, Whitby, Woods, Woollard; and 4 bushels were spent by our selvs; and 2 quarters 4 bushels sold at home at 2*s.* 6*d.* per bushell, for 2*l.* 10*s.* 0*d.*; and one peck lost in measuring, skreening, etc.

Of the 6th dressing, Friday February 19, there was in al 7 quarters; five quarters sold to Ashwel at 2s. 7d. per bushel, for 5l. 3s. 1d., and the remaining 2 quarters sold at home at 2s. 6d., for 2l. 0s. 0d.

Of the 7th dressing of rye, on April 16, there was in all 9 quarters and 2 bushels with 3 pecks of tayle wheate put to it, whereof 8 quarters 6 bushels and half were al sold at home at 2s. 6d. per bushel, for 8l. 16s. 3d.; and 3 bushels were spent by our selves; and yet there is half a bushel either lost in measuring or miscounted, but I suspect the latter because it did not lye a month. [p.90]

Of the 8th dressing of rye, May 14, there was in all 7 quarters 3 bushels and 3 pecks, whereof 2 pecks given to Rob: Macer; and all the remaining 7 quarters 3 bushels and one peck were sold at home readily, at 2s. 6d. per bushel, for 7l. 8s. 1d. ob.

Of the 9th dressing of rye, Munday May 23, there was in all even 7 quarters, whereof 5 quarters and 6 bushels and half were sold at 2s. 6d. per bushel, for 5l. 16s. 3d. 3 bushels and half were spent by our selvs, and the remaining 6 bushels were sold also at home at 8 groates per bushel, for 16s.; but still there is a bushel and half lost of tayle wheate that was put to it in the whole.

Notandum: Of the whole cropp of rye this yeare there was in al 56 quarters one bushel and half, whereof 7 quarters wanting a bushel was sowne upon about 17 acres in Southfield, whereof also I carryd in but 10 quarters to Royston and Ashwel, and I bought in 9 quarters in the roome of it againe; so that in effect my whole crop was sold at home.

An account of the barly gathered in in harvest that yeare [p.91]

Of the first dressing, by J: Casbourne on November 3, there was in all five quarters and 7 bushels, whereof 5 quarters and 6 bushels were sold to W: [sic; perhaps an error for M:] Thurgood at Royston at 11s. 10d. per quarter, for 3l. 8s. 0d., and the remaining bushel was spent with fowles etc.; a peck paide to Thrift.

Of the 2nd dressing, on Tuesday November 10 by Will: Thrift, there was also five quarters and 7 bushels, whereof one bushel was sold to Rich: Haycock, at 1s. 6d., and the remaining 5 quarters and 6 bushels were sold to Berkway at 11s. 9d. per quarter, for 3l. 7s. 6d.

Of the 3rd dressing, November 16, by J: Casbourne (which with the first dressing was all that lay in the middlesteade of the new barne), there was in all 6 quarters and one bushel, whereof a peck sold to Math: Payne at 4d. ob. and 2 pecks were sent to mill, and a peck tayle; and the remaining 6 quarters were sold to Berkway at 11s. 9d. per quarter, for 3l. 10s. 6d.

Of the 4rth dressing, by Will: Thrift, November 23, of which 5 quarters and five bushels were sold, one bushel to Henry Law, the rest to Mr. C[ook] at Chishill at 12s. 4d. per quarter, for 3l. 9s. 4d. ob. A peck spent, that was fanned up, and the remaining peck was sold to Math: Payne at 4d. ob.; and a peck paide Good: Dovy that was borrowed and spent before.

Of the 5th dressing, by Will: Thrift, December 4th, there was in all 7 quarters and 2 bushels, whereof 2 pecks spent by our selvs; and 2 pecks sold to W: Thrift for 10*d.*, and 6 quarters and half sold to Mr. Cook at 13*s.* 9*d.* per quarter, for 4*l.* 9*s.* 10*d.*, and the remaining 5 bushels sold at 15*s.* per quarter, for 9*s.* 4*d.*, with the 6th dressing.

Of the 6th dressing, on December 10 by John Casbourn, there was in all even five quarters; all sold to one at Shafnoe End at 15*s.* per quarter, for 3*l.* 15*s.*

Of the 7th dressing, by Will: Thrift on December 18, there was in all ten quarters and 3 bushels, of which 3 pecks sold at home at 1*s.* 7*d.*, a peck spent for horses' mash and 5 quarters and half sold to Mr. Cook at 17*s.* 4*d.* per quarter, for 4*l.* 15*s.* 4*d.*; and the remaining 4 quarters and 6 bushels carryd in to Rich: Sell's to be malted for me December 21 by Math: Payne and Thom: Barber.

Sum is 46 quarters 1 bushel 3 pecks [*apparently a later note*].

[*p.92*] Of the 8th dressing of barly, December 31 by Will: Thrift also, there was in all 12 quarters wanting a peck (besides tayle that would have made it up), whereof a peck sold to W: Thr: at 6*d. ob.*; and 2 pecks sent to mill for our selvs; and the 11 quarters 7 bushels remaining were carryed in to Rich. Sell's by the farm hands on Saturday January 2 to be malted for me.

Of the 9th dressing, by J: Casbourne on January 6, there was in all 6 quarters 3 bushels and 3 pecks, of which 3 pecks sold to Mat: P: and J: Casb. himself, at 1*s.* 7*d.*, and the remaining 6 quarters and 3 bushels were carryed in by the same men to Rich: Sell's on Friday January 8.

Of the tenth dressing, on Tuesday January 12, by J: Casb. also, there was in all five quarters and five bushels, of which 3 pecks spent; and a peck to Ant: Parish at 5*d. ob.*; and the remaining 5 quarters and half sold to Miles Thurgood at 14*s.* 4*d.* per quarter, for 3*l.* 18*s.* 10*d.*

Of the 11th dressing, on January 26, by John Casbourne also, there was in all 9 quarters and 4 bushels, whereof 2 pecks to Anton: Parish and one to Good: Payne, at 1*s.* 4*d. ob.*; and 4 quarters and 3 bushels were carryed to Rich: Sell's by M: Payne and Tho: Barber on January 27 to be malted; and five quarters sold to M: Thurgood at 13*s.* 9*d.* per quarter, for 3*l.* 8*s.* 9*d.*, and the remaining peck sold at home for a groat.

Of the 12th dressing, by Will: Thrift, February 4, there was in all 11 quarters and 7 bushels, whereof one bushel was sold at home to 3 persons, at 1*s.* 10*d.*, and the remaining 11 quarters and 6 bushels were all carryed in to Rich: Sell's to be malted for me on February 4 and 5 by Math: Payne and Thom: Barber.

Of the 13th dressing, on February 10 by W: Thrift, there was in all 5 quarters; all sold to Steven Hagger of Chishill at 15*s.* 2*d.* per quarter, for 3*l.* 15*s.* 10*d.*

Of the 14th dressing, on February 22 by Will: Thrift, there was in all 5 quarters and one bushel, whereof the odd bushel was sold to M: Payn, J: Casb: and Tho: W. for 2*s.*, and the remaining 5 quarters to M: Thurgood at 16*s.* per quarter and 1*s.* over in all, for 4*l.* 1*s.* 0*d.*

Of the 15th dressing, on February 27, by W: Thrift also, there was in all 5 quarters and 2 bushels, whereof 5 quarters sold to Ashwel at 16*s.* 3*d.* per quarter, for 4*l.* 1*s.* 3*d.*, and 3 pecks sent to mill for foules; and the remaining 5 peckes sold – to M: P. 3, to J: Casb: 1, to W: Th: 1 – for 2*s.* 6*d.*

Of the 16th dressing of barly, by John Casbourn, there was in all 6 quarters and five bushels, on Munday February 29, whereof 5 quarters and 5 bushels sold to James Good of Berkway at 15*s.* 3*d.* per quarter, for 4*l.* 5*s.* 8*d.*, and one remaining quarter was all sowne. [*p.93*]

Of the 17 dressing, on March 8, by John Casbourne also, there was in all 4 quarters and 6 bushels, of which one peck sold to Math: Payne at 6*d.* All the rest used for seed and sowne.

Of the 18th dressing, on March 15, by Jo: Casbourne also, there was in all five quarters and five bushels, which were all sowne upon about 12 acres in Southfield and 20 and an half in the Northfield – in all about 32 acres and half, which at 3 bushels an acre comes but to 12 quarters and a bushel; so that, besides 11 quarters and 3 bushels sowne of these 3 last dressings, there was some of the next dressing sowne also, quem vide.

Of the 19th dressing, March 23, there was in all six quarters and 6 bushels, whereof 9 bushels were saved for seed and five quarters were sold to M: Thurgood of Royston at 15*s.* 3*d.*, for 3*l.* 16*s.* 3*d.*; a bushel sold at home at 2*s.*; and the remaining 4 bushels were sold at 16*s.* 4*d.* with the next dressing, for 8*s.* 2*d.*; threshed by J: Casb:.

Of the 20th dressing, being the last of J: Casbourne, about March 31, there was in all 5 quarters 7 bushels and 2 pecks, whereof 2 pecks spent by our selvs and 5 quarters one bushel sold to M: Thurgood at 16*s.* 4*d.* per quarter, for 4*l.* 3*s.* 8*d.*; and 2 bushels were sold with the next dressing, at 15*s.* 4*d.*, for 3*s.* 10*d.*; a peck sold to Wil: Thrift at 6*d.*, and 2 bushels more sent to mil for sow and piggs; and the remaining 2 bushels sold at home for 4*s.*

Of the 21st dressing, April 6, by Will: Thrift, there was even five quarters, all sold to Mr. Hughes of Royston at 15*s.* 4*d.* per quarter, for 3*l.* 16*s.* 8*d.*

Of the 22nd dressing, on April 13, by Will: Thrift also, there was in all five quarters and 2 bushels, all sold to M: Thurgood of Royston at 14*s.* 9*d.* per quarter, for 3*l.* 17*s.* 5*d.*

Of the 23rd dressing, on April 22, by Wil: Thrift also, there was in all five quarters and half a bushel, of which the 2 pecks spent for fowles; and the five quarters all sold to one of Haydon at 14*s.* 6*d.* per quarter, for 3*l.* 12*s.* 6*d.*

Of the 24th dressing of barly, on April 28, by William Thrift also, there was in all five quarters and five bushels, whereof five quarters were sold to one Keitly of Royston at 14*s.* 8*d.* per quarter, for 3*l.* 13*s.* 4*d.*; and the remaining 5 bushels were saved for our owne expence, for fowls, piggs, etc. [*p.94*]

Of the 25th and last dressing, by Will: Thrift also, on May 6, there was in all 4 quarters and 3 bushels, whereof 4 quarters sold to one of Ashwel at 15*s.* per quarter, for 3*l.*; 2 bushels more sold at home at 3*s.* 10*d.*, and the remaining bushel saved for our owne expence.

Notandum: Of the whole crop of barly this yeare there was in all 162 quarters and one peck, whereof were sowne 12 quarters 3 bushels and 3 pecks; and one quarter and 6 bushels more were spent – that is in all 14 quarters one bushel and 3 pecks; and all the rest sold.

Of all sorts of corne this yeare there was as followeth,

in primis of wheate	27 quarters	6 bushels	0 pecks
item, of rye or miscellaine	56	1	2
item, of barly	162	0	1
item, of oates, in al	38	4	3
item, of pease, in al	28	1	3
summ total of all together is	312	6	1

[p. 95] **An account of the whole crop of oates gathered in in this yeare**

Of the first dressing of oates, on October 20, there was in all 7 bushels and 3 pecks, laide upon the hay at last and threshed in the malthouse after the wheate and rye; and were all laide in to malthouse chamber for our owne use.

Of the 2nd dressing, November 21, by J: Casbourne (out of the middlestead of the new barne, besides 12 quarters of barly), there was in all 5 quarters and half a bushel, of which one bushel was tayle carried into stable; and five quarters wanting half a bushel were laid into the first bin in the garner, whereof about one quarter and half were sowne upon the 4 acres on the Heath; and the remaining 3 quarters 3 bushels and half were al sold at home at 1s. 4d. per bushel, for 1l. 16s. 8d.

Of the 3rd dressing, by Will: Thrift and Math: Payne together on Tuesday January 16, there was in all 13 quarters, whereof 3 quarters and 3 bushels were sold at home to Thom: Nash and his brother Wedd etc. at 15d. per bushel, for 1l. 13s. 9d., and two quarters and 4 bushels more at 1s. 4d. per quarter [*sic, for* 'bushel'], for 1l. 6s. 8d.; and one quarter more sold also at home at 1s. 6d. per bushel, for 12s.; and the remaining 6 quarters and a bushel laide into garner and spent by our selvs.

Of the 4rth dressing, on Saturday January 23 by the same two men, there was [in] all ten quarters and six bushels, whereof 2 bushels were a kind of tayle, and Thom: had them into stable; and the remaining ten quarters and half laide into malthouse chamber for our owne use.

Of the 5th dressing, by Will: Thrift on Saturday May 14, there was in all 8 quarters and five bushels, of which Thom: had 2 bushels directly into stable; and one bushel sold at 1s. 8d. and 3 pecks more at 1s. 6d.; and the remaining 8 quarters and 5 pecks laide into malthouse chamber for our owne use.

Notandum: That the whole crop of oates this yeare is 38 quarters 4 bushels and 3 pecks.

[p. 96] **An account of the quarters of pease gathered in in harvest that yeare**

Of the first dressing of pease, being a few that lay in the bottom of the middlesteade, there was in all but five bushels, spent presently by our fatting hoggs, dressed up on November 24.

Of the 2nd dressing, on Thursday December 10, there was in all but 4 bushels and half, being the tithe of a few white pease, of which about 3 bushels and half sold to poor people at 6*d.* a peck and the rest eaten or spent.

Of the 3rd dressing, on Friday December 11, there was in all six quarters and 3 bushels of ordinary grey pease, whereof 2 quarters and 7 bushels sold at 23*d.* a bushel, for 2*l.* 4*s.* 1*d.*, and the 2 quarters and 7 bushels remaining were spent by our fatting hoggs.

Of the 4rth dressing, Friday January 8, there was in all 7 quarters 7 bushels and 3 pecks, whereof one bushel was tayle spent first by hoggs; and six quarters sold at home, some at 2*s.*, some at 2*s.* 2*d.*, some at 7 groats one with another at 2*s.* 2*d.*, for 5*l.* 4*s.* 0*d.*; and the remaining one quarter 6 bushels and 3 pecks spent by our selvs.

Of the 5th dressing, on Wednesday March 30, by Will Thrift, there was in all 3 quarters 5 bushels and half, whereof 4 bushels and 3 pecks were sold to Good: Nash, Good: Wedd, and J: Watson senior at 2*s.* 2*d.* per bushel, for 10*s.* 3*d. ob.*; and 3 bushels and half more sold to Good: Wedd, Good: Sell and J: Swann at 7 groats a bushel, for 8*s.* 2*d.*; and the remaining 2 quarters 5 bushels and a peck spent by hoggs, horses, etc.

Of the 6th dressing, on Tuesday April 19, there was in all even 9 quarters, all laide into granary for our owne use.

Notandum: That of the whole crop of pease this yeare there was in all 28 quarters one bushel and 3 pecks, whereof there were sold ten quarters 3 bushels and 3 pecks, for 8*l.* 13*s.* 6*d.*; and the rest, being 17 quarters and 6 bushels, spent.

An account of all sorts of corne or graine that I had of my [*p.97*] seventeenth cropp, which was gathered in in harvest 1692, when harvest began generally on Thursday the fourth or 4th of August (though there was some reaped on the Friday before and the Lordship and farme acre men had made a considerable entrance on the Tuesday before, though through the wetness of the season there was none carryed till Munday and Tuesday the 8 and 9 of August, and that which was carry'd in on those 2 dayes, the 6 and 7 dayes being very wett, was brought in but badly) and was compleatly ended on Tuesday the sixth of September, all but a loade or two left on purpose by the Lordship men for the next day; so harvest lasted 4 weekes and five days. The South Feild was tilth this yeare and the North Feild the brock, broake or eddish. All sorts of corne were brought in very well that came in after that 9th day of the month, unless any wheate were carryed in presently after reaping – for the straw was very greene – or unless any were carryed in on the 13 of August, because there was a very great storme of raine the night before; and we had no raine that was anything considerable all harvest longe after the 7th day but that, and that falling in the night, hindered no work the next day. I had 8 stetches of my owne this yeare in both fields, besides gleabe. Sheafe corne, that was inned on the 8 and 9 dayes aforesaide, which was not much, but the first that was brought in, was brought in the worst of any in this harvest. On the 12 day at night, or rather on the 13 day of August in the morning, it being about one or two of the clock, was a dreadful tempest of lightning, which was more terrible with us than the thunder, followed at last with a great puff of wind and raine; and was altogether very terrible; and a malting at Babran and an house at paper mills, by Sturbridge Faire, burned by that lightning: Laus Deo me minerentem et omnia mca protegenti.

[*p. 98*] **An account of the crop brought in in harvest that yeare, and first of the wheate, white or redd, id est Kentish**

Of the first dressing of wheate, finished on September 26, there was in all 3 bushels, whereof one bushel and a peck, being white wheat, was mingled with the rye that was sowne, and the 7 pecks remaining, being redd wheate and threshed on purpose to send to mill, at our neede was spent by our selvs.

Of the 2nd dressing, which was red wheate (threshed in the malthouse also, being all that was laide there), there was in all 2 quarters 2 bushels and 2 pecks, of which 11 bushels and half (together with 5 bushels that were bought at 5*s.* per bushel) were sowne upon betwixt 7 and 8 acres in Barr Feild, and 2 bushels and 3 pecks were sold at 4*s.* per bushel for 11*s.*, and the remaining 4 bushels and a peck were spent by ourselvs. The last of it was dressed on October 18.

Of the 3rd dressing, of red wheate, on Tuesday December 20, there was in al 4 quarters and 7 bushels, whereof 3 pecks were tayle put amongst the 6th dressing of rye and sold with it at 4*s.* per bushel, for 3*s.*; and the remaining 4 quarters 6 bushels and a peck (being made up out of the garner even 5 quarters) were sold to one Cornwel, a miller of Hormead, at 5*s.* 6*d.* per bushel, for 10*l.* 10*s.* 4*d.*

Of the 4rth dressing, of red wheate, on Munday February 13, there was in all 3 quarters 6 bushels and 3 pecks, whereof one peck was tayle put amongst the 4rth dressing of rye and sold with it at 10*d. ob.*, and 3 bushels and a peck more were sold at home, at 4*s.* 4*d.* per bushel, for 14*s.* 1*d.*; and 1 quarter 2 bushels and 3 pecks more at 4*s.* 8*d.* per bushel, for 2*l.* 10*s.* 2*d.*; and 3 bushels more at 5*s.* 4*d.*, for 16*s.* 0*d.*; and one bushels and 3 peck at 5*s.* 6*d.*, for 9*s.* 7*d. ob.*; and the remaining 1 quarter 3 bushels 3 pecks spent by our selvs, of which one bushel was borrowed to sow and paide again out of this to Edward Godfrey and Henry Thompson.

Of the 5th dressing, of red wheate, on Tuesday February 28, there was in all 5 quarters and 3 bushels, whereof one bushel tayle put amongst the 9 dressing of rye and sold for 4*s.* and five bushels sold at 5*s.* 6*d.*, for 1*l.* 7*s.* 6*d.*;

and 4 quarters more at 5s. 4d. at Royston, and 3 bushels at home at the same rate, for 9l. 6s. 8d.; and the remaining 2 bushels spent by our selvs.

Of the 6th dressing, of red wheate, on Thursday April 6, there was in all [p.99] five quarters and 2 bushels, whereof five quarters were sold to one Prist of Hitchin at 6s. 2d. per bushel and delivered there, for 12l. 6s. 8d., and one peck sold at home at 1s. 6d.; and the remaining 7 pecks were tayle, put afterwards to the 10 dressing of rye and sold with it at 4s. per bushel, for 7s.

Of the 7th dressing, of red wheate, on Thursday Aprill 13, there was in all five quarters and 2 bushels also, whereof one bushel was tayle, put to the 9th dressing of rye and sold with it at 4s. per bushel, for 4s.; and one quarter 2 bushels and half were sold at home at 5s. 4d. per bushel, for 2l. 16s. 0d.; and 2 bushels sold at Royston at 5s. 6d., for 11s.; and one quarter one bushel and a peck more were sold at 5s. 8d. per bushel, for 2l. 5s. 5[d.], and 2 bushels and half more at the same price, for 14s. 2d.; and 2 bushels and 3 pecks at 5s. 9d., for 15s. 9d.; and 3 bushels more sold at 6s. per bushel, for 18s.; and the remaining one quarter and 3 bushels were spent by our selvs.

Of the 8 dressing, of redd wheate, on Saturday Aprill 22, there was in all five quarters and 2 bushels, of which 3 pecks were tayle. I know not whither put to ye rye or no – it holds not out, see 14 dressing of rye; and 4 quarters were sold to Lydlington at 5s. 10d. per bushel, for 9l. 6s. 8d.; and 4 bushels sold at home at 5s. 8d. per bushel, for 1l. 2s. 8d.; and about five bushels and a peck were spent by our selvs.

Of the 9th dressing, on May 8, which was all white wheate, there was in all 2 quarters and 4 bushels and about half a peck, whereof one peck and half was tayle, put to the 9th dressing of rye and sold with it for 1s. 8d.; and 3 pecks were sold at 5s. 8d. per bushel, for 4s. 3d., and 7 pecks more were sold at 6s. per bushel, for 10s. 6d., and the remaining 2 quarters one bushel and a peck were all spent by our selvs, most of it in harvest.

Of the whole crop of wheat this yeare there was 35 quarters and a peck.

An account of the rye or miscellaine brought in in that harvest, viz 1692 [p.100]

Of the first dressing of rye, finished September 21, 1692, with the shalings and all, there was in all 4 quarters and six bushels, whereof 4 quarters and five bushels have been sowne on all the land in Short Shotts in Barrfield and other parcels on this side of it; and the remaining bushel sold to Mat: Payne at 8 groats per bushel, for 2s. 8d.

Of the 2nd dressing, Thursday October 6, there was in all 2 quarters 4 bushels and half, whereof was spent by our selvs 4 bushels; and sold at 8 groats six bushels and more, for 16s.; and 7 bushels and a peck more were sold at 3s. a bushel, for 1l. 1s. 9d., and the remaining 3 bushels and a peck were sowne (together with 6 bushels, that were bought at 10 groats per bushel, that were sowne upon the 2 White Acres against Lamps Acre); so that there is sowne now in all of rye in Barrfield about 19 acres.

Of the 3rd dressing of rye, on Munday November 14, there was in al 2 quarters and 3 bushels, being the first out of the barne, whereof 6 bushels sold at home at 10 groats, for 1*l.* 0*s.* 0*d.*, and 9 bushels and a peck more were sold at 3*s.* 6*d.* per bushel, for 1*l.* 12*s.* 4*d. ob.*, and the remaining 3 bushels and 3 pecks were spent by our selvs.

Of the 4rth dressing, on Thursday December 8, there was in all 3 quarters 5 bushels and half (besides a peck of tayle wheate put amongst it), whereof 2 bushels and half were spent by our selvs, and the remaining 3 quarters and 3 bushels were sold all at home at 3*s.* 6*d.* per bushel, for 4*l.* 14*s.* 6*d.*

Of the 5th dressing, on Wednesday January 1, there was in all six quarters one bushel and half, whereof a peck tayle laide into garner and five pecks spent by our selvs; one quarter sold at home and 5 quarters more at Royston, at 3*s.* 6*d.*, for 8*l.* 8*s.* 0*d.*

[*p.101*] Of the 6th dressing, on Thursday February 2, there was in all 8 quarters two bushels and 3 pecks, of [*p.101*] which two quarters and 6 bushels sold at home at 3*s.* 6*d.* per bushel, for 3*l.* 17*s.* 0*d.*; and 4 quarters and 4 bushels and 3 pecks more sold at 4*s.* per bushel at home also, for 7*l.* 7*s.* 0*d.*; and 3 bushels more sold at Royston at 4*s.* 1*d.*, for 12*s.* 3*d.*; and 5 bushels remaining were spent by our selvs.

Of the 7th dressing, on Tuesday March 7, there was in all 6 quarters, whereof 4 quarters and half were sold to Weston, miller in Hertfordshire, and delivered at Royston at 4*s.* per bushel abateing 1*s.* in five quarters, for 7*l.* 3*s.*; and 4 bushels and half sold at home at 4*s.*, for 0*l.* 18*s.* 0*d.*; and the remaining 7 bushels and half were given to the poor, viz to Widdows Dovy, Bourne, Ingry, George, Watson, etc.

Of the 8th dressing, on Tuesday March 14, there was in all 8 quarters and one bushel, whereof five quarters were sold to Sim: Beman of Shingy at 4*s.* per bushel, for 8*l.* 0*s.* 0*d.*; and 2 quarters 6 bushels and half sold at home at the same rate, for 4*l.* 2*s.* 0*d.*; and the remaining 2 bushels and half were given to the poor, viz to Widdow Reynolds, Joane Woods, R: Macer, Wil: Whitby, J: French.

Of the 9th dressing, on Thursday March 23, there was in all six quarters 5 bushels and half, whereof one quarter and a bushel were sold at Royston for 4*s.* wanting a penny, at 1*l.* 15*s.* 3*d.*; and 4 quarters 7 bushels and half more sold at home at full 4*s.* per bushel (only 5 bushels of it were sold to Ed: Payne at that rate), for 7*l.* 18*s.* 0*d.*, and the remaining five bushels were spent by our selvs.

Of the tenth dressing, on Friday May 19, there was in all 5 quarters 4 bushels and half, out of which were spent by our selvs 2 bushels and a peck; and 3 quarters 3 bushels and 3 pecks were sold at home at 4*s.* per bushel, for 5*l.* 10*s.* 0*d.*; and the remaining one quarter 6 bushels and 2 pecks were sold likewise at home at 11 groats per bushel, for 2*l.* 13*s.* 2*d.*

Of the 11th dressing, on Thursday May 25, there was in all 3 quarters and half a bushel, whereof 1 quarter 7 bushels and half were sold at home att 11 groats per bushel, for 2*l.* 16*s.* 10*d.*, and one bushel spent by our selvs; and the remaining one quarter sold at 10 groats per bushel, for 1*l.* 6*s.* 8*d.*

Of the 12th dressing, on Saturday June 3, there was in al 9 quarters and 2 bushels, whereof 3 quarters sold to Chrissol Grange at 11 groates, for 4*l.* 8*s.*

0*d.*, and 6 quarters more to Shepereth miller at ten groates, for 8*l.* 0*s.* 0*d.*, and one [*p.102*] bushel and half sold at home at 10 groats, for 5*s.* 0*d.*; and the remaining half bushel spent by piggs, it being partly the worst rye I had that yeare. [*p.102*]

Of the thirteenth dressing of rye, on Tuesday June 20, there was in all five quarters and seven bushels, whereof 4 bushels were sold at Royston and all the rest at home, all at 3*s.* 4*d.* per bushel, for 7*l.* 16*s.* 8*d.*

Of the 14th dressing of rye, on Saturday July 1, there was in all five quarters and half, whereof 2 quarters 7 bushels and half were sold at 3*s.* 4*d.* per bushel at home, for 3*l.* 18*s.* 4*d.*, and 4 bushels were sold at 4*s.* a bushel, for 16*s.*; and one quarter and 4 bushels were spent by our selvs; so that out of this and former dressings there is 4 bushels and half either not set downe, or lost, besides 3 pecks of tayle wheate of the 8th dressinge.

Of the whole crop of rye this yeare there was in all 77 quarters 7 bushels and 3 pecks.

An account of the barly brought in that harvest and yeare [*p.103*]

Of the first dressing of barly, on Tuesday November 1 by Will: Thrift, there was in all 6 quarters 2 bushels and half, whereof a bushel and half spent for fowles, and half a bushel for the sow that was borrowed before of Widdow Rand; and half a bushel sold at home for 1*s.* 3*d.*; and the remaining 6 quarters sold all to M: Thurgood at 19*s.* 8*d.* per quarter, for 5*l.* 18*s.* 0*d.* Note: that 6 bushels and half of it was threshed by J: Casbourn, the cuttings of the mow side.

Of the 2nd dressing of barly, by Will: Thrift on Tuesday November 15th, there was in all 7 quarters and 4 bushels, of which 3 pecks were paide to Widdow Rand that were spent before by fowles and hoggs; and 2 bushels sold to Good: Payne, Thrift, Casbourn, Law, for 4*s.* 7*d.*; and six quarters sold againe to M: Thurgood at 18*s.* 6*d.* per quarter, for 5*l.* 10*s.* 0*d.*; and 9 bushels and a peck remaine, of which one bushel and half were spent by fowles etc. – and now paide to G: Davy, Eastland, Rand, of whom they were borrowed before – and a bushel and 3 pecks more sent to mill for fowles, and in the wheate and rye; and the 6 bushels remaining all sold at home at 2*s.* 6*d.* per bushel, for 15*s.*

Of the 3rd dressing, by Will: Thrift also, November 22, there was in all 6 quarters and one peck, whereof ye peck was sold to Good: Swann for 7*d.* *ob.*, and the six quarters were all sold to a carpenter at Royston and [*sic for* 'at'] 2*s.* 6*d.* per bushel, for 6*l.* 0*s.* 0*d.*

Of the 4rth, by Will Thrift, Tuesday December 6, there was in all ten quarters and 6 bushels, whereof 2 bushels were sold to Good: Swann, J: Haycock, Thrift, Law, for 5*s.*, and the remaining ten quarters and half sold to Rich: Jackson of Royston at 19*s.* 11*d.*, for 10*l.* 9*s.* 1*d.*

Of the 5th dressing, by Will: Thrift on Friday December 16, there was in all ten quarters, all sold to M: Thurgood of Royston at 1*l.* per quarter, for 10*l.* 0*s.* 0*d.*

Of the 6th dressing of barly, Thursday December 22nd, there was in all 5 [*p.104*] quarters and 2 bushels, whereof one bushel sold at home to Goodm: Thrift and Payne, for 2s. 6d., and one bushel spent by our selvs, id est 2 pecks for the horses and 2 pecks put amongst the rye, and the remaining five quarters sold to one Val: Beldam of Royston at 19s. 9d. per quarter, for 4l. 18s. 9d., by W: Thrift also.

Of the 7th dressing, by W: Thrift also, December 29, there was in all six quarters and 4 bushels, of which five quarters sold againe to M: Thurgood for 5l. 1s., and one quarter and 2 pecks sold at home for 1l. 1s. 3d.; and 3 bushels and half sold at Royston for 8s. 9d.

Of the 8th dressing, by J: Casbourne on Tuesday January 2, there was in all 4 quarters 2 bushels and halfe very neare, whereof the 2 pecks were sold to him at 1s. 2d.; and the 4 quarters and 2 bushels sold at 1l. 0s. 3d. per quarter (being made up 5 quarters) to M. Thurgood, for 4l. 6s. 0d.

Of the 9th dressing, on Friday January 6 by Will Thrift, there was in all 8 quarters 2 bushels and half, whereof 7 quarters 6 bushels and half were sold to the same M: Thurgood at 1l. 0s. 3d. also, for 7l. 18s. 0d.; and the remaining 4 bushels were sold at home at 2s. 6d. (all but one peck put amongst our owne rye) for 9s. 4d. ob.

Of the 10th dressing, by J: Casbourne on Friday January 20, there was in all ten quarters, all sold to Mr. Pigg of Chrissol at 21s. per quarter abateing only 2s. in the ten quarters, for 10l. 8s. 0d.

Of the 11 dressing, Tuesday January 24, by J: Casbourne also, there was in all 2 quarters 4 bushels and half, whereof one quarter and half sold at Royston at 1l. 0s. 6d. per quarter, for 1l. 10s. 9d.; and one quarter and a peck sold at home to taskers and others, for 1l. 0s. 7d. ob., and the remaining peck spent by our selvs.

Note: That these 2 last dressings, and the 8th and 6 bushels and half of the first, making in all 17 quarters five bushels and half, were all the barly I had in the new barne.

Of the 12 dressing, Friday January 27, by W: Thrift, there was in all 2 quarters and 4 bushels, all sold to John Wadnam of Barly at 1l. 0s. 6d. per quarter, for 2l. 11s. 3d.

[p.105] Of the 13 dressing of barly, on Munday February 6 by W: Thrift, there was in all 10 quarters and 3 bushels, whereof 2 bushels were sold at home at 5s.; and the remaining 10 quarters and a bushel were sold to Urias Bowes of Barly at 1l. 0s. 3d. per quarter, for 10l. 5s. 0d.

Hitherto all the dressings by W: Thrift were threshed out of the old barne, making in all 72 quarters 5 bushels and 3 pecks; all that is yet to come were threshed out of the rickes.

Of the 14th dressing, by W: Thrift on Saturday February 18, there was in all 12 quarters and 2 bushels, whereof one bushell was sent to mill for our fowles etc.; and 4 bushels were sold at home, for 10s., and 7 bushels more sold at home at 8 groates, for 18s. 8d.; and the remaining ten quarters and six bushels were all sold to M: Thurgood at Royston at 22s. a quarter, for 11l. 15s. 6[d.]

Of the 15th dressing, by Will: Thrift on Tuesday February 28, there was in all ten quarters and 6 bushels, whereof 5 pecks were spent for fowles and amongst the rye; and one quarter and 3 pecks were sold at home (id est 4 bushels at 8 groats, for 10s. 8d., and 4 bushels and 3 pecks at 3s., for 14s. 3d.) and the remaining 9 quarters and half sold to Mr. Glenister of Royston at 1l. 4s. 0d. per quarter, for 11l. 8s. 0d.

Of the 16th dressing, on Tuesday March 21 by W: Thrift, there was in all 12 quarters and five bushels, whereof 3 bushels were sold at home for 9s., and the remaining 12 quarters and 2 bushels were carryed in on Thursday March 23 by Mathew Payne and Tho: Barber to Richard Sel's of Triplow to be malted for my owne use.

Note: That these 3 dressings, making in all 35 quarters and 5 bushels, came all from the little rick of barly that lay against my study window.

Of the 17 dressing, on Tuesday March 28, there was in all six quarters and half, whereof 6 bushels were sold at home at 22s. 8d. per quarter, for 0l. 17s. 0d., and ye remaining five quarters and 6 bushels, together with ten quarters bought, were sowne on about 26 acres and half in South Feild and 16 acres more in the Barrfeild, in all 42 acres and half, whereof 11 acres and a stetch in Waterden Shott is very great measure.

Of the 18 dressing of barly (by Will: Thrift, as the former was), on [p.106] Tuesday April 4, there was in all ten quarters six bushels and half, whereof five quarters were carryed in this day by Math: Payne and Tho: Barber to Rich: Sell's of Triplow to be malted for me; and on Saturday April 8 were carryed againe to the same person by the same men and for the same use 5 quarters and one bushel more; and the remaining 5 bushels and half sold at home at 2s. 10d. per bushel, for 15s. 7d.

Of the 19th dressing, by Will: Thrift, on April 8, there was but one quarter 4 bushels and half, whereof one peck spent by fowles; and one peck sold to Math: Payne at 8d. ob.; and the remaining 12 bushels were sowne, so that there is sowne now, with the ten quarters that were bought, in all 17 quarters and 2 bushels.

Of the 20 dressing, on Tuesday April 11, there was in all 8 quarters 6 bushels and half, whereof five quarters and 4 bushels were sold to Rich: Jackson at Royston, for 1l. 4s. 0d. per quarter, for 6l. 12s. 0d.; and 4 bushels were sold at home to divers persons, for 11s. 8d., and the remaining 2 quarters six bushels and half were sold with the next dressing to John Brand at Royston at 24s. 2d. per quarter, for 3l. 7s. 10d.

Of the 21 dressing, by Wil: Thrift also, on Tuesday April 28, there was in all 11 quarters, whereof 9 quarters five bushels and half were sold (with the 2 quarters of the last dressing etc.) at 24s. 2d. per quarter to the saide John Brand, for 11l. 14s. 2d. Five bushels were changed with him for malt; and one bushel made up the next dressing even measure; and 3 pecks more were spent by our selvs; and the remaining 3 bushels and 3 pecks sold at home at 9d. per peck, for 11s. 3d.

Of the 22nd dressing, by W: Thrift also, on Saturday April 29, there was in all 9 quarters and 3 bushels, which (with a bushel of the former dressing, whose price is not there mentioned) was made up 9 quarters 4 bushels and sold to Gregory Brand at 25s. per quarter and 6d. over in the whole, for 11l. 18s. 0d.

[p.107] Of the 23rd dressing of barly, by William Thrift also, on Tuesday May 9, there was in all eleven quarters and half, whereof five quarters and half sold to one of Ashwell at 24s. per quarter, for 6l. 12s. 0d., and five quarters and half more sold to Mr. Chapman of Ashwel the week after, at 1l. 3s. 4d. per quarter, for 6l. 8s. 4d.; and 2 bushels and 3 pecks were sold at home at 3s. a bushel – to Good: Swan 3 pecks, to Henry Law 2 pecks, to Widdow George 2 pecks, to W: Thrift 2 pecks, to J: Casb: 1 peck, to Jerem: Gooding a peck – for 8s. 3d.; and the remaining 5 pecks were spent by our selvs, id est, a peck with our rye, and a bushel sent to mill for sow and piggs.

Of the 24th dressing of barly, by Wil: Thrift on Tuesday May 16, there was in all ten quarters, whereof five quarters were sold to M: Thurgood at Royston, at 23s. 6d. per quarter, for 5l. 17s. 6d., and the remaining five quarters sold the week after to Mr. Morse of Baldock at 1l. 3s. 0d. per quarter, for 5l. 15s. 0d.

Of the 25 dressing, by Will: Thrift, May 19, there was in all three quarters one bushel and half, whereof one quarter 7 bushels and half were sold at 3s. a bushel, for 2l. 6s. 6d., at home by the peck, and the remaining one quarter 2 bushels were spent by ourselvs, with hoggs, fowles, etc.

Note: That the whole cropp of barly this yeare amounteth to the quantity of 198 quarters six bushels and one peck.

[p.108] **An account of the oates brought in in that harvest and yeare**

Of the first dressing of oates, on October 20, there [was] but 2 bushels and half, being the tithe of Antony Parishes close, laide upon the hay and threshed in the malthouse, used and eaten by horses presently.

Of the 2nd dressing, Munday November 28, by John Casbourne, out of the middlestead of the new barne, there was in all 6 quarters 5 bushels and half, all laide into garner for our owne present use.

Of the 3rd dressing, on Munday January 16 by Will Thrift, there was in all ten quarters and 3 pecks, whereof 7 quarters two bushels and 3 pecks were carryed directly into stable, and into garner for our owne use; and the remaining 2 quarters and 6 bushels of the best of them saved for seede, about 2 quarters sowne, the rest spent.

Of the 4th dressing, by Will: Thrift on Tuesday March 7, there was in all 14 quarters and 5 bushels, whereof five bushels being tayle were laide into garner; and the 14 quarters laide into malthouse chamber for our owne use, as was also the tayle.

Of the 5th dressing, on April 21, there was in all 2 quarters, all laide into garner for our owne use and spent before the other 14 quarters.

[*No summary of this crop*]

[p.109] **An account of the pease brought in in that yeare and harvest**

Of the 1rst dressing of pease, November 18, out of the middlestead of the new barne by J: Casbourn, there was in all 3 quarters 3 bushels and 3 pecks, all spent by our owne hoggs.

Of the 2nd dressing, by Wil: Thrift, from the rick January 2, there was in all 4 quarters and five bushels and half, whereof half a bushel was tayle yet spent at first by fatting hoggs; about 3 quarters sowne upon the five acres in the Rowze and half acre and stetch, and the remaining quarter and five bushels spent by fatted hoggs also.

Of the 3rd dressing, on Friday March 10, by W: Thrift, there was but 3 quarters one bushel and 3 pecks, whereof 3 bushels were a kind of tayle mixed with many oates because the oates lay upon them (but being good horse corne and spent by horses), and the remaining 2 quarters and 7 bushels neare upon remaine for fatting hoggs.

[*No summary of this crop*]

THE HOUSEHOLD ACCOUNTS

1705–1710

[*County Record Office, Cambridge, 603/A1*]

Tabulae vel codex sextus accepti et expensi, or an account of my receipts and disbursements, beginning with the month of Aprill in the yeare 1705, book the sixth; wherein the mony received is alwayes set downe in the left hand page, viz in page 1, 3, 5, 7, etc., and that which is disbursed or laide out in the right hand page, id est in the 2, 4, 6, etc.*, and the date of the yeare (that [always] † begins with Lady Day) is set on the top of both pages alike‡; and at the end of every month the whole summ of what hath been either received or expended is exactly cast up, and for the most part one against the other.

[*p.1*]	**April 1705 Receipts**	*l.*	*s.*	*d.*
	Received of Mrs. Peck for a bushel of wheat at 3*s.* per bushel; received of Athan: Carrington for 2 pecks	0	4	6
2	Received of one Mr. Ben or Pen of Puckeridge for 6 quarters and 2 bushels of the same and at the same price carryed in about 3 weeks agoe, and paid in part then, but nott al until now; in al	7	10	0
4	Received of Ann Thrift for a peck of barly	0	0	5
	Received of God: Harvey for 2 pecks	0	0	10
	Received of Mr. John Nichols of Royston for 5 quarters and 7 bushels of the last dressing of barly that I sold at 13*s.* 8*d.* per quarter, the summ of	4	0	0

---

* In this edition, for economy, the accounts are not printed in parallel, but consecutively, receipts then disbursements, for each month. The pagination of the original is shown in square brackets and the words 'Receipts' and 'Disbursements' have been added to the headings of each month for ease of reference.
† Hole in ms.
‡ Not printed in this edition. As before, the year has been modified elsewhere as necessary to that of the modern historical year, beginning on 1 January. Where it is changed from that in the manuscript it is printed in italics.

Day	Description	£	s	d
	Received of Edw: Godfry for wheat 3 pecks	0	2	3
9	Received of John Rayner for the tithe of his dovehouse 1703, at Michaelmas	0	9	0
	Item for his smal tithes in 1704, with 1s. 4d. for a yeare's quitt rents	0	15	2
10	Item for his smal tithes now due (and a yeare's quitt rents) but in part	0	7	0
	Received of Thom: Barber for 2 pecks of rye and a peck of wheat, in toto	0	1	9
12	Item for 3 or 4 bottles of rye straw	0	0	10
	Received of Rich: Haycock for wheat... [*hole in ms.*]	0	0	9
	Received of Jos: Wedd for a peck	[0]	0	9
	Received of Widdow Law for 2 pecks	[0]	1	6
	Item for a peck of rye then	0	0	6
14	Received of Mr. Will: Hurrel for half an hundred of willow setts, and he had 66 for his mony also — they were			
16	not so good as he might have had if they had been bespoaken in time — at	0	8	0
	Received of Wil: Fordham for a peck of barly	0	0	5
	Received of Jos: Wed for a peck of wheate	0	0	9
19	Received of Mrs. Westly for a peck and a bundle of rye straw 3d., in al	0	1	0
	Received of Edw: Parish for rye 2 pecks	0	1	0
	Received of James Swann for 4 pecks of malt	0	2	6
	Received of Mary Wallis for a peck of barly	0	0	5
20	Received of Edw: Parish for a peck	0	0	5
	Received of Jos: Wed for 3 pecks of rye	0	1	6
	Received of Will: Brock for six pecks	0	3	0
24	Received of Sam: Woods for 2 or 3 bottles of rye straw at divers times — or 4	0	0	5
	Item for 2 pecks of wheate against May	0	1	6
	Received of G: Woodward for a peck	0	0	9
26	Item for 2 pecks of rye the same time	0	1	0
	Received of Athan: Carrington for one peck of wheate at 3s. per bushel	0	0	9
	Item for 2 pecks of barly	0	0	10
27	Received of Widdow Law for wheat a peck	0	0	9
	Received of M: Waits for a peck of rye	0	0	6
30	Item for a peck of barly then also	0	0	5
	Received this month in al the summ of	15	1	2

[*p.2*]

April 1705 Disbursements

	£	s	d
Paide to Goodman John Rayner for the interest of ten pounds, due this month, for 15 months (and I now paide the principal and all), the sum of	0	15	0

4	Paide to Goodman Cock for mending the lock to the stable door	0	0	6
6	Item for mending the padds to one of the lock irons, wive's pattens, kettles, etc.	0	1	1
	Spent at Royston by Nat:; self and horse	0	0	6
9	Paide to Rich: Haycock for his children gathering stones for me 3 dayes	0	0	9
12	Paide for 3 quitt rents due to the Lord of the Mannor for 3 yeares ended at this last Lady Day 1705, at 13*s.* 2*d.* per annum for Cassanders, and 3*s.* 8*d.* per annum Hoyes, in all	2	10	6
14				
16	Spent by Nathaniel at market before, and unloading the 5 quarters and 7 bushels of barly at Mr. J: Nichols his house	0	0	10
	Paide for half a pound of whip cord bought of Will Numan of Barrington	0	2	0
	Spent by Nat: at Royston Markett	0	0	9
19	Paide to Sam: Woods by deduction of mony due to me for straw and corne, for 5 dayes going to plow and rowle	0	1	8
20	Spent by son Nathaniel, going to Walden being not well, with John Latimer to Dr. Harvy for bleeding of him 1*s.*, and spent besides for horse and self 7*d.*	0	1	7
23	For a couple of stones from the quarry at Orwel, at 4*d.* a piece there, for the cheese press, though half of them served our turne	0	0	8
24	Paide to Thom: Pinnock for 3 dayes work, in altering the cheese press and mending J: Bowellsworth's table broaken by us, at 1*s.* 2*d.* per day — he lives at Meldreth	0	3	6
26				
	For 3 pounds of pitch for the new hind wheeles of the wagon	0	0	9
27	Paide to Rich: Fairechild for 12 score and 7 pounds of the tire for them, at 3*d. qtr.* per pound; that is to say, ten score 5 pounds and half strakes and nayls, 23 pounds and half boxes, and 18 pounds hoops, etc. (abateing but 4*d.* 3*qtr.* in al), and so it comes to (vide September 19 last)	3	6	6
28				
30	To Marg: Cra: uxori for weaving of 4 or 25 yards of huggeback 10*s.*, for beife and smal meate of Malin and others about 1*l.* 1*s.* 6*d.* and all other occasions in this month, toto	2	16	0
	Expended this month in al the summe of	10	2	7

[*p.3*]

May 1705 Receipts

Received of Mrs. Peck for one bushel of wheate by self, at	0	3	0

		£	s	d
	Received of Athan: Carrington for three pecks of rye or miscellaine	0	1	6
8	Received of El: Bright for 2 pecks of barly	0	0	10
	Received of Rob: Macer for 3 pecks of wheat, 2 April 25 and one now	0	2	3
9	Item for 3 pecks of rye 1s. 6d.; received of Good: Gurner for a peck	0	2	0
	Received of Jos: Wed for 2 pecks	0	1	0
	Item for 2 pecks of wheate, one tayle; received of Good: Hankin for a peck	0	2	0
10	Received for the old Brock cow and calf	1	0	0
	Received of Edw: Phipp of Buntingford for 10 quarters of wheat, sold to him some time since (vide disbursments March 27 last)	11	8	0
11	Received of Ralph Carpenter and others for straw	0	0	5
	Received of Mrs. Westly for wheat a peck; received of Athan: Carrington for a peck	0	1	6
	Received of Wil: Wallis for 2 bushels of rye	0	4	0
12	Received of Lyd: Barber for 2 pecks	0	1	0
	Item for 2 pecks of wheate 1s. 6d.; received of M: Waites for a peck	0	2	3
14	Received of Edw: Godfry and others for rye straw, in al	0	0	5
	Item for the calf of the little black Welsh cow taken away now	0	8	0
16	Received of Edw: Phipp of Buntingford for 5 quarters of the 7th dressing of rye at 2s. per bushel	4	0	0
	Received of Thom: Nash for his small tithes in 704 and this yeare	0	6	7
17	Received of Mr. Taylour for his smal tithes in 1703	1	17	6
	Item for his smal tithes in 1704	2	1	4
	Item for 41 willow setts in 704	0	6	0
18	Received of Jos. Wedd, Hankin and Bright for 3 pecks of wheate	0	2	3
	Received of Widdow Gurner for rye a peck; received of Wallis of Chrissol for 2 bushels at 2s. per bushel — 4s.	0	4	6
19	Received of Ben: Bright for 3 pecks	0	1	6
	Received of Mrs. Peck for a bushel of wheate at 3s. by Thom: Watson	0	3	0
	Received of Mr. Norridge for 2 yeares quit rents ended at Lady Day last due to the Rectory	0	10	0
21	Received of Antony Parish for a peck of the last dressing of barly	0	0	5
	Received of Ann Thrift for wheate a peck	0	0	8
	Received of Athan: Carrington for a peck	0	0	8
	Item for 3 pecks of rye	0	1	6
22	Received of Widdow Haycock for a peck and 2 bundles of rye straw 6d., in al	0	1	0

	Received of Jos: Wed for 2 pecks of rye	0	1	0
	Received of Widdow Goode and Rich: Haycock for 1 peck and 2 pecks	0	1	6
23	Received of Rob: Macer for 2 pecks of wheate the 16th instant	0	1	6
	Received of Athanasius Carrington for a bundle of rye straw 3*d.*, and one of barly straw	0	0	7
[*p.*5]	Item for his smal tithes in 1704 and in this yeare 3*s.* 6*d.*	0	4	6
	Item for fetching a loade of barly straw from Triplow in November	0	2	0
24	Item for the tithe grass of his close, which he hires this next hay time	0	2	6
	Received of Eliz: Jude and Lyddy Barber for 3 pecks of rye	0	1	6
25	Received of Ralph Carpenter, Kate Reynolds, Widdow Ward, for 3 pecks of wheat	0	2	0
	Received of Edw: Phipp of Buntingford for 5 quarters and 4 bushels of it, at 13 shillings per loade	5	14	4
26	Received of Goody Goode and Brock for a bushel, at	0	2	8
	Item of her for a bushel of rye then	0	2	0
	Received of Thom: Barber for 2 bottles of straw	0	0	6
	Received of Goody Harvy for 2 pecks of barly	0	0	10
28	Received of Lydd: Barber for wheate 2 pecks; received of Widdow Law for a peck 8*d.*; received of Goody Bright for a peck 8*d.*; received of Ann Thrift for a peck 8*d.*; received of Jos: Wedd for 2 pecks 1*s.* 4*d.*	0	4	8
	Item of her for 3 pecks of rye 1*s.* 6*d.*; received of Widdow Gurner for a peck 6*d.*	0	2	0
	Received of Mr. Malin for the calf of the great black Welsh cow	0	10	0
29	Received of G: Brooks for a peck of wheat	0	0	8
	Received of Ath: Carr: for 2 pecks of barly	0	0	10
	Received for 3 pecks of rye	0	1	6
	Received of Rich: Haycock for wheat 2 pecks	0	1	4
30	Received of Mrs. Peck for a bushel	0	2	8
	Received of Edw: Parish, Good, Harvy, Good: Macer, for a bushel of barly	0	1	8
31	Received in part of Edw: Godfry for a peck of wheat July 17 1693 and paid in part December 93 and February *1696*	0	0	6
	Item for a bushel of barly he had on December 30, 1693, at 25 shillings per quarter	0	3	0
	Item for the calf of the littl Welch red cow at 3 weekes old just	0	7	0
	Received this month in al the full summ of	32	8	4

May 1705 Disbursements

[p.4]

	Paide to Mr. Wilson for a yeare's tenths due at Christmas last, and spent by a messenger I sent 2*d*.	2	19	11
8	Spent by Nathan: at Royston takeing mony, for toll, for the cow and calf, etc.	0	0	10
	Spent by Nat: for toll 2*d*., loading etc., when they fetched coales, Cambridg	0	0	10
9	For this paper book bought at London by James Rush			
10	because I disliked the other — this was bought before, though he came not to receive mony til now	0	1	2
11	For sugar candy for myself having a great cough and cold	0	0	4
12	Spent by Nat: and George carrying a load of rye to Edw: Phipp, unloading it there 3*d*., and so in al	0	0	9
14	Paide to Mr. Taylour for an overseers' rate for Mr[s] Norridge that I hire of her at 2*l*. per annum at 3*d*. per pound, vide March last, end	0	0	6
	Given to servants at Mr. Wedd's	0	0	6
	Pectorals for my cough at Royston	0	0	3
16	For horse there, Nat: spending nothing	0	0	2
	Paide to Mrs. Norridge for the use of the barne which I hired of her this yeare though my time is not ended till midsummer next, the sum of	2	0	0
17	For shooing an horse at Chrissol, going for a loade of wood there	0	0	5
	Paid to Will: Wallis of Triplow, for carrying up Hester's trunk, bringing downe Typtery the horse againe, etc.	0	3	0
18	For oranges bought at Cambridge	0	1	0
	Given to servant at Mr. Stevens: etc. at Hauxton	0	0	6
	For at cart whip 2*d*., lainers to it 2*d*.	0	0	4
19	Spent at Cambridge at the election of Knights of the Shire etc.	0	0	7
	To Goody Watson and her daughter for gathering up the stones out of the mudd in my close	0	0	2
21	Given to a seaman disabled and loosing his limbs about Christmas last and some others	0	0	5
	Spent by Nathan: at Royston Market	0	0	2
22	For 2 sticks of sealing wax, the one red and the other black, at 3*d*. each	0	0	6
	Given to a Scotchman born at Aberdeen and going to London in want etc.	0	0	4
23	Paide to Ralph Carpenter for 4 dayes work and half in filling dung, casting up dung, Thrift absent	0	3	9
	Paid to Sam: Woods for half a day about the horse pond, in fetching things for the engine etc.	0	0	2
[p.6] 24	Paid to Wallis of Chrissol for 100 of rodds 10*d*., for 200 withs 1*s*., and 6*d*. going to show my men theire wood, which they knew not	0	2	4

	Spent by men carrying the 5 quarters and half to Buntingf. and unloading	0	0	10
	Paide to Edw: Godfry for half a jagg of dung I had of him August 1, 1698	0	1	0
25	Item for 5 loades now, laid upon son Nathaniel's hired land in Northfield at 1s. 6d. per loade	0	7	6
26	Paide to Thom: Watson for 3 dayes work in March, lopping and other work	0	2	6
	Item for 11 dayes before April 15 tying up the old hedg in close, bringing them in, cutting out setts, lopping and plow etc.	0	9	6
28	Item for 12 dayes more before April 29 in like works and laying up the wood, in malthouse etc.	0	10	0
	Item for 2 dayes and better than half in odd times befor May 6	0	2	2
29	Item for 11 dayes before the 20 instant at plow, wood cart, and other works in the afternoon	0	9	2
	Item for 9 days work since at plow etc.	0	7	6
	Paide for a pint of sweet, eating sallade, oyle, at Royston, at	0	2	6
30	Paid to Thom: Dovy for a daye's worke about the great pond in the orchard, turning the engine I borrowed of Mr. Benning, vide June	0	1	3
31	To Marg: Cra: uxori for cheese about 15 pound and half 3s. 1d., 5 stone and 10 pound of beife 18s., 4 quarters of veale, one of lamb, about 12s. 6d., and al other occasions in this month besides, in al	3	3	0
	Expended this month in al the ful summ of	11	15	10

[p.5]

June 1705 Receipts

	Received of Widdow Nash, Good, Brock and Barber for 5 pecks of wheate	0	3	4
	Received of her and Good: Brock for rye 5 pecks	0	2	6
1	Received of J: Woods for a peck of wheat; received of Kate Reynolds 8d.; received of Jos. Wedd for a peck 8d.	0	2	0
	Received of them for 2 pecks of rye	0	2	6
2	Received of Widdow Law, Ward, Woodward, for a bushel of wheate; item of her and Mary Waites for 2 pecks more	0	4	0
	Received of Thom: Watson for 6 pecks, 4 at 3s. and 2 at 1s. 4d.	0	4	4
4	Item fetching him a loade of wood	0	5	0
	Item for 2 bushels of barly	0	3	4
	Item for 4 bushels of rye; received of Good: Gurner and Waites for 2 pecks	0	9	0

5	Received of R: Haycock, R: Macer and T: Barber for 6 pecks more		0	3	0
	Received of Ralph Carpenter for a bushel 2s., and 2 pecks of wheat		0	3	4
	Received of Hankin for a peck of wheate; received of Jos. Wedd and another for 3 pecks		0	2	8
6	Received of Ann Thrift for barly a peck		0	0	5
	Received of Thom: Dovy for rye straw the beginning of 703		0	0	5
[p.7]	Received of Mrs. Westly, Rob: Macer, Widdow Law, Mrs. Peck, for rye 2 bushels		0	4	0
7	Received of Wil: Jeeps for 3 pecks in part he had November 29 last, at 2s.		0	1	3
	Received of Rob: Macer for 2 pecks of... [omission] (May 25) and of Mrs. Westly a peck now, of wheate that is, at 2s. per bushel		0	2	0
	Received of Kate Reynolds for a peck		0	0	8
8	Item for a peck of rye then also		0	0	6
	Received of Goody Haycock for 2 pecks of tayl wheate, at		0	1	0
	Received of God: Parish for barly a peck		0	0	5
9	Received of Wil: Brock and Good: Gurner for 5 pecks of rye at 2s. per bushel		0	2	6
	Received of Wil: Brock for 2 pecks of wheat in part 1s., of Rob: Macer for a peck		0	1	8
	Received of Sam: Woods for a peck May 5		0	0	9
	Item for a peck of barly May 10 last		0	0	5
11	Received of G: Hankin and Widdow Nash for 3 pecks of wheate		0	2	0
	Received of Widdow Swann for a bushel		0	2	10
	Item for 4 bushels and 3 pecks of rye, and of Rob: Macer and E: Bright for 3 pecks		0	11	0
12	Received of M: Wallis for a peck		0	0	6
	Received for the Buck horse, of Thom: Ellis		7	15	0
	Received of Goody Bright and Barber for 2 pecks of miscellaine		0	1	0
	Received of her for 4 bundles of wheat straw		0	1	0
13	Item of them two and Rich: Haycock for a bushel of wheate, of Joane Woods for a peck, of Mrs. Peck for a bushel, toto		0	6	0
	Received of Widdow Law for a peck, and Antony Parish a peck, M: Waites a peck		0	2	0
	Item for a peck of rye of her		0	0	6
14	Received of Ed: Godfry for 2 pecks of wheate; received of Good: Brock for 2 pecks; received of Kate Reynolds for a peck		0	3	4
	Received of them two for 5 pecks of rye		0	2	6

HOUSEHOLD ACCOUNTS 1705

	Received of young J: Watson for a peck of barly	0	0	5
	Received for a bottle of straw of G: Gurner	0	0	1
15	Received of Edw: Godfry for 3 more, at	0	1	0
	Received of Good: Nash for 2 pecks of wheate	0	1	4
	Received of Rob: Macer and Widdow Goode for 2 pecks of the same	0	1	4
16	Received of Widdow Gurner and them two for 3 pecks of rye, and of Widdow Law a peck; received of Jos. Wed for 2 pecks 1s., toto	0	3	0
	Received of Good: Hankin and Law for 2 pecks of wheate	0	1	4
	Received of Widdow Gurner for the tithe grass of her orchard	0	0	11
	Received of Widdow Gooding for 2 pecks of rye; received of Eliz: Bright for 2 pecks	0	2	0
18	Received of Athan: Carrington for 6 pecks	0	3	0
	Item for the tithe of half an acre of St. Foine in Dod's Close	0	1	6
	Item for carrying it to his house	0	1	0
	Item for 3 pecks of wheate	0	2	0
19	Received of Mr. Wedd for the tithe of some St. Foine in Dod's Close	0	2	2
	Item for the tithe hay of his close	0	2	6
	Item for Mr. Latimer's tithe wooll	1	2	0
	Received of Will: Brock for a bushel of rye and 2 pecks of wheate	0	3	4
	Received of Jos: Wedd for a peck of wheat	0	0	8
20	Received of her and Mrs. Peck for rye a bushel, and for 3 bushels and a peck of Edw: Parishes wife	0	8	6
	Item for 2 pecks of wheate March 7 last	0	1	6
	Received for a peck of Goody Hankin	0	0	8
	Received of Ann Thrift for a peck of barly	0	0	5
	Received of Betty Jude for a peck of wheate	0	0	8
[p.9]	Received of Edith Wedd for 2 pecks of rye, or miscellaine	0	1	0
	Received of Mrs. Peck for 4 pecks of wheat	0	2	8
21	Received of Goody Swann for 2 pecks	0	1	4
	Item for 2 pecks of miscellaine; received of Widdow Haycock for a peck	0	1	6
	Received of Antony [*Parish*] for a peck of wheate; received of Widdow Gooding for a peck	0	1	4
22	Received of Athan: Carrington for a peck; received of Widdow Nash for 2 pecks; received of Peter Nash for 2 pecks	0	3	4
	Received for 4 bushels of Mr. Wedd, and a bundle of wheat straw 3d., toto	0	10	11
23	Received of Good: Gurner for a peck of rye; received of J: Woodward for 2 pecks	0	1	6

	Received of Edw: Phipp of Buntingford for 5 quarters of wheate at 14s. per loade of the last dressing	5	12	0
25	Received of J: Woodward's wife for 2 pecks	0	1	4
	Item for a bushel of rye the 4 instant	0	2	0
	Received of Good: Brock, Bright, Goode, for a bushel with a peck of tayl wheat in it	0	2	0
	Received of Edw: Parish for 2 pecks	0	1	0
	Item for the tith hay of his orchard	0	0	3
26	Item for 6 pecks of malt	0	3	6
	Received of G: Bright for 2 pecks of wheate	0	1	4
	Received of M: Wright for a peck of wheate	0	0	3
	Received of Eliz: Jude for a peck of rye	0	0	6
	Received of Will: Fordham and Goody Swan for a bushel of the same	0	2	0
27	Received of Widdow Goode for a peck of wheat	0	0	8
	Received of Widdow Gooding for a peck	0	0	9
	Received of Mrs. Peck for 2 pecks of rye	0	1	0
	Received of Lyd: Barber for a peck	0	0	6
28	Item for 2 pecks of wheate; received of Will: Brock for 2 pecks; received of Jos. Wedd for a peck	0	3	9
	Received of them 2 for 5 pecks of rye; received for 2 pecks of the miller	0	3	6
29	Received of Good: Gurner for a peck; received of R: Macer for 2 pecks	0	1	6
	Received of Good: Parish for a peck	0	0	6
	Received of Good: Swann for bushel	0	2	0
30	Received of Widdow Gurner and Robert Macer for 2 pecks of wheat 1s. 6d.; received of Widdow Law for a peck; received of Edw: Godfry for 2 pecks 1s. 4d.; received of Eliz: Bright for a peck 9d.	0	4	4
	Item for a peck of rye at that time	0	0	6
	Received of Widdow Nash for wheate 2 pecks; received of Athan: Carr: for a peck	0	2	3
	Received in this month in al the full summe of	24	0	5

[p.6]   **June 1705 Disbursements**

1	Paide to Will: Jeepes for a daye's work in turning the engine for foeing the great pond in the orchard, with Thom: Dovy, vide May	0	1	3
	Paide to Ralph Carpenter for a day in the pond, casting out the mudd	0	1	3
	Paide to Rob: Macer that came into pond about 4 a clock	0	0	7
2	Paide to Sam: Woods for a day in the same work	0	0	7

	Item for 2 days at plow, May 8 and 9	0	0	8
4	Item for part of a day in walking the horse about that was sick	0	0	3
	Item for another day at plow	0	0	4
	Paide to Rob: Macer for a day about the little pond, and casting water into the horse pond	0	1	3
5	Item, Thom. Fairchild for the same	0	1	3
	Item for the same work to Thomas Wootton, he coming 2 or 3 hours after the other	0	1	0
[p.8] 6	Given to a breife for the repaireing of the Church Minshal church in the county of Chester — the charges are computed at 1380*l.*, haveing laide out above 400*l.* already	0	1	0
	Spent by Nat: at Royston Market	0	0	6
	Spent by Will: Dovy at Storford Fayer for 2 horses, going to sel one, and urged by the buyer to spend 1*s.*	0	4	0
7	Item for his time and work	0	2	0
	Spent by Nat: at Royston Market	0	0	4
	Given to servant at Mr. W: 6*d.*; to the ringers the same day, for Hester Wedd	0	5	6
8	Paide to Ed: Godfry for his labour when they cast the little pond (vide 5 instant)	0	1	0
	For procurations at the archdeacon's visitation for 2 yeares	0	6	8
	To barber at Cambridge, and horse 3*d.*	0	0	9
9	Spent by Nat: going to London at his sister Hester's wedding 2 dayes	0	9	6
	Spent by him at Royston Markett	0	0	6
	Paide to Athan: Carrington for 9 loade of dung laide upon son Nat's hired land May 29 last at 1*s.* 10*d.*	0	16	6
11	Given to 2 poore people in want at door	0	0	2
	Spent by Nathan: going with Benj. Wedd to Duxworth	0	1	0
	Spent by him and George going for coals, and toll etc. 2*d.*, in al	0	0	10
12	For the little Bonny colt at Midsummer Faire, bought by Nat:	6	16	0
	For a whale-bone whipp at Royston	0	1	0
	Spent by Nat: going to sel	0	0	6
	Paide to Will: Morrice the sowgelder	0	0	7
13	Paide to Eliz: Gurner, widdow, for 5 loades of dung I had of her about this time at 1*s.* 8*d.* per loade, for Nat's land some, some my own	0	8	4
14	Spent by Nat: and George carrying the last dressing of wheate to Buntingford, theire stay occasioned by the raine	0	1	0
	Given to poor neighbours and strangers	0	0	4
15	Paide to J. Woodward's wife by deduction of mony due to me for corne for his herdman's wages due for 4 cowes for			

	the quarter ended the 24 day of this month	0	2	8
	Paide to the currier for dressing the bul hyde, for his work	0	4	0
16	Paide to Rich: Jackson for his yearly pay for keeping my pump, vide July 1, 1704, by Nat at Royston	0	1	0
	Spent by Nat: at Royston, and given 2*d*.	0	0	6
	Paide to Good: Swann for dressing a calve's skin delivered to her husband in his life time	0	0	6
18	Given to a bill of request for a loss hapening upon 7 inhabitants of Chatteris by fire March 30 last, to value of 817*l*., though the whole loss of rich men and al desiring no releif is 1787*l*.	0	1	0
	Given to poor at door	0	0	1
19	Given to poor again	0	0	1
	Paide for children; for son Sam since June last, vide for half a yeare's interest of 12 pound due April 23 last to Mrs: R:	0	7	2
[*p.10*]	Paide to son Ben's wife in gold, being with us the first time	1	3	6
20	Item for a yeare's interest due this Midsummer, for 50*l*. that month	3	0	0
	Paide for daughter towards 3 months board ended about July 6, 1704	2	10	0
	Item for her expenses coming down	0	10	0
21	Item four paire of shooes since 1704	0	10	0
	Item for a black petticoat, lace, cotton, etc.	0	16	6
22	Item holland 5*s*., muslin 5*s*. 6*d*., ghenting 2*s*. 6*d*., handkercheifs 3*s*., callico, ferrent cloth for aprons etc. 10*s*., linnen cloth and callico againe 11*s*. 10*d*., coloured aprons 3*s*. 6*d*., toto	2	0	6
23	Item going to London in May and comeing downe againe the 15 instant, to provide her self wedding cloaths, and in mony since she came downe 4*s*., in al	12	6	6
25	Item to her husband Ben: Wedd in Mr. Latimer's wooll 1*l*. 2*s*., and in Mr. Taylour's wooll 1*l*. 11*s*., toto	2	13	0
26	Paide also for son Nathaniel since June last, for 3 paire of new shooes for him, and mending his boots and shoos for him 2*s*. 6*d*. toto, a paire of buckles and combs 6*s*. 4*d*.	0	13	10
	Item for another paire of shooes and mending 3 paire for him 1*s*. 6*d*., toto	0	5	2
	Item for 2 paire of stockins for him, one yarne 2*s*., the other woosted 4*s*.	0	6	0
27	Item materials for a black wastcoate, viz, callimanco 5 yards ½ 12*s*. 6*d*., shaloun 2 yards ½ 4*s*, dimity 3 yards 3*s*., silk 1*s*. 8*d*., buttons 10*d*., and other things 9*d*.	1	2	9
28	Item 2 yards and 3 quarters of green plush for his best breeches 14*s*., 3 shammy skins 3*s*. 3*d*., 12 silver buttons 1*s*. 9*d*., silver thread 2*s*. 4*d*., pocket skins 6*d*., malling 4*s*.	1	5	10

HOUSEHOLD ACCOUNTS 1705

	Item materials for a new frock etc.	0	12	6
	Item materials for his best coate etc.	2	5	2
	Item mohair, buckram, silk, making of it etc.	0	5	2
29	Item shirts and handkercheifs for him 11s. 8d., 6 ells and half of doulas more 8s. 8d.	1	1	0
	Item 2 hatts for him 11s., 2 yards and half of fustian for him, buttons 6d., making it into breeches 2s., etc.	0	16	0
	Item for his old dow, to Widdow Goode	0	2	0
30	To Marg: Cra: uxori for 5 ston of beife 15s., veal, lamb, mutton, about 7s. 6d., cheese 20 pound about 4s., and al other occasions, in all	3	0	0
	Expended this month in all the ful summ of	48	8	10

July 1705 Receipts

[p.9]

	Received of Edw: Parishes wife for fetching a loade of hay from Molton in Orwel parish	0	2	6
9	Received from Mrs. Peck for a bushel of wheate and of M: Waites for a peck	0	3	9
	Received of Wil: Brock for 2 pecks of rye	0	1	0
	Received of Goody Hankin for wheat a peck	0	0	9
10	Received of Edw: Parish for a peck of barly	0	0	5
	Received of G: Gurner for a peck of rye; received of Mrs. Peck for 2 pecks	0	1	6
[p.11]	Received for 3 todd 11 pounds of Mr. Taylour's wooll at 9s. per todd	1	11	0
11	Received of Athan: Carrington for 2 pecks of miscellaine or rye	0	1	0
	Received of Widdow Swann for a parcel of hay out of the furthest close	0	6	0
12	Received of Will: Thrift for 2 sacks of chaff May 9, and now one	0	1	2
	Item for fetching him home a little wood that was left in Rofeway, with a cart and 2 horses, June 16	0	2	0
13	Item for a little peice of parchment he had of me for a register's bill	0	0	2
	Item for a peck of tayle wheate	0	0	6
	Item for a bushel of the best wheat, at	0	3	0
	Item for 4 pecks of the same, at	0	2	8
	Item for another peck of the same, at	0	0	9
14	Item for 7 bushels and 3 pecks of rye at 10 several times between March 26 last and this time at 2s.	0	15	6
	Received of Goody Swann for wheate 2 pecks	0	1	6
	Received of Thom: Eastland for 2 bushels and 3 pecks of the same at 3s.	0	8	3

133

16	Item for 2 bushels and a peck at 2s. 8d., but one peck at 9d.	0	6	1
	Item for 3 bottles of rye straw in April 23 and 6 instant	0	0	9
	Item for 13 willow setts, 6 paid, 7 gratis	0	0	6
17	Item for 2 bushels of barly at twice	0	3	4
	Item for fetching him a loade of wood in waggon and cart both	0	5	0
	Item for 2 bushels and half of rye March 1 last at 2s. per bushel	0	5	0
18	Received of Will: Fordham for 3 bundles of barly straw in March and April last, vide receipts of September last	0	1	5
	Item for 5 or 6 bundles of rye straw in April and May last	0	2	2
	Item for 4 bundles of wheate straw in March and June last	0	1	7
	Item for 2 bushels of the bought pease	0	4	4
19	Item for 2 bushels of wheate	0	5	4
	Item for the tithe of an acre and half of St. Foine out of Dod's Close	0	4	4
	Item for 4 bushels of rye betwixt September 29 last and this time	0	8	0
20	Received of R: Macer for 5 pecks on June 25 and now	0	2	6
	Item for a peck of wheate then	0	0	8
	Received of Jos: Wedd for a peck now	0	0	9
	Received of Goody Brock for 2 pecks of rye	0	1	0
21	Received of Widdow Haycock for a peck	0	0	6
	Received of Athan: Carrington for a peck of wheate	0	0	9
	Received of Eliz: Bright for rye a peck; received of Mary Waites for 2 pecks; received of Edith Wedd for 2 pecks	0	2	6
23	Received of Edw: Parish for a peck of barly	0	0	5
	Received of Widdow Gurner for a peck of rye; received of Widdow Reynolds for a peck	0	1	0
	Item for a peck of wheate then also; received of Mary Waits for 2 pecks; received of Peter Nash for 2 pecks	0	3	9
24	Received of Mr. Bowes for 2 pecks that he had sometimes since	0	1	4
	Item for a bushel of rye at twice	0	2	0
	Received of Athan: Carrington for 2 pecks of wheat, by miller	0	1	6
[p.13]	Received of Widdow Reynolds by the hands of her son Thom. for a peck of wheat	0	0	9
26	Received of Edith Wedd for a peck of rye	0	0	6
	Received of Widdow Swann for 2 pecks	0	1	0
	Item for 2 pecks of wheat, by self	0	1	6
	Received of Edith Wedd for a peck	0	0	9
27	Item for a peck of rye then	0	0	6
	Received of Ed: Parishes wife for a peck of barly laide by for expenc	0	0	5

	Received of Widdow Swan for 2 pecks of rye	0	1	0
	Item for 2 pecks of wheate also	0	1	6
30	Received of Jos: Wedd for a peck of rye	0	0	6
	Received of Widdow Gurner for a peck	0	0	6
	Received of Mary Weights for a peck of wheate, by self	0	0	9
31	Received of Ben: Brightman for a peck of rye or miscellain	0	0	6
	Received this month in al the full summe of	8	0	1

[p.10]   **July 1705 Disbursements**

	Given to poore in want at door	0	0	2
4	Spent by Nathan: at Royston	0	0	4
7	For the Duke colt, exchanged for the black nagg at Royston Fair, that cost 7*l.* 8*s.* 6*d.* June 29 1703, and in mony now, with a shilling to horse keeper and 1*s.* 8*d.* spent and for toll etc. 2*l.* 12*s.* 8*d.*, in al	10	1	2
[p.12]	For linsed oile and red leade to mark my sacks withal	0	0	2
	For coach hire from Buntingford to Royston for my wife, not wel	0	1	0
9	To Captaine Hitches man, borrowing his calash to bring my wife from Royston	0	1	0
	Spent by Nathaniel at Buntingford and Royston going for his mother	0	0	6
11	Given to 3 souldiers taken and kept prisoners by French 9 months	0	0	3
	For a pound of ten and 8 penny nailes	0	0	4½
	Paide to Will Numan the coller maker for 4 dayes work, he and his brother	0	8	0
12	Item for a calve's skin 2*s.* 4*d.*, half a bullock hyde 3*s.* and etc.	0	5	4
	Item 2 croopers out of his leather 5*s.*, 2 headstals of the same 2*s.* 8*d.*	0	7	8
13	Item 3 pound and 3 quarters of woosted fring at 3*s.* per pound	0	11	3
	Item for divers other things, vide his bill in the beginning of this book, dated the 16 instant, in fold	0	4	3
14	For another pound of 6 and 8 penny nailes	0	0	4½
	Paide to Will: Thrift for threshing 20 quarters and 2 bushels of the four last dressings of barly	0	16	0
	Item for 3 days in April, 12, 13, 14, in hedging the roundabout	0	2	6
16	Item for 12 days between April 20 and May 7, cutting up trees, throwing up dung in lane and yard	0	10	0
	Item for one day in foeing the orchard ponds, turning the engine, see disbursements in June 1 to 5 etc.	0	1	3

17	Item for 16 dayes before June 24	0	12	7
	Item for 15 dayes more til this time	0	12	6
	Item for his part in mowing the closes	0	2	1
	Paid to Thom: Eastland for his part in the same also	0	2	1
18	Item for threshing 5 quarters and 6 bushels and half of his last barly at 9*d*.	0	4	4
	Item for threshing 22 quarters and 7 bushels of his last dressings of rye	1	8	7
	Item for threshing 50 quarters of the 7 last dressings of wheate at 1*s*. 3*d*.	3	2	6
19	Item for 12 days in February, March, April, in sowing, spreading dung, hedging, fagotting, remoovein the straw etc.	0	10	0
	Item 2 dayes and half in cleansing the horse pond and orchard ponds, se disbursments in June 1 and 5	0	3	0
20	Item for 23 dayes and half in spreading dung, filling it, and hay, daubing	0	19	7
	Item for his work Munday 16 instant	0	0	10
21	Paide to Will: Fordham for 6 dayes of his son Henry in October and November last, mending Mrs. Norridge's barne doors 8*d*., wal in kitchin garden, mending cow racks, ladder, oak bin in stall, a pent house for J: Watson's oven 1*s*. 2*d*., and other things, in al	0	8	0
	Item for 2 dayes and half mending my seats in the chancel	0	3	9
23	Item for 3 dayes now, in mending my carts, ladders, tressels, etc., against harvest, and rack in stable, vide disbursments in September last, the end	0	4	6
24	For gloves against harvest	0	6	0
	For a couple of ridgetrees	0	1	0
	For a new pitchfork	0	1	6
[*p.14*]	Spent by son Nathaniel at Royston	0	0	6
	Item by George going for some Wedd wool	0	0	6
25	For a new cart rope at Cambridge	0	2	0
	Paide to 5 or six acre men for reaping an acre and half of wheate of mine and as much of Nat's	0	10	0
26	Paide to 2 acre men for reaping 3 acres of wheate for me 9*s*., and 3 half acres of rye for Nat: 4*s*., one of them Porter of Clavering, in al	0	13	0
	Paide to 3 other men that came by the day, 12 dayes amongst them	0	12	0
	Paide for a letter brought from Royston	0	0	8
30	To Marg: Cra: for smal meate in this month about 13*s*. 6*d*., cheese 2 stone 4 pound at Royston at 3*s. ob.* about 9*s*. 6*d*., and 6 ston 2 pounds and half at Ickleton at 2*s*. 8*d*. 17			
31	shillings, sugar 9 pound 4*s*., malagas etc. 25 pounds 8*s*. 3*d*.,			

	candles 18 pounds 5s. 6d., salt a bushel 5s. 6d., and al other occasions	5	0	0
	Expended this month in al the full summ of	29	13	1

[p.13] **August 1705 Receipts**

	Received of Edward Godfrey for 3 pecks of wheate that he had on July 12 last at 3s.	0	2	3
8	Received of Jos: Wed's wife for a peck	0	0	9
	Item for 2 pecks of rye of her	0	1	0
9	Received of Widdow Gurner for a peck	0	0	6
16	Received of Mr. John Nichols of Royston for ten quarters and half of mault, sold to him at 17s. per quarter and just skreened	8	18	6
18	Received of Will: Brooks for 2 acres of haume August 14 last	0	6	0
	Item for 5 pecks of rye at several times between June 14 and July 18	0	2	6
20	Item for a bushel of wheate at divers times between June 5 and July 19 last, 3 [*pecks*] at 2s. 8d., one at 9d.	0	2	9
	Received of Ralph Carpenter for 2 pecks of wheate July 20 last, at	0	1	6
21	Item for 3 pecks more since	0	2	3
	Item for 5 pecks of peares, at	0	0	11
	Item for a bushel of rye now	0	2	0
22	Item for a bushel of malt	0	2	4
	Received of Widdow Swan for 2 pecks of wheat	0	1	6
	Received of Ben: Bright for a peck	0	0	9
	Item for a peck of rye then	0	0	6
24	Received of Mary Weights for a peck of the old wheate at 3s.	0	0	9
	Received of Athanasius Carrington for half a bushel of malt	0	1	2
	Received of B: Bright for 3 pecks	0	1	9
27	Received of Jos. Wedd for 2 pecks of rye	0	1	0
	Received of Rob: Macer for a peck	0	0	6
	Received of Eliz: Bright for 2 pecks	0	1	0
	Item for a peck of wheate then	0	0	9
28	Received of Rob: Macer for 2 pecks of rye or miscellaine	0	1	0
	Received of Mary Waits for a peck of the same, at	0	0	6
[p.15]	Received of Athanasius Carrington for a bushel of rye	0	2	0
	Received of Jos: Wed's wife for a peck of barly that was left of the old	0	0	5
29	Item for a peck of rye by Tho: Watson	0	0	6
	Received of Antony Parish for wintering his cow 7 weekes			

30	last winter, viz from November 29 til January 18 besides what she was in March following and after	0	3	0
	Item for half a bushel of rye in October last	0	1	0
	Item for smal tithes for several years last past to this present yeare	0	8	0
31	Received of Sam: Woods for 3 bottls of wheat straw about June 18	0	0	6
	Received this month in al the summe of	11	9	10

[p.14] August 1705 Disbursements

	Paide for a spring lock for my study door of Mr. Finch at Cambridge	0	1	4
11	Paide to Thom: Davies, one of tythe men from Hare Street in Hertfordshire	1	10	0
14	Paide to Thomas Wallis and Benjamin Ingrey, two other of tythe men	3	0	0
16	Paide for the use of two horses hired of Edward Lunn of Great Shelford	1	0	0
	Paide to Antonyes wife and Mary Waites for gathering my barly and oates 3 dayes a peice at 2s. per day	0	6	0
18	Given for largesses in the whole towne this yeare	0	13	0
	For unloading malt at Royston	0	0	4
20	Spent by Nathan: selling it, and sack bands etc.	0	0	7
	Paide to Will: Numan for 6 hempen reignes, and a pair of plow traise 1s., laying them in and work 3d., toto	0	2	3
21	Paide to Will: Brooks for his wages last harvest	1	10	0
22	Paide to Ralp Carpenter for 5 loades of dung laid upon son Nat's land July 18 at 1s. 8d. per loade	0	8	4
	Item for 3 dayes and half serving the thatcher for my reeke	0	3	6
24	Spent by Nathan: at market 4d., and for pounding my hogs at Foxton 4d.	0	0	8
	For a pound of 6-penny nayles	0	0	4
	For a whole box of Lockier's pills	0	4	0
27	Paide to Rich. Haycock for putting on new heeles to my old shooes	0	0	8
	Spent by Nathan: at Royston	0	0	4
	Given to a souldier in want	0	0	2
28	Spent at Cambridge with son Sam and daughter Mary, and her 2 sisters, seeing the colledges, King's Colledge Chappel, and at the inn for victuals, horses and al, being six etc.	0	8	8
[p.16] 29	Paide to Widdow Gurner for her daughter [p.16] Rose, for picking rye sheaves 5 dayes at 2s. 3d.; item to herself for 5			

	dayes and half at the same work at 2*s.* 9*d.*, toto	0	5	0
	Given to a souldier, an officer, and a scholler in want	0	0	6
30	Spent by son Nathaniel at Finchingfeild, going to meet daughter Mary, and at Mr. Martin's at Newport that day 8*d.*	0	2	0
31	To Marg: Cra: uxori (besides things that were bought in July) for sugar 18 pound about 8*s.* 4*d.*, malagoes 6 pound 2 shillings, currans 10 pound 5*s.*, butter 6 pounds 3*s.*, rise and carryway 6*s.*, tobacco a pound and half 3*s.*, 6 yards of damask 2 yards and a quarter wide at 7*s.* 6*d.*, 2*l.* 5*s.* 0*d.*, and al other occasions this month.	4	16	0
	Expended this month in al the summ of	14	13	8

[*p.15*]

September 1705 Receipts

	Received of Sam: Woods againe for a bushel of malt July 18 last	0	2	4
1	Item for 2 pecks of rye the next day	0	1	0
	Item for 2 pecks of wheate that day, and againe August 3rd	0	1	6
3	Received for 4 quarters 7 bushels of the first dressing of this new wheate, out of the malthouse, with a bushel of the old put to it, to make it 8 loade, and soald to Edward Phip of Buntingford at 11*s.* per loade, for	4	8	0
4	Received of Thom: and Will: Ward (for they came to reckon together) for a bottle of wheate straw	0	0	2
	Item for a bushel of barly March 19 last	0	1	8
5	Item for his smal tithes, viz 2 cowes, 2 calves, one lamb, wool, etc.	0	6	2
	Item for a bushel of wheate of him at two several times, at	0	2	8
6	Item for 6 bushels and an half of rye at several times also between March 19 and June 26 last at 2*s.* per bushel	0	13	0
7	Received of Will: Brock what remained due in part for 18 bushels and half of rye ever since March 28 last, at 2*s.* a bushel	0	2	7
	Item for what remained for 2 pecks of wheate ever since June last, vide both those places aforesaide	0	0	4
8	Item for fetching him 2 loades of moor straw both in one day	0	3	0
	Item for 6 pecks of rye April 25 last	0	3	0
	Item for 2 pecks of wheat that month	0	1	6
	Item for one bushel of wheat at several times in August last	0	3	0
10	Item for 6 pecks of rye at twice in the month of July last	0	3	0

	Item for 2 bushels and 3 pecks in August and this instant September	0	5	6
	Received of Rob: Macer for 3 pecks	0	1	6
11	Received of Edw: Parish by the hande of his wife in part of his half year's rent ended at Lady 1705, the sum of	1	0	0
	Item in part for 6 pecks of malt they had on August 29, at 3s. 6d.	0	2	0
[p.17]	Received of John Thompson for the tithe hay of Bridg Close this last hay time	0	3	0
12	Received of John Bowelsworth for half a bushel of malt	0	1	2
	Received for a bottle of rye straw	0	0	2
	Received of Urias Bowes for smal tithes for 2 yeares last past, one cow, one calf	0	2	4
13	Item for 6 pecks of wheate att divers times	0	4	4
	Item for 2 bushels and half of rye since July 2 last at divers times	0	4	10
14	Received of Thom: Watson for 3 pecks of mault betwixt June 7 and July 28	0	1	9
	For fetching him a loade of moor straw about July 19	0	1	6
	Item for bringing him a loade of turf with Eastland, at	0	2	0
17	Item for 3 pecks of wheat at 2s. 8d.	0	2	0
	Item for a bushel more of the same	0	3	0
	Item for 5 pecks of the same at 3s 4d.	0	4	2
	Item for 4 bushels and half of rye since June 4 last at several times	0	9	0
18	Received for 3 pecks of the new rye of R. Macer	0	1	3
	Received of Thom: Eastland for carriage of turf with Thomas Watson from the faire, at	0	2	0
	Item for 6 pecks of rye, 3 of them new rye at 1s. 3d., and old at 1s. 6d.	0	2	9
19	Item for 2 bushels and 3 pecks of wheat	0	8	2
	Item for 2 bushels of mault	0	4	8
	Received of William Thrift for 2 bushels and a peck of the same	0	5	3
	Item for 2 pecks of barly in August	0	0	10
20	Item for wintering 2 cowes last winter 19 weekes a piece at 5d. per weeke	0	15	4
	Item for 3 bushels of rye	0	5	8
	Item for a bushel of wheate	0	3	0
21	Received of Will: Fordham senior for 3 bushels and a peck of rye that he hath had at several times, at 2s. per bushel	0	6	6
22	Received of Goodman Thomas Barber for 3 stretches of hame straw in Waterden Shot October last, at	0	3	0
	Item for 7 pecks of rye at several times in November last	0	3	6

HOUSEHOLD ACCOUNTS 1705

	Item for 6 pecks of wheate on November 8, at 2s. 8d., that is	0	4	0
24	Item for 2 pecks of rye on January 11	0	1	0
	Item for a peck of wheate February 21	0	0	9
	Item for a bushel of rye at that time	0	2	0
25	Item for a bushel of rye April 13 last	0	2	0
	Item for 2 pecks of wheat in that month also at twice	0	1	6
	Item for a peck of wheat May 4	0	0	9
26	Item for 3 pecks of wheate in the month of June last, 25 and 28	0	2	0
	Item for 2 pecks of rye then	0	1	0
27	Item for 3 pecks in July last	0	1	6
	Item for 2 pecks of wheate that month	0	1	6
	Item for a peck of barly August 3	0	0	5
28	Item for 3 pecks of wheate in that August 9 and 13 at 3s.	0	2	3
[p.19]	Item for 3 pecks of the same on June 4, promised to be paide that weeke but it was not, at	0	1	4
29	Item for 6 pecks of the new wheate since harvest at 2s. 4d.	0	3	6
	Received this month in al the sum of	14	8	7
	Received in the last six months, viz since March last, the full summ of	105	8	5

[p.16] **September 1705 Disbursements**

	Spent by son Nathaniel at Royston Market	0	0	6
	Paide to Sam: Woods for 2 dayes at cart and other things before June 20	0	0	8
1	Item for his whole time from June 20 until Michaelmas next, he wanting his mony to buy him some cloaths at Sturbridge Fair, and so had it beforehand, in al by the great [?] foraliftime	0	14	0
3	Paide to Br. Spa: for the interest of 20 pound which should have been paide June 24, when the half yeare ended	0	12	0
4	Paide for the first quarter of the land tax granted for this present year for Hoyes, or Ed: Parishes	0	2	6
	Item for Cassanders to the same	0	9	0
	Item for parsonadge to the same tax	7	0	0
5	Spent at Buntingford carrying the wheat to Ed: Phips, and unloading 3d., toto	0	0	9
	For an hand or riding whip at the faire	0	1	8
6	Paide to Mr. Hanchet for letting my wife blood, and phisick for her when she returned from London very ill	0	6	6

		Paide to Goodman Thom: Ward for a quarter herdman's wages due to him at Christmas last for 4 cowes	0	2	8
7		Item for 4 loades of dung about February 25, and 8 more about June 10 last, partly upon my owne and partly upon son Nathan: hired land	1	1	0
		Paide to Will: Fordham for putting a new purloyne to the hogestyes, and given to a man in want at door 2*d*., toto	0	0	5
8		Paide to Will: Brock for trimming myself 3 quarters and half at 29 [?]instant	0	10	6
		Item trimming my sons Nathan:, Sam:	0	5	10
		Item for his wages this harvest	1	12	0
10		Item for 15 dayes work threshing the wheate in malthouse and barly in Nat's barn	0	12	6
		Item for 3 loads of dung since harvest	0	5	6
		Item for an odd job helping Eastland etc.	0	0	6
		Paide to Edw: Parish for 2 dayes in the latter end of harvest	0	2	0
11		Paide to John Thomson for 13 loads of dung we had of him before harvest, partly upon Nat's and partly upon my owne land, at 1*s*. 5*d*. a piece	0	19	6
		Spent by Nathan: at Royston Market	0	0	5
		Paide to Thom: Watson for 2 dayes in casting ponds in orchard, see June	0	2	6
[*p.18*]		Item for 15 dayes and a part of one before June 30, the part at 8*d*.	0	13	2
12		Item for 18 dayes more in July before we began harvest, in the hay and filling and spreading dung, etc.	0	15	0
		Item for his wages this last harvest	1	10	0
13		Item for 10 dayes more after harvest, at plow, haume cart, dung cart, and gathering peares and such	0	8	4
		Item for 22 dayes in this month at sowing rye, haume cart, dung cart, gathering apples, etc.	0	18	4
14		Spent by son Nathaniel at Royston	0	0	2
		For lyquorish for myself, having a cold	0	0	2
		Paide to Thomas Eastland for 5 dayes and half before harvest in hay and other things	0	4	7
17		Item for 2 dayes and half since harvest at haume cart, going to grange, etc.	0	2	1
		Item for a loade of dung in the latter end of August last	0	1	8
		Item for his wages last harvest	1	12	0
18		Paide to Will: Thrift for 3 dayes before harvest in hay and dung cart etc.	0	2	6
19		Item for 13 dayes since harvest, and for part of a journey at plow, haume cart, throwing up dung, making room in his own barne, etc.	0	11	2
		Item for his wages last harvest, he being hurt and lame 5 dayes before we made an end, vide November	1	7	0

20	Paide to John Waites for his wages this last harvest	1	10	0
	Paide to Athan: Carrington junior for his help 4 dayes and an half this harvest	0	4	0
21	Paide to Will: Fordham senior for his wages this last harvest	1	12	0
	Item for lodging Thom: Wallis in the time of harvest, Nat: being sick	0	1	0
22	Item for half a day setting pales about the reeke, etc., 9*d*., and 2 dayes of his son Will mending boardes in stable chamber, horse block, etc., vide December following	0	3	5
	Paide to Will: Fordham junior for his wages last harvest also	1	10	0
24	Paide to Goodman Thomas Barber for his wages last harvest	1	10	0
25	Paide to Mr. Thomas Malin for 61 stone and 6 pound of beife, mutton and lamb, that we had of him this harvest at 2*s*. 8*d*. per stone, besides divers parcels that wee had since harvest in this month	8	3	8
26	To Marg: Cra: uxori for mutton, veale, etc., about 15*s*., pidgons, oisters, wine, about 10*s*., to Widdow Haycock for			
27	nursing 9 dayes in harvest at 7*s*. 2*d* and a week afterwards 2*s*., daughter Hester and Rhoda being both very sick, and other things	2	16	0
28	Item for Elizabeth Fere, for her help at harvest 12*s*., and afterwards til Michaelmas, that is 6 weeks, toto	0	18	0
[*p.20*]	Item for Rhoda Phipp her maidservant [*p.20*] for her whole yeare's wages now due	3	0	0
29	Paide to George Carpenter for his yeare's wages now due also	5	0	0
	Expended this month in all the sum of	49	16	2
	Disbursed in the last six months, viz since March last, the full summ of	164	10	2

[*p.19*]

October 1705 Receipts

	Received of Margaret Swann, widdow, for a bushel of wheate in September 26 and 15 last, at 3*s*. and 2*s*. 8*d*.	0	2	10
1	Item for the carriage of a loade of goods to Sturbridge Faire	0	3	0
	Item for her half yeare's rent ended on September 29 now past	1	11	6
3	Received of Widdow Haycock for a peck of wheat on August 24	0	0	9
	Item for 2 pecks of rye then and now	0	0	11

	Item for a loade of rye straw laid in	0	5	6
4	Received of Jos: Wedd for a peck of rye	0	0	5
	Item for a peck of barly then	0	0	5
	Received of Tho: Reynolds for a peck of rye	0	0	5
6	Received of Edith Wedd for a peck 5*d*.; item for a peck of new wheate 7*d*.	0	1	0
	Received of Rich: Haycock for rye 2 pecks	0	0	10
12	Received of Edw: Phip of Buntingford for 5 quarters of the 2nd dressing, at about 19 pence per bushel	3	3	4
	Received of Ben: Bright for 2 pecks	0	0	10
	Item for a bushel of mault	0	2	4
16	Received of John Watson for a bushel	0	2	4
	Received of Widdow Goode for a peck of rye 5*d*.; received of Rob: Macer for 4 pecks 1*s*. 8*d*.	0	2	1
	Item for a peck of new wheat 7*d*.; received of Thomas Barber for 2 pecks	0	1	9
17	Received of John Watson what remained due of his half year's rent ended Lady Day 1704 since the October 5 or 6, 1704	0	3	3
18	Item for his half yeare's rent ended on September 29, 1704	0	19	0
	Item for his half yeare's rent ended at Lady Day 1705	0	19	0
19	Item for 2 bushels of malt November last	0	4	6
	Item for 2 pecks of wheat December last	0	1	6
	Item for a bushel of barly in April last	0	1	8
	Item for 2 pecks of rye September last	0	1	0
20	Item in part of his half yeare's rent ended on September 29 last	0	4	3
22	Received of Wil: Brock for a bushel of new rye 1*s*. 8*d*.; received of Joan Woods for a peck 5*d*.	0	2	1
	Received of Rob: Macer for malt 2 pecks	0	1	2
	Item for a peck of wheate	0	0	7
23	Received of Ed: Parish for a peck of barly	0	0	5
	Received of J. Boweles for a peck of wheat	0	0	7
	Item for 2 pecks of the new rye	0	0	10
[*p.21*]	Received of Will Dovy for wintering his cow 1704 5*s*., and for smal tithes last year 2*s*. 6*d*.	0	7	6
25	Received of the coller maker for 4 pound of old bells	0	2	0
	Received of Tho: Barber, Rob: Macer, Widdow Swann, for 6 pecks of wheate	0	3	6
27	Received of Mr. Selby of Cambridge for 6 quarters 2 bushels of rye at 1*s*. 6*d*. per bushel	3	15	0
30	Received of Good: Godfry, Woods, Gurner, Bright, Waits, Macer, Mrs. Westly, for 4 bushels and half at 2*s*. and 1*s*. 8*d*. per bushel	0	7	9

31	Received of Rob: Harvy for 6 pecks of malt, at	0	3	6
	Received this month in al the full summe of	13	19	4

[*p.20*]

October 1705 Disbursements

	Paide to Will: Brock for his yeare's wages as clerk now due	0	6	8
1	Paide to Widdow Swann for 9 loads of dung June 26 and 27 last, five of them but indifferent at 7*s.* 6*d.*, the other 4 at 8 shillings, upon Nat's land	0	15	6
	For sowing silk, hooks and eyes, for an undercoate for myself	0	1	0
3	For 4 bushels of lime for the seede wheat from Eversden at 5*d.*	0	1	8
	For a pound and half of 8*d.* and 6*d.* nayls	0	0	7
	Given to a poor man at door	0	0	2
6	Paide to Goodm: Botterel of Chrissol for coming over twice to the Duk horse, and drinks for him	0	5	0
	Spent by men carrying rye to Buntingford and unloading sacks	0	0	8
12	Paide to Ben: Bright for a loade of dung laide upon son Nat's land	0	1	11
	Spent by son Nathaniel at market	0	0	4
16	Paide to John Watson junior for 2 dayes and half thatching the rest of the parsonadge stable October 1704 — part of it done the summer befor	0	3	9
17	Item for 2 jaggs of dung in October last, viz at 3*s.*, and 3 bushels of hen dung March 7, 1704, 1*s.* 6*d.*	0	4	6
	Item for 2 dayes work on June 6 and 12, casting out the mudd in the 2 ponds in the orchard, vide June 1, [170]5	0	2	6
18	Item for 3 dayes and half thatching my reeke August 28 last	0	5	3
	Item for a load of dung about June 24 last, carryed with 2 horses	0	1	8
19	Item for 5 dayes this last September 25 thatching the hoggs' stye and mending the old barn over the doors	0	7	6
	To Featherstone for making a new bottome to the chaff sieve	0	0	6
20	Given to a breif for John Bainton and 2 others by name of Kirton in Lyndsey, Lincolnshire, looseing by fire on			
21	Saturday December 16, 1704, 1,000*l.*, collected on July 1 last and brought in now	0	1	0
	Spent by Nathaniel at Royston on 2 several market dayes	0	0	6

22	Item at Cambridge Market	0	0	6
	For 2 pound of 6*d.* and 8*d.* nayles	0	0	9
	Paide to Will: Dovy for 2 loads of dung upon Nat's land between February 18 and 25 – he saith three	0	5	0
[*p.22*] 26	Paide to Will Numan the coller maker for half an horse hyde 4*s.* 6*d.*, half a calve's skin 1*s.* 6*d.*, and work 6*d.* in al	0	6	6
	For tol at Cambridge 2*d.*, spent 9*d.*, and to barber for Nathaniel	0	1	2
27	Given to a poor blind man at door	0	0	2
	For 700 sprindles or spitts of Wallis	0	1	8
	Spent by Nathaniel at Royston Market	0	0	4
	For a new halter at Cambridge	0	0	3
30	For 3 pounds of grease for wagon	0	1	0
	For 7 dressing broomes at Cambridge	0	1	0
31	To Marg: Cra: uxori for cortex peruviana for her ague of Mr. Wright 3*s.*, beife and mutton about 1*l.* 6*s.* 0*d.*, and al other occasions this month	2	14	0
	Expended this month in al the summe of	6	13	0

[*p.21*] **November 1705 Receipts**

	Received for Mr. Selby of Cambridge for 6 quarters of rye more at 1*s.* 6*d.*	3	12	0
3	Received of Widdow Gurner for 2 pecks of malt at 2*s.* 4*d.* per bushel	0	1	2
6	Item for a peck of rye; received of Mr. Greenhil for 2 bushels	0	3	9
	Received of Cas: Woods for wheate a peck	0	0	6
	Received for 2 pecks of Jos. Wed — wheat, best	0	1	2
7	Received of Ralph Carpenter for rye 6 pecks, Widdow Law 2 pecks, Benjamin Bright 2 pecks, Wil Brock 6 pecks,			
9	Jos. Wedd 3 pecks, Cass. Woods a peck	0	8	4
	Received for a bottle of straw of Jos. Wed	0	0	2
	Received for 2 pecks of wheat of R. Macer	0	1	2
12	Received of Thom: Barber for a peck	0	0	7
	Received of Edw: Parish for a peck of barly, and of Mr. Greenhil for a bushel	0	2	1
16	Received of J. Waites for 2 pecks of rye	0	0	10
	Received of T. Barber for 2 pecks of wheat	0	1	2
19	Item for half a bushel of rye 10*d.*; received of Edw: Godfrey for a bushel; received for 3 pecks of Widdow Gurner, Hankin, Mrs. Westly	0	3	9
	Received of Widdow Swan and Cass: Woods for 3 pecks of wheate at 2*s.* 9*d.*	0	1	9

HOUSEHOLD ACCOUNTS 1705

20	Received for a bottle of straw of Ward	0	0	1
	Received of Ben: Bright for 3 pecks of rye, Widdow Gooding 2 pecks, Wil Brock 6 pecks, Jos: Wedd 3 pecks	0	5	10
21	Item for 2 pecks of old wheat	0	1	4
	Received of Rob: Macer for a peck; received of Kate Reynolds for 2 pecks; received of J: Bowles for a peck; received of Thom Barber for 2 pecks	0	3	6
22	Received of Widdow Gurner for a peck of rye; received of John Waites for 3 pecks; received of Cass: Woods for 3 pecks	0	2	11
24	Received of Mrs. Norridge by the hands of son Nathaniel for 4 bundles of willow lopps for the straw house, to thatch upon in the room of reede	0	1	6
26	Received of Widdow Goode for a peck of barly	0	0	5
	Received of K: Reynolds for a peck of rye	0	0	5
	Item for a peck of wheate then	0	0	7
[p.23]	Received of Widdow Nash for 2 pecks of old wheat	0	1	4
27	Received of Edw. Godfry for a peck of rye	0	0	5
	Received of Will Brock for six pecks	0	2	6
	Received of G: Hankin for a peck of wheat	0	0	7
28	Received of Rob: Macer for 2 bushels of rye or miscellaine	0	3	4
	Received of Peter Nash for 2 pecks of the new wheate at 2s. 4d.	0	1	2
29	Received of Widdow Law for 2 pecks of rye; received of Widdow Gurner for a peck	0	1	3
30	Received of Edward Phipp of Buntingford for five quarters of the 6th dressing of the same at 13s. per quarter	3	5	0
	Received this month in al the summ of	9	10	7

[p.22] **November 1705 Disbursements**

	Spent by men at Cambridg, carrying 6 quarters of rye whereof the unloading 6d., tol 2d., spent 6d.	0	1	2
3	Paide to Will: Thrift for threshing 9 quarters and half of oates, which my horse had though the oates were son Nathaniel's, and al his crop	0	6	4
6	Paide for the 2nd quarter of this yeare's tax for Hoyes, or Ned Parishes	0	2	6
	For Cassanders to that tax	0	9	0
	For the same tax of parsonadge	7	0	0
9	For the half yeare's duty upon windowes ended at Michaelmas last	0	5	0
	Given to a gentleman, an officer, disabled at the taking of Namur etc.	0	1	0

12	Paide to Goodman Soale for mending up a plow or two	0	1	0
	Paide to Rich: Carter for a ringe of wood bought of him in Rofeway	0	16	0
14	Given to a breife for Sam: Allen and 9 other familyes of Rolleston in County Stafford, loosing by fire (on November 30, 1703) 113*l*. 14*s*. 4*d*., collected October 21, handed now	0	1	0
15	For a letter from Mr. Mitchel	0	0	3
	Spent by son Nathan: at Royston	0	0	6
	Paide for a pound of 10*d*. nayles etc.	0	0	5
	Item for a pound of 6*d*. and 8*d*. nayles	0	0	4
20	Spent by son Nathan: at Royston	0	0	6
21	Given to one James Sleigh, living at Durham, going for London with his family, taken by the French, retaken by Dutch and loosing 200*l*., as he saith	0	0	2
24 25	Given to a breife for Al Saints Church in Oxford, the steeple falling March 8, 1698, and beating downe a great part of the church, which they were forced to take down, the charges of rebuilding being computed to amount to 1,800*l*. and upwards	0	1	0
	For 3 pounds more of 6 penny and 8 penny nayles used about malthouse, door-pales at Cassanders, etc.	0	0	9
[*p.24*]	Paide to George Rule for a new cutting knife	0	2	0
28	For a pound of 6*d*. nayles used about pales and hogtub style, and some other little nayles for pailes	0	0	5
	Spent by Nathaniel at Royston Market	0	0	6
29	Item at Buntingford carrying the 5 quarters of rye and unloading	0	0	9
30	To Marg: Cra: uxori for 2 dozen of candles 10*s*., for 12 stone of beif, besides mutton, veale and lamb, about 1*l*. 15*s*., and al other occasions in this month	3	2	6
	Expended in this month in all the full summ of	12	13	1

[*p.23*]
December 1705 Receipts

	Received of Edw: Parish for 2 bushels of the new wheate at 2*s*. 4*d*.	0	4	8
3	Received of Cass: Woods for a peck	0	0	7
	Item for a peck of rye 5*d*.; received of Good: Hankin for a peck 5*d*.; received of Edw: Godfry for a bushel	0	2	6
4	Received of Wil Thrift for half a bushel of wheate, at	0	1	2
	Item for 2 bushels of rye at 1*s*. 8*d*.; received of Mary Waites for 2 pecks	0	4	2
5	Received of Mr. Urias Bowes of Barly for 6 quarters and 2			

	bushels of the 2nd dressing of barly at 15s. per quarter	4	13	0
6	Received of Lyd: Barber for wheat a peck; received of Jos: Wedd for 2 pecks, old	0	1	11
	Item for 2 pecks of rye 10d.; received of Wil. Brock for six pecks	0	3	4
7	Received of Ben: Bright for 2 pecks 10d.; received of Widdow Gurner for a peck 5d.; received of Mr. Urias Bowes for 6 bushels	0	11	3
	Item for 2 bushels of wheat of him	0	5	4
8	Item for 11 quarters and 7 bushels of the 3 and 4 dressings of barly, sold to him at 15s. 8d. per quarter	9	6	0
	Received of Widdow Gooding for 2 pecks of rye	0	0	10
10	Received of Mr. Greenhil for one bushel of barly	0	1	10
	Received of Ben: Bright for 4 pecks of malt	0	2	6
	Received of Widdow Nash for wheat 2 pecks	0	1	2
12	Received of Athanasius Carrington for 2 pecks of rye and a peck of barly	0	1	4
13	Received of Widdow Gurner for wheat a peck; received of Widdow Goode for a peck; received of Rob: Harvy for 2 pecks, old	0	2	6
	Received of Mr. Urias Bowes for five quarters 3 bushels of the 5th dressing of barly sold at 15s. 6d.	4	3	0
14	Received of J: Waits for a peck of barly	0	0	6
	Received of Peter Nash for wheat 2 pecks	0	1	4
	Received of Widdow Law for 2 pecks of rye	0	0	10
17	Received of Urias Bowes for 4 bushels	0	6	8
	Item for a bushel of old wheate	0	2	8
	Item for 6 quarters of the sixth dressing of barly at 15s. 6d.	4	13	0
18	Received of Mrs. Westly and Widdow Law for half a bushel of rye	0	0	10
[p.25]	Received of Widdow Swan for half a bushel of wheate, old wheate	0	1	4
19	Received of Mr. Wedd for six bushels	0	16	0
	Received of Jos. Wedd for 2 pecks	0	1	4
20	Item for a bushel of rye at that time; for a peck of Widdow Gurner 5d.; received of Edward Godfry for a peck 5d.	0	2	6
21	Received of J: Woods for 2 pecks of wheat; received of Thom: Barber for 2 pecks [?] in part; received of Kate Reynolds for a peck	0	3	2
	Item for 2 pecks of rye at 10d.; received of M: Waites for 3 pecks 1s. 3d.; received of Widdow Nash for 2 pecks 10d.	0	2	11
22	Received of Edw: Parish for one bushel of barly, at	0	1	11
26	Received of Athanasius Carrington for five pecks of rye; received of Widdow Goode for 2 pecks; received for a bushel of wheat of Widdow Nash	0	2	11
	Received of the miller for 2 bushels	0	8	0

	Received of Thom: Eastland for a bushel of mault	0	2	6
27	Item for one of the 2 furlong stetches laide in by my cart and men	0	2	0
	Item for 2 bushels and a peck of rye since October 13 to this time	0	3	9
28	Item for 3 bushels and 3 pecks of wheate in the same time	0	10	0
	Received of Mrs. Westly for rye 3 bushels	0	5	0
	Item for a bushel of barly	0	2	0
	Item for a bundle of rye straw	0	0	3
29	Received of Mr. Urias Bowes of Barly for 15 quarters and half of the 7 and 8 dressing at 15*s.* 3*d.*	11	6	0
	Received of Peter Nash and Jos: Wedd for 6 pecks of rye, and of the last for 2 bottles of rye straw 4*d.*	0	2	10
31	Received of Jos. Wedd and Rob: Harvy and Widdow Goode for 3 pecks of wheat	0	2	0
	Received of Edw: Parish for 1 bushel and half of malt at 2*s.* 6*d.*	0	3	9
	Received this month in al the full summe of	40	13	1

[*p.24*]
December 1705 Disbursements

	Paide to Will: Fordham senior for 3 dayes mending up cow rack, stable rack, barne doors	0	4	6
3	Item making a new door to malthous	0	1	6
	Item mending pales at Cassanders	0	1	4
4	Item for 6 dayes between November 19 and 30 making a style to hog tub yard, a cow rack, an head to the wagon, mending orchard and street gates, a new post to orchard	0	8	0
5	Paide to Will: Thrift for threshing 4 quarters 1 bushel 2 pecks of pease	0	4	2
	Item for threshing 9 quarters and half of oates out of middlestead	0	6	4
6	Item for threshing 13 quarters of the 2nd and 3rd dressing of barly	0	9	9
	Spent by son Nathan: at Royston	0	0	6
7	To Thom: Fairechild helping to in my barly reek, part of it into Eastland's	0	0	9
	Paide to Mrs. Norridge for 5 loades of dung laide upon Nat's land	0	5	0
8	For 150 rodds for thatching	0	1	6
	Paide to Will: Strett, yelming about half a day, the straw being blowne off the reek the 2nd time – John Watson had nothing	0	0	3
10	Paide to Wallis for 1,000 spitts	0	2	6

HOUSEHOLD ACCOUNTS 1705

	Paide to Will: Numan for 3 collers, 3s. 6d. each of them, in al	0	10	6
12	Item for cloth 1s. 2d., halters 6d., a pair of bitts 3d., work 6d., toto	0	2	3
	Spent by son Nathan: at Royston	0	0	6
	Item for unloading the 5 loade of barly	0	0	4
13	For 2 pound 5 ounces of tarr	0	0	7
	For a letter and a bundle for London	0	0	6
14	Paide by son Nathan: to Mr. Jackson for mending chancel windowes blowne down	0	2	3
	Spent by son Nat that day	0	0	2
17	For unloading the 6 loade of barly, and to a man that helped George when he hurt his hand so much	0	0	5
	Paide to Thom: Eastland for cutting 30 acres and one roode of haume at 1s. 9d.	2	0	4
[p.26]	Item for threshing 4 quarters and 2 bushels of pease from the reek	0	4	3
18	Item from threshing 3 quarters and 4 bushels of oates from reek	0	2	4
19	Item for five dayes work in October at haume cart, making roome in the old barnes and such like	0	4	2
20	Item for 3 days and half in this present month, 2 and an half of them in helping to get in the oates and pease and barly, the other at plow and other things	0	2	8
21	Item for threshing 6 quarters and 2 bushels of the 2nd dressing of wheate at 1s. 3d. per quarter	0	7	10
22	Item for threshing 21 quarters and 6 bushels and half of the 4, 6, 7 dressings of barly at 9d. a quarter	0	16	4
	Item for threshing 18 quarters and 4 bushels of the 1, 2 and 3rd dressings of rye at 1s. 3d. per quarter	1	3	1
26	Item for dressing 28 quarters and 2 bushels more of the 4, 5, 6, 7th dressings of rye also at 1—3 also	1	15	5
27	For unloading the 15 quarters and half of barly at Mr. Urias Bowes'	0	0	7
	For a sheet almanack and one bound for the ensuing yeare	0	0	7
28	To Marg: Cra: uxori for black woosted 2s., yarn besides 2s., and for cloathes that were bought at London and not			
31	paide for til now 1l. 10s. 0d., and other things for cloathes and taylour, etc., since, viz in this month about 8s., a side of mutton 4s. 6d., and for al occasions in this month	3	8	0
	Expended this month in al the full summe of	13	9	2

151

REV. JOHN CRAKANTHORP

[p.25]   **January 1706 Receipts**

	Received of Widdow Goode for 2 pecks of barly of the 8 dressing, at	0	0	11
2	Received of Good: Hankin for wheat a peck	0	0	8
	Received of Peter Nash for 2 pecks	0	1	4
	Received of Rich: Haycock for a good bundle of rye straw	0	0	3
	Received of Widdow Ward for a bundle	0	0	3
3	Received of Widdow Gooding for three pecks of rye	0	1	3
	Received of Widdow Gurner for a peck	0	0	5
	Received of Jone Woods for a peck	0	0	5
	Item for a peck of wheate	0	0	8
4	Received of Edw: Parish what remained due for 6 pecks of malt ever since September 11 last	0	1	6
	Item for 7 pecks of rye	0	2	11
	Item for 7 bushels of barly of the 2nd and 3rd dressings and 2 pecks now at 11 pence, in al	0	13	7
[p.27]	Received of Widdow Nash for half a bushel of rye or miscellaine	0	0	10
	Received of Ben: Bright and J. Thompson for 3 bundles of rye straw	0	0	8
5	Received of Thom: Watson for 2 bushels of wheat at divers times since October 4 last, by agreement for an whole yeare, at 3s. 4d.	0	6	8
7	Item for five bushels and an half of rye in the same space of time	0	11	0
	Received of John Waites for 3 pecks	0	1	3
	Received of Will: Brock for 6 pecks	0	2	6
	Item for a peck of wheate	0	0	8
8	Received of Widdow Gooding for a peck	0	0	8
	Item for 2 pecks of rye 10d.; received of Jos: Wedd for a bushel 20d.; received of Widdow Law for 2 pecks 10d.	0	3	4
9	Item of her and Cass. Woods and John Waites for each of them for a bottle of rye straw	0	0	4
	Received of Jos: Wedd for 3 bottles	0	0	6
	Received of Widdow Swann for half a bushel of wheate, old wheate	0	1	4
	Received of Widdow Law for a peck	0	0	8
10	Item for a peck of rye that day	0	0	5
	Received of Widdow Goode for 2 pecks	0	0	10
	Received of Widdow Nash for half a bushel of wheate to miller	0	1	4
11	Received of R: Macer for 2 pecks in part 1s. 2d.; received of Peter Nash for 2 pecks 1s. 4d.	0	2	6
	Item for 2 pecks of rye 10d.; received of Athan. Carrington for a peck	0	1	3
12	Item for 2 pecks of barly at 11d.; received of Mr. Greenhil			

HOUSEHOLD ACCOUNTS 1706

		£	s	d
	for a bushel; received of Widdow Gooding for a peck	0	3	3
	Received of Mr. Greenhil for a parcel of wheat straw by the lump	0	0	9
14	Received of Rob: Harvey for wheat 2 pecks; received of Will: Brock for a peck	0	2	0
	Received of him for six pecks of rye; received of Widdow Gurner for a peck	0	2	11
15	Received of Ed: Phip for 10 bushels, carryed with Nat's at 1s. 10d.	0	18	4
	Received of J: Woodward by discount with his wife for 3 bushels	0	5	0
	Received of Widdow Eversden for a loade of rye straw laide in	0	6	0
16	Item for 400 spitts for thatching	0	1	0
	Received of Thom: Reynolds for a peck of wheate, at	0	0	8
	Item for a peck of rye 5d.; received of Widdow Gooding for 3 pecks	0	1	8
17	Received of Will: Dovy for a parcel of wheate chaff, at	0	1	0
	Received of Mr. Urias Bowes of Barly for 12 quarters of the 8 and 9 dressings of barly at 15s. per quarter	9	0	0
18	Received of Widdow Goode for a peck	0	0	6
	Received of Cass Woods for wheate a peck; received of Thom: Barber for a peck	0	1	4
	Received of J. Thompson and others for straw	0	0	6
22	Received of Widdow Gooding for rye a peck; received of Wil: Brock and Jos. Wedd for 2 bushels and half and a little wheate	0	4	9
	Received of Will: Thrift for 4 fannfuls of wheat, rye and barly chaff	0	1	10
[p.29]	Item for a bushel of malt on September the seventh last at 1l. per quarter	0	2	6
23	Item for a peck of wheate at 8d.; received of 2 pecks of Lyddy Barber; received of Widdow Law for 2 pecks tayle 10d.	0	2	10
	Received of them two and Widdow Nash for a bushel and 3 pecks of rye, at	0	2	11
24	Received of Rich: Fairechild for the carriage of 20 bushels of coales when my wagon was at Cambridge	0	2	6
	Received of Thom: Barber in part for the 4 home stetches of haume in Short Spotts, vide end of September	0	0	6
25	Received of Widdow Ward for 2 pecks of rye; received of J: Woodward for 3 pecks	0	2	1
	Item for a peck of wheate 8d.; received of Widdow Gooding for a peck of tayle	0	1	1
	Item for a peck of barly 5d.; received of Widdow Swann for a peck	0	0	11
	Received for a bundle of straw	0	0	4

26	Received of Macer for a peck of barly	0	0	6
	Received of John Bowelesworth for a peck of wheate 8*d*., and a peck of Cass Woods	0	1	4
	Received of J: Wates for 3 pecks, of Eliz: Bright for a peck of rye, toto	0	1	8
28	Received of Widdow Nash for wheat 2 pecks	0	1	4
	Received of P: Nash for 2 pecks of rye; received of Widdow Nash and another for 3 pecks of the same	0	2	1
	Received for a bundle of wheate straw	0	0	4
29	Received of Widdow Goode for a peck of barly	0	0	5
	Item for a peck of wheate 8*d*.; received of Rob: Harvey for 2 pecks 16; received of Athan: Carrington for 2 pecks	0	3	4
30	Received of Mrs. Westly for a peck 8*d*.; received of Kate Reynolds for a peck 8*d*.	0	1	4
	Item for 2 pecks of rye 10*d*.; received for a peck of Widdow Gurner	0	1	3
31	Item for a peck of wheate 8*d*.; received for a peck of W: Brock, tayle 5*d*.	0	1	1
	Item for 6 pecks of rye 2*s*. 6*d*; received of Widdow Goode for 2 pecks	0	3	4
	Received of G: Godfry for a peck of barly	0	0	6
	Received this month in al the sum of	16	6	7

[*p.26*]   **January 1706 Disbursements**

2	Paide to Cozen Ned Sparhawke for half a yeare's interest of 20*l*. now due — I sent 2 guineas by son Nathaniel October 29, but he being not at home, nor Mr. Landy, he never came from his horse but left them with the maide — that is 1*s*. more than due	0	13	0
	Spent by son Nathaniel at Royston, with his brother and sister and horse etc.	0	0	10
3	Paide to Thom: Watson for 26 dayes work in the month of October, at plow, sowing, haume carting, etc., at 10*d*. per day	1	1	8
4	Item for 8 dayes in November before the 11 of that month at 9*d*. per day in the like works	0	6	0
	Item for 16 dayes more in that month before December 1, at 9*d*., per day also	0	12	0
[*p.28*]	Item for half a day inning the barly reek December 1, when they were beaten out with wett etc.	0	0	4
5	Item for 17 dayes and half before December 23 at 9*d*. per day	0	13	2
	Item for 4 dayes that month besides	0	3	0
7	Item for five dayes this month and December 31 —			

		£	s	d
	Saturday at the wood for himself	0	3	9
8	Paide to Rich: Carter by Thom: Watson for stub-mony for a ringe of wood in Chrissol Park 2s., and a groate he spent then.	0	2	4
	Paide to him for hiring a man that helped him to bring back my bul — strayed almost as farr as Kneesworth or Bassingborne	0	0	2
9	Paide to Will: Numan for lineing a collar for one of my horses	0	1	2
	For a new whale-bone whipp	0	1	2
	For a new curry comb	0	0	7
11	For an hundred of 4d. nayles	0	0	4
	Spent by son Nat: at Royston	0	0	6
12	Spent at Buntingford carrying 5 quarters of rye, whereof but 10 bushels for me as my part	0	0	3
14	Given to one Mr. Danvers of the County of Berks., loosing his house and goods by fire in the great wind August 11 last, etc., at Hempsted Ash betwixt Newbury and Reading, a mile out of the roade on one side	0	0	6
15	Paide to Henry Wallis of Crissol for 350 withs for thatching	0	1	9
16	Paide to J: Woodward's wife by discount for 2 quarters herdman's wages ended at Christmas last, for 4 cowes, 2s. 8d. per quarter, vide June 14 last	0	5	4
	Paide the 3rd quarterly payment of this yeare's tax for Hoyes, now in the occupation of Edw: Parish	0	2	6
17	Item for Cassanders to the same tax	0	9	0
	Item for parsonadge to the same tax	7	0	0
18	Paide to Will: Thrift for threshing 12 quarters and 3 bushels of the 5th and 9th dressings at 9d. per quarter	0	9	3
	Item for threshing 23 quarters and one bushel of the tenth and 11th dressings of barly also as the two former dressings were, at 9d. per quarter	0	17	4
19	Paide to Rich: Fairchild for his constable's rate, for Cassanders, what I had of my owne, at 2d. per pound	0	0	9
21	Item to the same rate for parsonadge at 137 pound per annum	1	2	10
22	Paide to him for work done upon son Nathaniel's tally since March 19, 1704, that is 10 months and about 3 dayes	1	18	0
[p.30]	Item for remooving the horses shooes, forty and one in the same space of time at 1d. per remoove	0	3	5
23	Item for 82 new shooes in the same space of time at 4d. per shooe	1	7	4
	Given to a poor man in want	0	0	1
24	For a paire of hames and traise of John Thompson, second hand	0	1	6

REV. JOHN CRAKANTHORP

	For unloading the 12 quarters of barly received 17 instant	0	0	6
	Spent by son Nat at Royston, going to sel barly, as he did	0	0	2
25	Paide to Thomas Fairchild for making a new latch to the parlour door and mending the pin for the sweepe of the pump, the iron in most being mine	0	0	6
26	Spent by Nathaniel at Royston Market	0	0	4
	Paide to Rob: Harvey for 3 dayes sowing of pease at 7d. per journey	0	1	9
	For a little pad lock for the littl door in the old barne, for Thrift's safes	0	0	7
28	For a bottome of pack thread	0	0	3
	Spent by son Nat at Royston	0	0	6
29	Given to Richard Harrison, Sam: Harper and John Ballesson, inhabitants in Merriden in Warwicksh., looseing by fire January 15, *1705* the sum of 478*l*. and upwards, collected the 20th of this instant	0	0	6
	To the glazier putting in 2 quarryes glass in kitchin and buttery	0	0	2
30	To Marg: Cra: uxori for salt 5*s*., cheese at 3*d*. per pound 1*s*. 2*d*., beife 2 ston 5 pounds 7*s*. 6*d*., hen feathers for a bed 42 pounds 14 shillings, materials for a riding hood and making about 19*s*., and al other occasions this month	3	18	0
	Expended this month in al the sum of	22	3	1

[*p.29*]
February 1706 Receipts

2	Received of Mr. Urias Bowes for six quarters and 2 bushels of the 12 dressing of barly, at 15 shillings wanting one penny etc.	4	13	3
	Item for a bushel of rye 1*s*. 8*d*.; received of Kate Reynolds for 2 pecks	0	2	6
4	Received of Widdows Gooding and Law for 3 pecks of tayle wheate, at	0	1	3
	Received of Thom: Barber for 2 pecks; received for 2 pecks of Widdow Gooding, tayle	0	2	2
	Received of Widdow Nash for 2 pecks	0	1	4
5	Received of Eliz: Bright for 2 pecks of rye; received of Jos: Wedd for a bushel	0	2	6
	Received of Widdow Law for 2 pecks; received of Lyddy Barber for a bushel	0	2	6
6	Item for a peck of barly of 13 dressing	0	0	5
	Received of Athan: Carrington for 2 pecks; received of Widdow Gooding for a peck	0	1	3

7	Item for a peck of wheate 8*d.*; received of Widdow Goode for a peck	0	1	4
[*p.31*]	Received of Mary Waites for three pecks of rye 1*s.* 3*d.*, of Widdow Nash for 2 pecks, Widdow Ward 2 pecks, and Edward Godfry 2 pecks, toto	0	3	9
8	Item for 2 pecks of wheat 1*s.* 4*d.*; received of Good: Haycock for a peck	0	2	0
	Received of Rob: Macer and Widdow Strett for 2 pecks of barly, 14 dressing	0	0	10
9	Received of Mary Parish for what remained due of her half yeare's rent ended at Lady Day 1705	0	12	6
	Item in part of her half yeare's rent ended September 29 last	1	1	6
11	Received of Mr. Urias Bowes for 7 quarters of the 13 and 14 dressings of barly at 15*s.* per quarter	5	5	0
	Received of Widdow Goode for 2 pecks	0	0	10
	Recieved of Widdow Gooding and Wil Brock for 6 pecks of rye	0	2	6
12	Item of the last for a peck of wheate; received for 2 pecks of Good: Harvey	0	2	0
	Received for 2 pecks of Widdow Law; received for a peck of Mary Waites	0	2	0
	Received of J: Bowels and Widdow Gurner for 3 pecks of rye 1*s.* 3*d.*	0	1	3
13	Received of Edw: Parish for barly 2 pecks	0	0	10
	Received for a peck of malt of Geo: Kettle	0	0	7½
	Received for 3 pecks of rye of B: Bright; received of Jos: Wedd for a bushel	0	2	11
14	Received of Kate Reynolds for a peck	0	0	5
	Received of God: Hankin for wheat a peck, and M: Waites for a peck	0	2	8
16	Received of Good: Woodward for 2 pecks; item for a peck of barly then	0	0	5
	Received of Rob: Macer for 2 pecks 11*d.*; received of Edw: Godfry for a peck 5*d*	0	1	4
	Item for a peck of rye 5*d.*; received of Rob: Macer for 2 pecks	0	1	3
19	Received of Good: Gurner for a peck	0	0	5
	Received of Mr. Greenhil for a bushel of barly of the 15 dressing	0	1	8
	Received of Mary Waites for malt 2 pecks	0	1	3
20	Received of Widdow Goode for a peck of tayle wheat of the 6 dressing 5*d.*; received of Good: Kettle for 3 pecks of the same, and a peck April 16 last 9*d.*; received of Joan Woods for a peck now 7*d.*	0	3	0
22	Item for a peck of rye 5*d.*; received for 2 pecks of Widdow Nash; received for 6 pecks of Will: Brock	0	3	9

	Received of Widdow Swan for a peck of barly of the 16 dressing	0	0	6
23	Item for a peck of wheate 7*d*.; received of Widdow Gooding for a peck	0	1	2
	Received of Eliz. Bright for a peck; received of Widdow Nash for 2 pecks	0	1	9
26	Received for 5 quarters and half sold to Cambridge at 2*s*. 5*d*.	5	6	4
	Received of Eliz: Bright for rye 2 pecks; received of Mary Waites for 2 pecks	0	1	8
	Received of Widdow Gooding for 2 pecks; received of Kate Reynolds for a peck	0	1	3
27	Received of Widdow Gurner for 9 weekes wintering her bullock to this day, at 6*d*. per week	0	3	6
	Received this month in al	19	19	4½

February 1706 Disbursements

[*p.30*]

	Paide for unloading the 6 quarters 2 bushels at Mr. Ur: Bowes	0	0	3
2	Given to a briefe for 12 inhabitants of Bradmore in Nottinghamshire loosing by fire (hapning on July 2, 1704, beweene 2 and 3 in the afternoon) the summ of 2,400*l*. etc.	0	1	0
3				
4	Spent by son Nat: at Royston	0	0	6
	Paide to Thom. Fairchild for putting on the bayle of a paile	0	0	3
6	Paide to William Numan the colleremaker for three new pipes against the horses' sides, at 5 shillings a peice, and for work the same day he brought them	0	15	6
7	For half a box of Lockier's pills	0	2	0
	Given to a wooman out of Ireland having had great losses, and amongst the rest her husband	0	0	2
[*p.32*]	To son Nathaniel what he spent at Cambridg Market, selling	0	0	6
8	For unloading at Barly 7 quarters of barly of the 13 and 14 dressings	0	0	4
	Paide to Rob: Harvey for 2 journeyes sowing of oates, by the hands of his mother	0	1	0
9	For 30 square pavings at Cambridge, for my oven that must be new made, at 2*d*. a peice	0	5	0
	For 6 or 7 dressing broomes, bought at Cambridge also	0	1	0
11	For toll at Cambridg 2*d*., and spent by George and Eastland 9*d*., toto	0	0	11
14	Paide for a colt bought at Oundle Faire 7*l*. 10*s*., and			

	expences 3 nights at Gidding at Mr. C: 4s. 6d., toto	7	14	6
	Item besides al this to the horsekeeper	0	1	0
15	Paide to Rob: Harvey and Will: Stiffen for making me a new oven, besides a day of Tho: Eastland to help them	0	4	0
16	Paide to Rich: Kettle by the hands of his wife for 2 loades of dung at this time laide upon son Nathaniel's hired land in Northfeild	0	3	0
	Spent by Nathan: at Royston	0	0	4
	For a sample bagg that he kept of Urias Bowes, that he sent farthings in to us before hand	0	0	3
19	Paide to Mr. Harris for tanning a bul hyde sent to him November 9, 1704, and brought home and workt out about July 12 last, vide that from the currier June 15 (vide) 4s.	0	6	0
20	Spent by son Nath: at Cambridg at that market, going for mony for 44 bushels of wheat left there a fortnight before	0	0	6
22	Given to a breife for Rich: Davyes and 12 other inhabitants of the parish of St. Saviour's in Southwark at the Bankside in the county of Surry, looseing by fire			
24	(happening on Friday March 30, 1705) the sum of 1,131 pounds	0	1	0
	For a new wanty bought by Nathaniel	0	0	4
26	Spent by him at Royston Market	0	0	6
27	Paide to Dr. Harvey for comeing over to my daughter, being sick about harvest last, and phisick, for her and some for her husband Ben: W:	0	14	0
	Paide to Wil: Clerk for his peck	0	0	6
	Paide for a new seed lepp	0	1	6
28	To Marg: Cra: uxori for 34 pounds of cheese 8s. 7d., candles 38 pound 12s. 9d., salt 5 pecks 8s., 8 wooden dishes at 3d. per dish and a wooden bowle 1s. 11d., black lace for uxor's taylour's work, for Nathan:, etc., about 12s. 6d., and al other occasions	3	12	0
	Expended this month in al the summ of	14	7	10

[p.33] **March 1706 Receipts**

	Received of Widdows Reynolds, Law, Good, and Edw: Godfry for 6 pecks of wheat	0	3	6
1	Received of Widdows Rand, Goode, and Edw. Godfry for 5 pecks of rye	0	2	1
	Received for 100 spitts of Mrs. Norridge	0	0	3
	Received of G: Hankin for wheat a peck	0	0	7
2	Received of Pet: Nash for 2 pecks January 30; received of Good: Macer, Brock, Barber, for 7 pecks more at 3s. 4d.	0	5	5

		£	s	d
	Received of Jos: Wed, Macer, Gurner, Brock, for 2 bushels and half of rye at 1s. 8d.	0	4	2
4	Received for 5 quarters of the 17 dressing of barly of Mr. Bowes at 15s. and 3d. per quarter	3	16	0
	Received of Goody Macer for 2 pecks of it	0	0	10
	Received of Good: Barber and Bright for a bushel of wheate equally	0	2	4
5	Received of Athan: Carrington for half a bushel of barly; received of Will: Thrift for a peck 6d.	0	1	6
6	Item for 2 fannfuls of barly chaff 1s., and a parcel of wheate chaff 3d.	0	1	3
	Item for a bushel of malt, at	0	2	6
	Item for fetching a ring of wood from Essex	0	5	0
7	Received of Good: Hankin for wheat a peck	0	0	7
	Received of G: Woodward and Wil: Dovy for 2 pecks of barly of the 17 dressing	0	1	0
8	Received of Jos: Wedd and Widdow Goodin for 3 pecks of tayle wheate 1s. 2d.; received of Kate Reynolds for 2 pecks of the best 1s. 2d.; item of Widdow Goode, Good: Godfry, Peter Nash, for 5 pecks	0	5	3
9	Received of Goody Waites, Reynolds, Hankin, Gurner, Barber, Brock, Fordham, Goode, Godfry, for 6 bushels of rye	0	10	0
	Received of him and Widdow Swann for 2 pecks of barly of the 18 dressing	0	1	0
11	Received of Wil: Thrift for 4 bushels	0	7	8
	Received of his son Will: for wheate straw	0	0	3
	Received of Widdow Gooding, Barber, Bright, Bowlesworth, for 4 pecks of wheate	0	2	4
12	Received for a peck of Good: Hankin	0	0	7
	Received of Widdow Goodin for rye 2 pecks	0	0	10
	Received of G: Woodward for a peck of barly	0	0	10
	Received of Good: Parish for 2 pecks, best	0	1	0
13	Received of Widdows Goode, Waites, Gooding, and another for 6 pecks of wheate	0	3	6
	Received of Wil: Thrift junior, Godfry, Gurner, Goode, Barber, for six pecks	0	3	6
14	Received of Good: Gurner, Godfry, Hankin, Wedd, Macer, for 2 bushels of rye	0	3	4
	Received of Mrs. Westly for 10 weekes wintering 2 cowes at 6d. per weekes	0	10	0
16	Item for 2 pecks of wheate	0	1	2
	Received of Good: Hankin, Waites, Woodward, for a bushel more	0	2	4
21	Received of Good: Gooding, Bright, Waites, Gurner, for 7 pecks of rye	0	2	11
	Received of Mr. Bowes of Barley for 5 quarters of the 19			

23	dressing of barly at 15s. 6d.	3	17	6
	Received of Goody Haycock for 2 pecks of tayle wheate 9d., 2 pecks best 1s. 2d.	0	1	11
[p.35]	Received of one Mr. Body, a miller of Buntingford, for 5 quarters of the 9th dressing of wheate at 2s. 4d.	4	13	4
26	Received of Athanasius Carrington for a peck of barly of the 19 dressing	0	0	6
	Received of him and Jos: Wedd for rye 2 pecks	0	0	10
	Received of Thom: Eastland for 2 bushels	0	3	4
27	Received of him for carriage of 1,000 of turfe from Cambr: — wayes very bad	0	2	6
	Item for 2 bushels of malt February 13	0	5	0
	Item for 2 bushels and a peck of wheate at 8 groates per bushels, and a bushel and half more at 2s. 4d. per bushel, al since December 30 at several times	0	9	6
	Received of Athan: Carr: for 2 pecks	0	1	2
28	Item for a peck of rye 5d.	0	0	5
	Received of Will Thrift for 3 bushels of the 19 dressing of barly	0	5	9
	Received of Mrs. Greenhil for a peck of the same	0	0	6
	Received of young Wil Thrift for a peck	0	0	6
	Received of Widdow Gooding for a peck	0	0	6
29	Received of Will: Dovy for an oaken staddle out of Essex, at	0	0	6
	Received of Good: Carrington, Gooding and Jos: Wedd for 6 pecks of rye	0	2	6
	Received of Widdow Gooding, Barber, Wedd, Brock, for 6 pecks of wheate	0	3	6
30	Received of her and Good: Parish for 2 bushels of tayle wheat of the 8th dressing, or rye, for they were mingld	0	3	4
	Received of Widdow Nash and Widdow Gooding for 3 pecks of wheate at 2s. 4d.	0	1	9
	Received this month in all the summe of	18	18	4
	Received in the last six months, viz since Michaelmas last, the full summe of	119	7	3½

[p.34]	**March 1706 Disbursements**			
1	For unlading the 5 quarters of barly at Mr. Bowes', and spent otherwise, in al	0	0	6
	Paide to Will: Thrift for threshing 20 quarters of the 12 and 13 dressings of barly at 9d. per quarter	0	15	0
2	Item for threshing 20 quarters and 6 bushels of the 14, 15, 16, 17 dressings of barly at 9d. per quarter	0	15	6

4	Item for a day getting in my reeke on Saturday February 23		0	0	10
	Paide to Will: Numan for a paire of plow trayse, hames 6*d*., leather and work laying them in etc.		0	3	6
5	Spent by Nathaniel at Royston Market		0	0	5
	Paide to John Thompson for 6 loade of good dung in October last upon Nat's 2 acres by Harberlow Hil		0	12	0
6	Item for 7 loades more, whereof 2 upon my owne and five upon Nat's land, at 1*s*. 8*d*. per loade		0	11	8
	Item for his help a day inning reeke		0	0	10
7	Item for his farrier's work, for Ball being very bad of the quinsey		0	7	6
	Item for drinks for the other horses and helping theire eyes etc.		0	6	6
9	Paide to Miles Manning for cleaning my jack and clock in part etc.		0	1	0
10	Given to a breife for the repairing the church of St. John's in Beverly in Yorkshire much decayed, the charges of repairing computed at 3,500*l*.		0	1	0
11	Paide to Mr. Love for 40 bushels of coales fetched now at 10*d*. per bushel 1*l*. 13*s*. 4*d*., toll 2*d*., toto		1	13	6
	Spent by men then 5*d*., and by Nat: the market day before at Cambridg		0	0	9
12	Paide to John Boweles for neare 2 bushels of pitch brands carryed out with hen dungs on to my 2 acres part of the 12 acres		0	0	10
	Spent by Nathan: at Royston Market		0	0	5
13	For 30 red pavins for the bottom of the oven, the other, February 9, being white and flying when they came to neal the oven, at 1*d*. *ob*.		0	3	9
14	Paide for a pound of 6*d*. nayles, and given to a distracted wooman 2*d*.		0	0	6
	Paide to Edw: Godfry for neare 3 bushels of hen dung upon my 2 acres, part of the 12 abutting on Michel's path		0	1	4
16	Paide to Miles Manning for mending the stable lock 4*d*., and a padlock 3*d*., toto		0	0	7
	Spent by son Nathan: at Royston Market		0	0	6
19	Spent by him againe at Walden Market		0	0	6
22	Given to a bill of request for William Fuller of Dry Drayton in this county, loosing by fire happening upon him lately an 150*l*. and upwards		0	1	0
[*p*.36]	Paide to Rich: Wallis of Chrissol for fagotting up a ringe of wood for me in Chrissol Park		0	1	6
26	Paide to Widdow Gurner for 4 bushels of hen dung upon my 2 acres, part of the 12 on the north side of Michel's path		0	2	0

	Spent and unloading at Mr. Bodye's	0	0	9
	Paide to Thom: Eastland for threshing 5 quarters 5 bushels 2 pecks of the 8 dressing of barly	0	4	3
27	Item for 6 journys at plow 23 instant, the last	0	4	0
	Item for 4 dayes griping, spreading dung, serving mason, plow, setting trees	0	3	4
	Item for threshing 13 quarters 2 bushels and half of the 8 and 9 dressings of rye	0	16	7
28	Item for threshing 50 quarters 7 bushels of the 3, 4, 5, 6, 7, 8 dressings of wheate at 1s. 3d. per quarter as the rye also	3	3	7
	Paide to young Will: Fordham a journy at plow	0	0	6
	Paide to Goodman Soale for a new paire of hind wheels for the waggon, made last April 26 or 27, vide	1	16	0
29	Item for a new paire of shafts — waggon	0	5	6
	Item for making a new plow, at	0	1	6
	Item for a new felloe and spoke in the tumbrel wheeles, broaken,.... [one or two words illegible]	0	2	0
	Item for an horse tree or wipple tree	0	0	6
30	Spent at Royston by son Nathan:, at market	0	0	6
	To Marg: Cra: uxori to pay Mary Burton 6d., beife and mutton of Th: Malin about 4 stone 11s. 6d., cheese 50 pounds about 13s., cloths, shoos, silk, galoon, shaloone for uxor about 13s., and al other occasions in this month, toto	3	10	0
	Expended this month in al the summe of	16	12	5
	Expended in the last six months, viz since September 29 last, the full and entire summe of	85	18	7

[p.35]

April 1706 Receipts

	Received of Robert Harvey for six pecks of wheate his mother had in March last, viz the 4, 15 and 29	0	3	6
1	Received of Widdow Swann for 3 pecks	0	1	9
	Item for her half yeare's rent ended at Lady Day last 1706	1	11	6
2	Received of Edith Wedd, Widdow Gooding, Widdow Goode, Ben: Bright, for 2 bushels of rye or tayle wheate	0	3	4
	Received of him for a peck of wheat; received of Lydy Barber for a peck 7d.; received of Rob: Macer for 2 pecks	0	2	4
3	Received of them two for 4 pecks of rye	0	1	8
	Received of Widdow Gurner and Widdow Law for half a bushel 10d.; received of Athan: Car: for 2 pecks	0	1	8

		£	s	d
	Item for a peck of barly of the 20 dressing	0	0	6
4	Item for 2 pecks of wheate 1s. 2d.; received of Richard Haycock for 2 pecks	0	2	4
	Received of Goody Brock and Jos: Wedd for 2 bushel of rye	0	3	4
5	Received of Rob: Macer for a peck of barly	0	0	6
	Received of Goody Brock for 2 pecks of wheate in part	0	0	10
	Received of the miller for 2 tithe piggs lately, that is Fraunces Hales	0	2	6
[p.37] 6	Received of one Mr. Hodge, bayly to the Lady Miller, for 5 quarters and half of the 8th dressing of wheate at 11s. 6d. per loades, delivered at Buntingford	5	1	3
	Received of Will: Thrift junior, Goody Goodin, Hankin, for 3 pecks	0	1	9
	Received of Ed: Godfry for a peck of barly	0	0	6
8	Received of Wil: Brooks and Lyd: Barber for a bushel of rye	0	1	8
	Received of Good: Hales for 10 weeks wintering of 2 calves — yearlings	0	6	8
9	Received of Goody Law, Barber, Jos: Wedd, Widdow Goode, for 9 pecks of rye or tayle wheate, mingled together, at 1s. 8d.	0	3	9
	Received of Good: Barber, Goode, Jos: Wedd, for 4 pecks of wheate	0	2	4
10	Received of Mr. Bowes of Barly for 7 quarters 2 bushels of the 20 dressing of barly at 15s. 4d. per quarter	5	11	0
	Received for a peck of Mary Waites	0	0	6
	Item of her and G: Hankin for 2 pecks of rye	0	0	10
	Item of her and Good: Godfry and Gooding, for 5 pecks of wheate at 2s. 4d.	0	2	11
12	Received of J: Woodward and Widdow Gurner, Good: Bright, Boweles, Mr. Wedd, for 2 bushels and half at 2s. 4d.	0	5	10
13	Received of J: Woodward and Widdow Gurner for a bushel of rye; received of Rob: Macer for 2 pecks	0	2	6
	Received of him and Tho: Barber, Rich: Haycock, for a bushel of wheate; received of Wallis of Chrissol for a bushel	0	4	8
	Received of Good: Brock, Goode, Hankin, Law, for 6 pecks of wheate	0	3	6
16	Received of her and Athan: Carrington, Ben: Bright, Mr. Greenhil, for 2 bushells of rye	0	3	4
	Received of Thom: Watson for 4 bushels more since January 9 last at 2s.	0	8	0
	Item for 7 pecks of barly at that time	0	3	4
	Item for a bushel of malt, at	0	2	7
17	Item for 2 bushels and a peck of wheate in the same space of time	0	7	6

HOUSEHOLD ACCOUNTS 1706

	Received of Goodman Parish for a bushel of barly of the 21 dressing	0	1	10
18	Received of Rob: Macer and William Brock for 2 pecks of it	0	1	0
	Received of Will: Thrift for 5 pecks of wheate since March 7 last	0	2	11
	Item for a bushel of malt	0	2	7
	Item for 3 bushels of rye at 1s. 8d.	0	5	0
19	Item for 3 bushels and a peck of barly, and 2d. a bottle of rye straw	0	6	5
	Received of Athan: Carrington, Jos: Wedd, Widdow Gurner, Widdow Haycock, Brock, for 2 bushels and a peck of rye	0	3	9
20	Received of Athan: Car., Mary Waites, for 3 pecks of wheate and 2 pecks more of Rob: Macer	0	2	11
	Received of Mr. Taylour for his smal tithes in the yeare 1705	1	19	4
22	Received of Athan: Carrington, Edward Godfry, for 3 pecks of barly	0	1	6
	Received of Good: Hankin, Macer, Woodward, for 6 pecks of rye or tayle wheat	0	2	6
	Received of Edw: Parish for 6 pecks	0	2	6
23	Item for 6 pecks of malt	0	3	11
	Received of Widdow Strett for a peck	0	0	8
	Received of Jos: Wedd and Ben Bright for 3 pecks of miscellaine	0	1	3
	Received of Jos: Wedd and Rob: Harvey for 5 pecks of wheate	0	2	11
[p.39] 24	Received of Urias Bowes of Barly for 9 quarters and 3 bushels of the 21 dressing of barly at 14s. 4d.	6	14	0
	Received of Good: Law and Macer for 2 pecks	0	1	0
	Received of Widdow Law and Gurner for rye 3 pecks	0	1	3
26	Received of Good: Woodward, Woods, Nash, Law, Hankin, Barber, and 2 others for 4 bushels of wheate at 2s. 4d.	0	9	4
	Received of Rob: Macer for 2 pecks	0	1	2
	Received of Good: Brock for 3 pecks and Widdow Goode for a peck 7d., toto	0	2	4
27	Received of Mary Waites, young J. Watson, Widdow Nash, for six pecks	0	3	6
	Received of Mr. Latimer for a bushel in April 1703, at	0	2	6
29	Received of Ath[an]asius Carrington and Robert Macer for 2 pecks of rye	0	0	10
	Received of Good: Harvey for a bushel of malt by son Nat:	0	2	7
30	Received of Woodward for a peck of the last dressing of barly	0	0	6

	Received of Thomas Hope for 4 short pieces of a willow tree, at	0	0	6
	Received in this month in al the summ of	28	10	8

[*p.36*]   **April 1706 Disbursements**

	Paide to Robert Harvey for 16 journys in sowing barly at 6*d.* per journey	0	8	3
	Item for half a day laying bottom of the oven, mending copper, etc.	0	0	9
1	Paide to Goody Swan for a loade of dung out of the limeyard upon Nat's 2 acres at Harberlow Hil October 20	0	2	6
	Item for 16 bushels of hen dung and pitch brands upon 2 acres, part of the 12 on the right hand of Michel's path	0	8	0
2	Item for 700 of white bricks wanting eleven, for oven, chancel floor, etc., at 2*s.* 6*d.* at home, id est 6*d.* carriage	0	17	3
	Paide to Goodman Soale for a plow beame, I having none of my owne	0	1	6
	Item for making a new plow now	0	1	6
4	Paide to young Goody Haycock for her children gathering stones 6 dayes	0	1	6
	Paide to Mr. Wright for coming over to my wife, and phisick 5*s.* 2*d.*	0	7	6
	Spent at market by son Nathan:	0	0	6
5	Given to George towards the healing of his hand, being very much hurt with the wagon at Barly	0	1	0
	Given to 3 lame souldiers at door	0	0	3
[*p.38*] 6	Spent by son Nathan and George unloading the 5 quarter and half of wheate at the George Inn, Buntingford, whereof for unloading 4*d.*	0	1	0
	For unloading the 7 quarters and the loade before at Mr. Uri: Bowes	0	0	6
8	For a letter from Mr. Michel, brought on purpose from Royston by a messenger	0	0	6
9	Paide to Will: Numan the collermaker for half an horse hyde 4*s.* 6*d.*, a calve's skin 1*s.* 6*d.*, work before, part of a day 6*d.*, part to-day 8*d.*	0	7	2
	Spent by Nathaniel at Royston Market	0	0	4
	Item for nayling a couple of shoos on the Buoy horse, going to the wood	0	0	2
10	Given to one Richardson, a tenant of Sir G: Cotton's at Biggleswade, forced to abscond by surety for his wive's brother	0	0	6
	Paide for the half yeare's duty upon houses and windowes ended at Lady last	0	5	0

11	Paide for the last quarter of the land tax for 1705, for Hoyes or Ed: Parishes		0	2	6
	Item for Cassanders to the same tax		0	9	0
	Item for parsonadge to the same tax		7	0	0
12	Paide to Thom: Watson for 24 dayes between January 6 and February 3 at 9*d.* a day		0	18	0
	Item for 23 dayes and half more from February 3 to March 3 at 10*d.* per day		0	19	7
13	Item for 23 dayes more between March 3 and 31 at plowing, sowing and other things, and works about the yard; and given him back put of Bartholomew's contract, etc.		1	3	5
16	Item for 5 dayes and the greatest part of a day before April 7, the sixth being the last included in this reckoning		0	4	9
17	Paide to William Thrift for threshing 36 quarters and 2 bushels of the 17, 18, 19, 20, 21 dressings of barly at 9*d.* per quarter since March 6 last, when he also discounted 4*s.* for 2 years quitt rents ended at Lady Day last		1	7	2
18	Spent by Nathaniel at market		0	0	4
	Given to 2 lame souldiers at door		0	0	2
	Paide to Mr. Taylour for an overseers' rate at 2*d.* per pound for what I hold of my owne		0	0	9
19	Item to the same rate for parsonadge		1	2	10
	Paide for 2 pound of 6, 8, 10 penny nayles of Mrs. Peck for wheelebarrow, court yard gate, pales, etc.		0	0	9
20	Paide for the yeare's quit rents ended at Lady last, that is for Cassanders 13*s.* 2*d.*, Hoyes 3*s.* 8*d.*, toto		0	16	10
	Spent by Nat: at market, Royston		0	0	6
22	For a pound of pitch at Triplow		0	0	3
23	Given to John Derby and John Arnold of Newton, looseing by fire, hapning the 2nd of this instant April at 10 in the forenoone, the sum of 60 pounds 16*s.* 0*d.*, whereof Arnold's house, valued at 25 pounds and Derby's goods at 24*l.* and Arnold's at 11*l.* 16*s.*, in all 60*l.* 16*s.*, one house being Mr. S:'s, another John Pale's, the 3rd Arnold's		0	6	8
[*p.40*] 24	Given to one Captain Timothy Groves, having been a souldier under Charles 1st and 2nd, now 88 yeares old, his son taken by the French and looseing 10,000 pounds		0	0	6
26	Paide to Robert Hills, a gardner of Linton, for seedes 6*d.*, and ten dayes in both the gardens at 1*s.* 6*d.* per diem		0	15	6
27	Paide to Rob: Harvy for 7 dayes helping him, digging and fetching and carrying into garden 25 loads of clunch and five of gravel at 10*d.* per day		0	5	10
	Paide to Jonathan Peck for 2 dayes and a part in the same work helping him		0	2	0
29	Paide to Eastland for 3 dayes and half in digging, fetching gravel, etc., for the same		0	2	11

30	Paide to Will: Fordham for 2 dayes bushing up the garden pales, and nayles 6*d*.	0	3	6
	To Marg: Cra: uxori for 45 pounds etc. of cheese 12*s*. 2*d*., and aprons etc. and other things above 1*l*. 0*s*. 0*d*., weaver for 42 yards of cloth, and warping 9*d*., toto 13*s*., and biefe and smal meate of Malin, etc., about 1*l*. 3*s*., and all other things this month, toto	4	0	0
	Expended this month in all the sum of	23	9	5

[*p.39*]

May 1706 Receipts

1	Received of Mr. Urias Bowes of Barly for six quarters and 2 bushels, being the last of my barly that I have to carry out this yeare	4	4	0
	Received of Thom: Barber and Will Thrift junior for 2 pecks of the same	0	1	0
2	Received of Tho: Barber and Jos. Wed: for a bushel of miscellaine	0	1	8
	Received from young Will Thrift for a bottle of barly straw	0	0	3
	Item for 3 pecks of wheate 1*s*. 9*d*.; received of Tho: Barber for 2 pecks 1*s*. 2*d*.	0	2	11
3	Received of Rob: Macer for a peck	0	0	7
	Item of him for 2 pecks of rye 10*d*.; received of Woodward for a bushel	0	2	6
	Received of Will: Brock, Rob: Macer, Edw. Godfry, for 3 pecks of barly	0	1	6
4	Received of Mr. Taylour for an 100 of willow setts this last spring	0	17	6
	Received of Widdow Swan and Athan: Carrington for 3 pecks of wheate at 7*d*. a peck; received of Joan Woods for a peck	0	2	4
6	Received of Widdow Gurner for the use of my churchyard, being very much spoiled with the hoggs (abateing 6*d*.)	0	2	0
	Item for a peck of miscellaine at 5*d*.; received for a peck of Widdow Goode 5*d*.; received for a peck of Widdow Law 5*d*.	0	1	3
7	Item for a peck of wheate 7*d*.; received of Jos: Wedd for a peck 7*d*.	0	1	2
	Received of Mr. Wedd for a bushel of barly	0	2	0
	Received for a peck of Rob: Macer	0	0	6
8	Item of him and J. Bowles, and J: Waites and Will: Brock, for 5 pecks of wheate	0	2	11
	Received of Brock for 4 pecks of rye	0	1	8
	Received of Jos: Wedd, Thom: Barber and Edw: Parish, for 2 bushels and a peck more	0	3	9

HOUSEHOLD ACCOUNTS 1706

	Item of the 2 last for 2 pecks of wheate	0	1	2
9	Received for the Ball horse bought at St. Neots about June 14, 1704, at 9*l*. 13*s*. 6*d*. by son Nath: and Will	13	0	0
[*p.41*] 10	Received of Rob: Macer, Athan: Car: Widdow Gurner, Eliz: Bright, Mary Law, Mr. Greenhil, for 9 pecks of rye	0	3	9
	Received of Good: Carrington, Goode, Macer, Hankin, Law, Barber, for 7 pecks of wheate at 2*s*. 4*d*.	0	4	1
11	Received for the Bonny colt bought last Midsummer Faire at 6*l*. 16*s*. 0*d*.	4	0	0
16	Received of Mary Brooks, Good: Jeepes, Macer, Bright, Godfry, young W: Thrift, Thompson, for 2 bushels of barly	0	4	0
	Received of Good: Waites and Goode for half a bushel of rye	0	0	10
17	Received of Good: Hankin and Wates for half a bushel of wheate 1*s*. 2*d*., straw 2*d*.; received of Jos: Wedd for 2 pecks 1*s*. 2*d*.	0	2	6
	Received of her and Widdow Gurner for 3 pecks of miscellaine at 5 per peck	0	1	3
18	Received of Good: Brock, Barber, Carrington, Widdow Haycock, for 7 pecks of it	0	2	11
20	Received of Good: Brock, Barber, Carrington, Widdow Swan, Haycock and the miller for 2 bushels and half of wheate at 2*s*. 4 per bushel	0	5	10
	Received of Good: Jeeps, Brooks and Brock, Macer, Woodward, Carrington, Strett, for 2 bushels of barly	0	4	0
22	Received of Jos: Wedd and others for the straw	0	0	4
	Received of Mrs. Westly for a bushel of rye some time since	0	1	8
	Item for a bushel of barly	0	2	0
	Item for smal tithes this yeare	0	3	0
23	Item for another bushel of barly	0	2	0
	Item for a bottle of straw	0	0	2
	Received for a peck of barly of M: Waites	0	0	6
24	Received of her for wheat a peck 7*d*.; received of Good: Hankin for a peck 7*d*.; received of Widdow Law for a peck 7*d*.	0	1	9
	Item of her for a peck of rye 5*d*.; received of Eliz: Bright for a peck 5*d*.	0	0	10
25	Item for a peck of wheat of her 7*d*.; received of Mary Waites, Law, Hankin, Bowelesworth, for 4 pecks	0	2	11
	Received of John Rayner for his tithe pigeons and one calf in 1705	0	9	6
27	Item for his smal tithes this yeare, excepting pigeons and orchard etc.	0	6	6
	Item for a jag of rye straw last summer	[0]	5	0

28	Received of Rich: Haycock's wife and Edw: Parishes for 2 pecks of wheate 1s. 2d.; received of Joane Woodds for a peck 7d., and a peck of some other 7, toto	0	2	4
	Received of Clem: Norman for a bushel of malt by Nat, at	0	2	7
29	Received of Mr. Malin for 5 quarters and 2 bushels of rye, carryed in about November 12, at 1s. 8d. per bushel	3	10	0
	Item for a calf of the little cow	0	7	0
30	Received of Will: Thrif junior for malt 2 pecks	0	1	4
	Received of Good: Macer, Edith Wedd, Lyd: Barber, for barly 2 pecks and half	0	1	3
31	Received of Good: Gurner, Carrington, Bright, Wedd, Barber and the miller for 4 bushels 2 pecks of rye	0	7	0
	Received of Good: Carrington, Barber, Macer, for 4 pecks of wheate	0	2	4
	Received this month in al the full summ of	31	9	10

[*p.40*]

May 1706 Disbursements

	Paide to Will: Frinkel for making 25 quarters of malt for me at 18d.	1	17	6
	For the tax upon the saide malt	4	16	1
1	Paide to son Nathaniel for 15 bushel and 3 pecks of rye, being without	1	6	3
	Item for 3 pound and a quarter of binding of him that was left of Mrs. Norridge's	0	1	0
	Item for his expenses at Reach Fair etc.	0	1	2
2	Paide to Rob: Harvey for mending up the backhouse chimney that was fired (but wee were mercifully saved from great harm), a day and half	0	2	3
3	Paide to Thom: Holt for setting up a stud in the roome of the burnt ones, and reiving splints, etc., and Hester's chest 4d.	0	1	10
	For a dozen on church chatechisms	0	0	10
	Paide to Rich: Jackson for his yearly pay in keeping up my pump, vide June 15, 1705, and July 1, 1704	0	1	0
4	Paide to Miles Manning mending my clock and causing it to strike	0	1	6
	Paide to Mr. Wilson for tenths due last Christmas — the bishop's secretary	2	19	9
	Paide to Wallis of Crissol for 400 spitts	0	0	11
	Spent by Nathan at Royston	0	0	4
5	Paide to Thom: Fairechild for mending the fire fork to the copper 3d., and 3 plates to my peck 3d., in al	0	0	6
	Given to a breife for Iniskilling in Ireland, loosing by fire,			

6	happening on June 2, 1705, to the ruine of 100 families, and the loss of 8,166*l*. in monyes and goods	0	2	6
	To Will: Morrice the gelder cutting the old boar 4*d*., and 5 boare piggs, toto	0	0	6
	For a pound of 6 penny nayles	0	0	5
7	Spent by son Nathaniel 6*d*., and given to Mr. Clerk's man that bought the horse at Royston 1*s*., and an halter bought to leade him away in 3*d*.	0	1	9
[*p.42*] 8	Given to one George Hanson of Chrisleston in the parish of Sandal Magna both deafe and dumb	0	0	2
	Paide to Goodman Soale for putting 2 new felloes into the tumbrel wheeles	0	3	0
9	Item for a new shaft to the wagon	0	2	3
	Item for heading a plow since	0	1	0
	Spent by Nath: at Royston Market	0	0	5
16	For a colt of St. Neot's Fair 9*l*. 17*s*. 6*d*., to the horse keeper 1*s*., spent there 2*s*. — they call him Jolly, of Browning — toto	10	0	6
	Given to a scholler in great want	0	0	6
	For a new pocket knife at Cambridg	0	0	5
	For procurations at the archdeacon's visitation	0	3	4
17	Spent there and to barber	0	1	2
	Paide to Goodman Rayner for 3 bushels of barly son Nat: had of him for seede in 1705 at 2*s*. 2*d*.	0	6	6
18	Spent by son Nathaniel at market	0	0	4
	For 2 paire of white bugle cuffs for self	0	1	0
20	For 6 quarters and 6 bushels of oates bought at Cambridge at 11 shillings per quarter, and fetched there	3	14	3
	Item spent by men then, and tol 2*d*.	0	1	2
	For a parcel of whipcord, 3 quarters of a pound or thereaboutes	0	1	6
23	Spent by Nat at Royston Market	0	0	6
	Paide to Mr. Jackson for new ledding 2 casements, one at Edw: Parishes, the other at Widdow Swann's, and a little matter in the chancel	0	1	11
24	Paide to Rob: Hills the gardner a daye's work againe, in rectifying the garden againe, vide April 24 last	0	1	6
	Paide for a militia tax for Cassanders, that is for drums and colours	0	0	9
25	For parsonadge to the same tax	0	11	5
27	Given to the wife of Christopher Hunt of Hinxton, to releive him out of prison, being thrown by surety for one John King of Duxworth for 30 pound, who is dead, and the debt lieth upon her husband, who paide 20*l*. of it in one weeke	0	0	6
28	Paid for a beaver hatt or a beaveret for myself, bought at London by son Benjamin, at	1	0	0

	Spent by son Nathaniel and George for toll at Cambridge, and loading, and for an earthen pot broake by our horses 3*d*., etc.	0	1	3
29	Item for a cart whipp and lainer	0	0	4
	For a couple of fork stales	0	0	11
	For a pound of nailes for the quern	0	0	4
30	To Marg: Cra: uxori for linnen for shirts, shifts for selves and children, and handkerchiefs etc. about 1*l*. 15*s*. 0*d*., cheese 18 pounds and half 4*s*. 6*d*., for 7 ston of beife,			
31	besides lamb and veale, of Malin and Fidlin, about 1*l*. 10*s*. 0*d*., and 18 ston 6 pounds in September, February, March, April last, besides veale, mutton, lamb, about 3*l*. 1*s*. 6*d*., and al other occasions, in al	8	10	0
	Expended this month in al the ful summ of	37	3	0

[*p.43*] **June 1706 Receipts**

1	Received of Thomas Nash for the tithe of an acre of Saint Foine this yeare	0	3	0
	Item for the tithe of an acre and half in Dod's Close also, in 1705	0	4	6
	Item for 200 spitts in November 1703	0	0	6
	Item for his smal tithes this yeare	0	5	4
3	Received of Mr. Wells of Baldock for 5 quarters of the tenth dressing of rye	2	13	4
4	Received of Wil: Brock, J: Woodward, Widdow Goode, Athan: Carr., Mary Waites, for 2 bushels and a peck of rye	0	3	9
	Received of Wil: Brock, Woodward, Law, Wedd, Carrington, Waites, for 2 bushels and half of wheate, at	0	5	10
6	Received of Widdow Jeepes, Law, M: Brock, for 2 pecks and half of barly	0	1	3
	Received of Mrs. Wedd and Mr. Greenhil for 6 pecks more, at	0	3	0
7	Received of Widdow Gurner and the miller for 3 pecks of rye	0	1	3
	Received of him and Mrs. Westly for a bushel and 3 pecks of wheate	0	4	1
8	Received of miller, Goody Hankin, Haycock, Bowels, Gurner, Goode, Swan and Barber for 2 bushels and half more	0	5	10
	Received of Good: Brooks, Bright, Haycock, Barber, for 5 pecks of rye	0	1	10
10	Received of Good: Gurner and Barber for a bushel of the same	0	1	4

HOUSEHOLD ACCOUNTS 1706

	Received of Edw: Parish, Macer, Waites, Peter Nash, Barber, young W: Thrift, for 7 pecks of barly	0	3	6
11	Received of Wil: Brock for 2 pecks of wheat and 4 of rye on Friday last	0	2	6
	Received of Goody Hankin and miller for 5 pecks of wheate, at	0	2	11
	Item for 2 pecks of rye	0	0	8
	Received of Widdow Goode, Hankin, Mr. Greenhil, for 7 pecks of rye at 4*d.* a peck	0	2	4
12	Item of him for a parcel of straw at twice	0	1	1
	Received of Rich: Haycock and Mary Waites for 2 pecks of wheate	0	1	2
	Received of Athan: Carrington for half a bushel of the same	0	1	2
	Item of him and Mary Waites for rye 2 pecks	0	0	8
13	Received of Widdow Haycock for malt a peck	0	0	8
	Received of Good: Brock for barly a peck	0	0	6
	Item for 2 pecks of wheat 1*s.* 2, and for a bushel of rye 1*s.* 4*d.*	0	2	6
	Received of Mrs. Norridge by son Nat for fetching a loade of clay	0	1	0
14	Item for 200 withs for thatching	0	1	0
	Item for 900 spitts or sprindles	0	2	3
	Item for 4 loades of rye straw laide in	1	2	9
	Received of Thom Eastland for a bottle of wheate straw in April last	0	0	2
15	Item for 8 good willow setts in April	0	0	8
	Item for a bushel of rye	0	1	4
	Item for a bushel of malt, at	0	2	7
	Item for fetching a cant of wood from Langly with 2 teemes	0	6	0
17	Item a loade of rye straw laide in	0	6	0
	Item for 4 bushels and 3 pecks of wheate since March 26 last at 2*s.* 4*d.*	0	11	1
	Received of Edw: Godfry for 2 pecks the 5th instant, and 4*d.* in part for a bushel now at 2*s.* 4*d.*, id est in al	0	1	6
	Item for 2 stetches of haum at the Buts	0	1	6
	Received of J: Woodward for wheat 3 pecks	0	1	9
[*p.45*]	Item for six pecks of rye at twice, all in this month at 1*s.* 4*d.*	0	2	0
18	Received of Goodman Harvey for a peck of barly	0	0	6
	Received of Widdow Gurner, Lyd: Barber, J: Aspinal, for 7 pecks of rye	0	2	4
19	Received of Lyd: Barber and Widdow Goode for 2 pecks of wheate, at	0	1	2
	Received of Good: Hankin, Swann, Barber, Waites, Peck, Widdow Haycock, for 2 bushels and a peck of the same	0	5	3

173

	Received of Good: Bright, Brooks and miller for 5 pecks of rye at 4*d*.	0	1	8
20	Received of Good: Brooks, Bright, Waites, for 3 pecks of barly, at	0	1	6
	Received of J: Woodward and Mary Thrift for 2 pecks more of the same	0	1	0
21	Received of Good: Macer, Hankin, Wedd, Thrift junior, miller, for rye 3 bushels	0	4	0
	Item of Macer, Hankin, Carrington, for a bushel of wheate, at	0	2	4
	Received of Widdow Nash for 2 pecks	0	1	2
	Received of the miller, Good: Barber, Waites, for 2 bushels and 3 pecks rye	0	3	8
22	Received of Mrs. Westly, miller, Aspinal, Widdow Haycock, Gurner, R: Carpenter, for 4 bushels and half of rye at 1*s*. 4*d*.	0	6	0
	Received of one Body, at Buntingford, for 6 quarters and 2 bushels of wheat, hardly 2 shillings a bushel	4	18	0
24	Received of young Wil. Thrift and Widdow Ward for 5 bushels and half of rye	0	8	2
	Received of the miller and Wil: Brock for 7 pecks of the same	0	2	4
	Item of him and Rob: Macer for 3 pecks of wheate at 7*d*. per peck	0	1	9
25	Received of Peter Nash, Widdow Haycock, Jos: Wedd, for 2 pecks and half of barly	0	1	3
	Received for a peck of Goody Brock	0	0	6
	Received of Wil: Thrift for 7 pecks since May 1 last, at	0	3	6
	Item for 21 weekes of his cow, going in my yard last winter, at 6*d*. per week	0	10	6
26	Item for his wennel the same time at 3*d*.	0	5	0
	Item for 2 pecks of malt, at	0	1	4
	Item for 6 pecks of wheate	0	3	6
	Item for 6 bushels of rye, at	0	9	6
	Received of J: Woodward and Eliz: Bright for a bushel of the same	0	1	4
	Item of the first for a peck of wheate	0	0	7
27	Received of Ann Gooding, widdow, for 6 pecks of rye her husband had in *1704*, February 19, at	0	4	5
	Item for a flooring of pease straw in Lady Day 1704 to her husband	0	0	6
	Item in part for a bushel of tayle wheat February 14 last in her widdowhood	0	1	1
28	Received of Mr. Taylour for a deal board 11 foot long, etc.	0	1	8
	Received of the miller, Mr. Greenhil, Mary Waites, Widdow Goode, for 7 pecks of rye	0	2	4

	Item of the 2 last for 4 pecks wheate	0	2	4
	Received of Rich: Fairechild for a bushel of malt	0	2	7
29	Received of Jos: Wedd for a peck of wheate, at	0	0	7
	Received this month in all the summe of	18	9	5

[p.44]
June 1706 Disbursements

	Paide to Thom: Nash for his overseers' rate for what I hold of my owne	0	0	9
3	Item to the same rate for parsonadge	1	2	10
	Spent by men carrying rye to Baldock, and unloading there 4d.	0	0	10
4	Spent with daughter and neighbours going to see theire house at Foxton	0	1	0
	Given to Mr. Wedd's maide Anne	0	0	6
5	Spent by Nathan: at Royston Market	0	0	6
6	Given to a breife for the burrough of Great Torrington in Devon, loosing by fire (happening on June 21, 1705, in the house of Thomas Dingle, and 23 other families) 1,600l. and upwards	0	1	0
	For little nayles for the hopper and fatt of the querne, having a new spout etc.	0	0	5
7	To Goodman Boman of Royston for neare 2 dayes work of himself and his son, dressing the querne, making a new extra box, sides, etc.	0	6	0
8	Paide to Thom: Eastland for a journey spreading dung for son Nat: in Northfeild on March 28 last	0	0	8
	Item for a journey this month, filling dung cart – Nat at Royston	0	0	8
10	Item for 12 dayes and half in April at wood cart into Essex, fagotting, hedging in closes and in the orchard, etc., vide April 27 last	0	10	5
	Item for 4 dayes more in this month and May, filling dung, watering the rye straw for barne, etc., and an odd job at rowle while Nathaniel came home	0	3	6
11	Item for threshing 13 quarters and 2 bushels of wheate in April and this month at 1s. 3d. per quarter	0	16	6
	Item for threshing 28 quarters and six bushels of rye dressed up April 8 and May 7, 23, 31 and June 11, at 1s. 3d. per quarter	1	15	11
12	Paide to Edw: Godfry for a loade and half of dung upon Nathaniel's land, now	0	3	0
13	Paide to J: Woodward for a quarter herdman's wages, due for 4 cowes the 24 of this month, deducted with his wife for corn she hath had before and now, viz rye 6 pecks, wheate 3	0	2	8

		£	s	d
	Given to poor at door in want	0	0	1
	Spent by Nath: at Royston Market	0	0	4
	Given to poor at door againe	0	0	1
14	For a plow cord bought at Royston	0	1	0
	Spent by Nat: at market againe	0	0	4
	Spent by men carrying six quarters and 2 bushels of wheate to one Body, a miller at Buntingford, unloading	0	1	0
15	Paide to Mary Parish, making hay in my close part of a day	0	0	4
	Given to a breife for the repairing the parish church at Basford in Nottinghamshire, the charges whereof are computed at 1,482*l.* and upwards	0	1	0
17	Paide to Will: Thrift for threshing 12 quarters of the last dressing of barly at 9*d.* per quarter	0	9	0
	Item for threshing 9 bushels of ... [omission] at 1*s.* and 10 bushels of oates at 8*d.*	0	2	0
[*p.46*]	Item for 2 dayes, May 3 and 4, hedging in the courtyard, at 10*d.* per day	0	1	8
18	Item for 23 dayes more to the 2nd of this instant June	0	19	2
	Item for 26 dayes since that in filling, spreading dung, mowing and making hay, etc.	1	1	8
19	Item for mowing the close next to the Linch Lane, etc.	0	1	8
	Item for a single day since in hay	0	0	10
	Paide for wantyes and ridgetrees, a couple of each, at Royston	0	1	6
20	Paide to Ann Gooding, widdow, for 4 loades of dung, February 18 and one about this time, at 18 pence, upon son Nat's hired land	0	6	0
21	Paide to Rich: Fairechild for 5 journeyes at plow in pease and barly seede, at 6*d.* per journy	0	2	6
	Item for a day inning my reek, March	0	0	10
	Item for a day now, and dung cart, and spreading dung in afternoone	0	0	10
22	Given to 2 lame men, old, at door	0	0	2
	To little Will Brock gathering chipps	0	0	1
24	Paide to or for children since June last, vide, inprimis for son Samuel to Mrs. Richardson for the interest of 12*l.* which he hath of hers	0	14	4
25	Item for son Ben for a yeare's interest of 50*l.* that he oweth, due this Midsummer etc., and 10*s.* to Mrs. Rich.	3	10	0
	Item to daughter Hester that I lent son Sam and he hath now paide her	100	0	0
	Item in ten guineas laid downe before the other could be paide	10	15	0
	Item in broade gold given her	1	3	6
26	Item in 5 todd and half tithe wool this yeare, 3 lordship, 2 and an half farme	2	10	0

HOUSEHOLD ACCOUNTS 1706

	Item for son Nathaniel for a great coate, with the exchange of his old one, the sum of	0	14	6
27	Item for dimity, silk, shooes mending and such like, besides al his dung paide for these 2 yeares for his farme, not here mentioned	0	3	6
	For 9 paire of gloves against harvest for harvest men, at 6*d.* a pair	0	4	6
28	Spent by son Nathaniel at Royston that day, buying cheese and other things	0	0	6
29	To Marg: Cra: uxori for about 26 pounds of cheese 6*s.* 6*d.*, and for 5 stone of beife and mutton, veale and lamb 1*l.* 11*s.* 6*d.*, and al other occasions in this month	3	3	0
	Expended this month in al the ful sum of	131	18	1

[*p.47*]

July 1706 Receipts

	Received of Eliz: Bright and the miller for a bushel of wheate	0	2	4
2	Received of Widdow Haycock, Jos: Wed, Mr. Greenhil, Widdow Gurner, miller, J: Aspinal, for 4 bushels and half of rye or miscellaine	0	6	0
3	Received of Mary Waites, J: Woodward, for a bushel of the same	0	1	4
	Received of him, Good: Harvy and Hankin for 5 pecks of wheate, at	0	2	11
4	Received of Widdow Swan for a peck	0	0	7
	Received of Mrs. Westly, miller, Widdow Nash, for 7 pecks of rye, at	0	2	4
	Received of Widdow Strett, Wil Dovy, Edw. Parish, for 3 pecks of barly	0	1	6
6	Received of Widdow Haycock, Jos: Wed, Hankin, Widdow Nash, for 5 pecks of wheate	0	2	11
	Received of Jos: Wedd, Widdow Goode, Widdow Nash, Mr. Greenhil, for 6 pecks of rye	0	2	0
8	Received of Widdow Gurner, Atha: Carrington, Widdow Nash, Hankin, for 4 pecks more	0	1	4
	Received of Rob: Macer, Mrs. Westly, Athan. Carrington, Widdow Nash, for 6 pecks of wheate at 7 groates	0	3	6
9	Received of young Good: Thrift for barly a peck	0	0	6
	Received of Jos: Wedd, Widdow Swan, Widdow Carrington, Good: Harvy, for 5 pecks of wheate at 7 groats	0	2	11
12	Received of Jos: Wedd, Widdow Gurner, Widdow Nash, Good: Macer, Carrington, Mr. Greenhil, for one bushel and 3 pecks of rye at 1*s.* 4*d.* per bushel	0	2	4
13	Received of Clemt: Norman for 2 bushels of malt on			

	March 26, at	0	5	2
	Item for 2 bushels of malt more now	0	5	2
	Item for a peck of rye soone after	0	0	4
16	Item for 2 pecks of wheate then	0	1	2
	Received of Will: Brooks for one bushell of rye towards the beginning of this instant, at	0	1	4
17	Item for 3 pecks of wheat the same time at 7 groats	0	1	9
	Item for 2 pecks of wheate afterwards in this month also	0	1	2
19	Item for 2 pecks of rye that time	0	0	8
	Received of Joan Woods for a peck	0	0	4
23	Received of Widdow Gurner for a peck; received of Thom: Barber for 2 pecks; received of Eliz: Bright for a peck	0	1	4
24	Received of Widdow Ward for 2 pecks of wheate by her daughter Betty	0	1	2
	Received of Jos: Wedd for rye a peck	0	0	4
	Received of Widdow Haycock for a peck	0	0	4
26	Received of Widdow Gurner for a peck	0	0	4
30	Received of Athan. Carrington for a peck; received of Good: Harvey for a peck; received of Rob: Macer for 3 pecks	0	1	8
	Received of Widdow Nash for 3 pecks	0	1	0
	Received this month in all	2	15	9

[p.48]

July 1706 Disbursements

2	Paide to a man of Shepereth for comeing to the red cow, being hurt, and leaving oyle to dress her withal againe, if neede require	0	1	0
	For an hatt band to the hatt sent me from London by son Ben about the 27 May last, quem vide	0	1	6
3	Spent by Nathan: at Royston	0	0	3
	For a paire of gloves for George, there being not enough of the former June 28, at	0	1	0
4	For a new cart rope at Royston	0	3	0
	Spent by son Nathaniel at Ickleton Faire, going to buy cheese	0	0	6
6	Given to one Elizabeth Browne of Wiggen in Lancashire, left with 3 smal children and 40*l*. debt and kept in prison a yeare after her husband's death etc.	0	0	3
8	Given to 3 other poore lame etc. at door	0	0	3
	For a pound of 6 and 8*d*. nayles to mend the stable walls broaken down	0	0	5
9	Paide to little Dick the taylour for 5 dayes upon the mow this harvest at 6*d*. per diem	0	2	6

12	For a letter sent to Good: Davyes, in or before harvest began, which I promised to repay againe	0	0	3
	Given to a couple of seamen wounded and maimed in the warrs, and taken and kept prisoners in Fraunce	0	0	2
16	To the same man mentioned in the first instant, one Marmduke Feaks, comeing over to the red cow again and to rowel Typtery being hurt	0	1	0
17	Paide to Clement Norman for reaping the 2 acres of rye in Brook Shot in Northfield this harvest	0	6	0
	Given to a West India merchant undone by the French in those parts	0	0	3
23	Paide to Goody Swann for 5 loades of dung, laid upon Nat's land, and an half load besides, at	0	9	0
24	Item for the use of her horse in harvest and 14 or 20 dayes before	0	12	0
	Paide to old William Fordham for 3 dayes work in February and January last, making and mending cow racks, lyeletts, pales in kitchin garden	0	4	0
26	Item half a day mending cow rack againe February 12, at (vide August 26)	0	0	9
30	To Marg: Cra: uxori for mutton of R: Fid: 9s., salt 2 bushels 10s. 8d., and cheese 79 pounds, whereof 5 stone at Ikleton at 2s. 4d. ob., id est 18 cheeses about 14s. 2d., rise 9d., sugar 18 pound 6s. 5d., currans 14 pound 6s. 9d., malagas 56 pound 16s. 4d. at 3d. ob., out of which 5 pounds dross, 5 more left after harvest, and al occasions besides, in al	4	0	0
	Expended this month in al	6	4	1

August 1706 Receipts

[p.49]

	Received of Good: Hankin for one peck of wheate at 7 groates	0	0	7
8	Received of Widdow Haycock for a peck of rye or miscellaine	0	0	4
	Received of the miller for 2 pecks	0	0	8
	Received of Jos: Wed for 6 pecks of malt that they had July 15	0	3	11
9	Item for 2 pecks of rye they paide for now	0	0	8
	Received of Ben: Bright for a peck	0	0	4
12	Received of Widdow Carrington for a peck 4d.; item for a peck of wheat	0	0	11
	Received for a peck of Ben: Bright	0	0	7
	Received of Widdow Gurner for a peck of rye	0	0	4
14	Received of Widdow Nash for 2 pecks; received of J: Woodward for a peck	0	1	0

	Received of Good: Hankin for wheate a peck	0	0	7
	Received of Widdow Gurner for a peck of rye	0	0	4
16	Received of Widdow Goode for a peck more; received of Mr. Greenhil for a peck; received of M: Waites for 2 pecks; received of Eliz: Bright for a peck	0	1	8
17	Received of Widdow Carrington for 2 pecks of wheate, and received of Edw: Godfry for 2 pecks	0	2	4
	Received of Edw: Parish for 2 bushels and half of malt at 2s. 4d.	0	5	10
22	Received of Jos: Wedd for a parcel of rye straw, about 3 bundles	0	0	9
	Received of Eliz: Bright for a peck of wheate at 2s. 4d.	0	0	7
	Received of Jos: Wedd for a peck of rye	0	0	4
23	Received of Jone Woods for a peck	0	0	4
	Received of Eliz: Fere for 3 pecks	0	1	0
	Received of Widdow Goode for a peck	0	0	4
	Received of Widdow Nash for 3 pecks	0	1	0
24	Received of miller and Mr. Greenhil for 3 pecks of the same	0	1	0
	Received of Will: Brooks for 2 acres and one rood of haume straw in Short Spotts, after harvest 1705	0	5	0
26	Item for a bundle of straw June 19	0	0	3
	Item for a peck of barly July 26	0	0	6
	Item for 2 pecks of wheate	0	1	2
	Item for 2 pecks of rye that time	0	0	8
27	Received of Rich: Fairchild for one peck of malt at 2s. 4d.	0	0	7
	Received of W: Brooks for rye a peck	0	0	4
29	Received of Thom: Barber for his smal tithes in 1704, 5, 6, cow, etc.	0	4	0
	Item for what remained due for 4 stetches of haume in Short Spotts ever since January *1706*	0	2	6
30	Item for wintering his calf 11 weeks, viz from January 24 to April 13 at 3d. per week, and I reckon but 9 weeks	0	2	3
	Item in part for 2 pecks of wheate remaining since December 20 last, vide December	0	0	2
	Received this month in al	2	2	10

[*p.50*]

August 1706 Disbursements

	Paide to James Miles of Foxton, as tithe man in the roome of Thom: Wallis, about 14 dayes	1	5	0
8	Paide to Thomas Davyes my old tithe man of Great Hormead	1	10	0
9	Paide to his son David for tithe man in the roome of Ben: Ingry, dead	1	10	0

	Paide to Ben: Wedd the son of Jos:, for his harvest this yeare	0	15	0
12	Spent by Nathaniel at Royston on several market dayes	0	0	10
	To barber at Braintre	0	0	6
14	For 2 new straps to the saddle, they being broke in going thither	0	0	4
	Given to servants at son Sam's	0	2	0
	For a baite at Old Samford, self and sons and 3 horses	0	2	4
16	For new heeles to my shooes of R: Hay.	0	0	4
	Paide to Will: Brooks by deduction of mony due to me in part for his wages this last harvest	1	10	0
17	For materials for a coate for myself, viz quater of an ounce of silk 5*d*., 2 ounces of mohaire 1*s*. 5*d*., buckram and			
22	canvas 9*d*., three dozen of buttons 1*s*. 6*d*., 4 yards of shaloone 7*s*. 4*d*., 2 yards and half of broadecloth 1 pound,			
23	for the making, 2 taylours 2 dayes, 2*s*. 8*d*., besides victuals, which may be reckoned worth 2*s*. more	1	14	1
	Item for 3 dozen of buttons to repair my waistcoate, at	0	0	7
24	Given to Captain Hitches man, borrowing his calash to Braintree	0	1	0
	Given to one in want and lame	0	0	1
26	Paide to Thom: Barber for his wages this last harvest, at	1	10	0
	Paide to Will: Fordham senior for his help in inning my reeke on February 23 last, vide July 24	0	0	10
27	Item for a day and half March 15, April 20, making a new wheele barrow for the garden, mending waggon shafts, etc., vide April 29	0	2	3
29	Item for 2 dayes, June 11 and 12, rectifying hay house grunsel, making cart-ladder, tressels, and mending rakes against harvest etc.	0	3	0
	Item for half a day of his son Wil: and Thom: Holt, each, mending carts	0	1	6
30	Item mending the little rack in the stable 12 instant (vide September)	0	1	6
	To Marg: Cra: uxori for mutton 9*s*. 2*d*., beife 33 pound 5*s*. 6*d*., tobacco of Thom: Davyes 1*s*. 3*d*., taylour's making and mendin, Nat, Hester 4*s*. 3*d*., toto	2	0	0
	Expended this month in al the summ of	12	10	2

[*p.51*]

September 1706 Receipts

Received of Thom: Barber for 2 bottles of wheat straw January last	0	0	4

	Item for a peck of malt in June	0	0	8
2	Item for 2 bushels and 3 pecks of rye since February 18 last at divers times and prices, viz 1*s.* 8*d.* and 1*s.* 4*d.*	0	4	3
	Item for 3 bushels and 3 pecks of wheate, 5 pecks at 2*s.* 8*d.*, 10 pecks at 7 groats	0	9	2
3	Received of Widdow Nash for 2 pecks	0	1	2
	Received of Will: Fordham senior for 2 parcels of chaff, March 2 and 12	0	0	5
	Item for 3 good bundles of haum straw	0	0	6
4	Item for 4 bundles of rye straw, and an oaken stick 2*d.*, in al	0	1	4
	Item for 2 pecks of barly, June 11	0	1	0
	Item for carriage of 6 quarters of barly in December last to Barly at 8*d.* per quarter	0	4	0
6	Item wintering 2 cowes for him 16 weekes a peice, at 6*d.* per week	0	16	0
	Item for 12 bushels and 3 pecks of rye since February 6 last til now	0	19	11
	Received of Eliz: Bright for a peck	0	0	4
	Item for a peck of wheate then	0	0	7
7	Received of Mr. Latimer for 6 bushels of malt he had August 29 last	0	14	0
	Received of Mary Parish for 14 weekes wintering of her cow last winter	0	7	0
9	Received of Lyd: Barber and Mary Brooks for a bushel of rye, 2 pecks each	0	1	4
	Received of Good: Soale for 3 old cart spoks	0	0	3
	Received of Thom: Watson for 2 pecks of malt at 2*s.* 8*d.*	0	1	4
10	Item for 6 pecks of barly at 2*s.* per bushel	0	3	0
	Item fetching a cant of wood out of Chrisal Park	0	5	0
	Item for 3 bushels and 3 pecks of rye, 2 bushels at 1*s.* 8*d.*, 7 pecks at 1*s.* 4*d.*, in al	0	5	8
11	Item for 4 bushels and half of wheate since April 18 last at 7 groates	0	10	9
	Received of Widdow Carrington and Robert Macer for half a bushel of the same	0	1	2
12	Received of Widdow Nash for 3 pecks of rye; received of Edw: Godfry for a bushel	0	2	4
	Received of W: Brooks for wheat 2 pecks	0	1	0
	Received of Widdow Swan for 2 pecks	0	1	0
	Received of John Bowels for a peck; received of Rob: Macer for a peck; received of the miller for 3 pecks	0	2	6
13	Received of Eliz: Bright for 2 pecks of rye, of Widdow Gurner for a peck	0	1	0
	Received of Mrs. Westly for 2 bushels	0	2	8
	Received of Mary Waites, Widdow Nash, Edw: Godfry, Rob: Macer, for 2 bushels	0	2	8

		£	s	d
	Received of Jos: Wedd for a peck	0	0	4
14	Received of Mr. Greenhil for half a bushel of wheate, at	0	1	0
	Received of Widdow Gooding for a peck; received of Ben: Bright for 2 pecks	0	1	6
16[sic]	Received of him for 2 pecks of rye; received of Good: Nash for 2 pecks	0	1	4
	Received of Thom: Reynolds for a peck of wheate; received of Rob: Macer for 2 pecks, one before and one now	0	1	6
[p.53]	Received of Edward Godfry for 3 pecks of wheate	0	1	6
16	Received of Will: Ward for wintering a cow in 1704, 7 weeks at 5d.	0	2	11
	Item for wintering a cow 12 weeks in 1705 at 6d., and wintering another 5 weeks, the same, 1705	0	8	6
17	Item for a bushel of malt June 14	0	2	7
	Item for a bushel of rye that month	0	1	4
	Item for small tithes this yeare	0	7	8
18	Received of Will: Brock for the haume straw of 2 acres and 3 roods laide in by my men – hauming and carriage comes to 8s. 2d., and so it is but 6d. per acre	0	9	8
	Item for a bushel of tayle wheat in January and February last, at	0	1	8
19	Item for 5 bushels and half of wheate between November 20 and August 15 last	0	13	7
	Item for 17 bushels and an half of rye since October 25 last to this time, ten bushels at 1s. 8d., the rest at 1s. 4d.	1	6	8
20	Received of Will: Thrift senior for 3 pecks of barly in July and August	0	1	6
	Item for 1 bushel and half of malt July 20 and August 20 last, at	0	3	11
21	Item for a bushel and 3 pecks of wheate between July 10 and the 11 instant	0	4	1
	Item for 3 bushels of rye in the same space of time at 1s. 4d.	0	4	0
23	Received of Thomas Eastland for 2 bushels of rye betwixt July 12 and the 10 of this instant, at 1s. 4d.	0	2	8
	Item for 2 bushels and half of wheate in the same space of time, at 5 several times, all at 2s. 4d. per bushel	0	5	10
24	Item for 2 bushels of malt June 25 last at 2s. 7d., and 2 pecks about August 29 at 2s. 4d., in al	0	6	4
26	Item for 3 pecks of wheate in this month besides, at 2s. per bushel	0	1	6
	Item for 100 of spitts or sprindls for thatching July 11 last	0	0	3
27	Item for 3 bushels more of rye in this month at twice and the same price with the former	0	4	0

	Received of John Boweles for a bushel	0	1	6
28	Received of Mr. Greenhil for 3 pecks of the same, being new rye; received of Rob: Macer for a bushel; received of Ben: Bright for a bushel	0	4	1
	Received of the miller for a bushel	0	1	6
30	Received of Widdow Nash for a bushel; received of Joane Woods for a peck; received of Mary Waites for a peck	0	2	3
	Received of Widdow Gurner for 2 pecks and a peck June 26 last at 4*d*.	0	1	1
	Received of Widdow Ward for 2 pecks	0	0	9
	Received of Eliz. Bright for 3 pecks	0	1	2
	Received this month in al the summ of	12	10	6
	Received in the last six months, viz since March last, the ful summ of	95	19	0

[*p.52*] **September 1706 Disbursements**

1	Given to a breife for Jacob Bell, Will: Johnson, John Dodgen, John Bridgar, Vincent Bishop, Thom: Gibson, and several other sufferers by fire, hapning on August 1, 1705, to the loss of 2,706*l*. and upwards in the parish of St.			
2	Olave, Southwark, in the county of Surry, the sum of	0	1	0
3	Paide to Will: Fordham senior for his wages this last harvest (for the rest of his reckoning see July 24 and August 26 and April 29)	1	12	0
4	For 600 spitts or sprindles for thatching, of Wallis of Chrissol	0	1	6
6	Paide to Good: Soale for a new felloe put into the tumbrel wheeles a little while before harvest	0	1	6
	Item for heading a plow now, at	0	1	0
7	Paide to Thom: Watson for 22 dayes between April 7 and May 5, and one half day, in all	0	18	9
	Item for 21 dayes betwixt May 5 and June 2 following at 10 per day	0	17	6
	Item for 20 dayes and half more before June 30 at the same rate	0	17	1
9	Item for 11 dayes and half before July 15	0	9	7
	Item for 19 dayes in August after we ended our harvest	0	15	10
	Item for harvest wages now due	1	10	0
10	Item for 11 dayes and half befor the 15 day of this instant	0	9	7
	Paide to Thom: Eastland for his wages this last harvest also	1	12	0
	Spent at the faire with wife, daughters, Mrs. W:, etc., in wine, etc.	0	5	0

HOUSEHOLD ACCOUNTS 1706

11	For 3 paire of black spectacles	0	1	6
	Paide for the first quarterly payment of the land tax granted for this present yeare for Hoys, id est Parishes	0	2	6
12	Item for Cassanders to that tax	0	9	0
	Item for parsonadge to the same tax	6	18	0
	Paide to Wallis of Chrissol for an hundred rodds 1s., an 100 spitts 3d., an 150 withs at 9d., in al	0	2	0
	Given to a maimed souldier at door	0	0	1
13	Spent at Mrs. Winstanleye's with son Ben and his wife 4s., at 1s. a peice, and at the White Lion 1s. 10d., in al	0	5	10
	Paide to Ben: Bright for a loade of dung laide upon my owne land	0	1	8
14	Paide to Will: Ward for six load of dung November 1 and 2 last upon son Nat's hired land at 20 pence	0	10	0
16[sic]	Item for 10 loads about Midsummer, partly upon Nat's and partly upon my land (besides 14 shillings deducted by Nat for straw)	0	4	8
	Item for 5 loades more this September upon my land, al of it at 1s. 8d. per loade also	0	8	4
[p.54]	Given to a seaman whose ship was cast away by stormy weather, loosing 24 men at that time	0	0	1
16	Paide to Will: Brock for trimming myself 4 ful quarters since our last reckoning, at 3 quarter	0	12	0
17	Item trimming son Nathan: and I think son Sam 29 times in al, at 2d.	0	4	10
	Item for his wages last harvest, though I told him withal that I would give him but 1l. 8s. if we lived another harvest together	1	12	0
18	Paide to old Will: Thrift for 2 dayes before harvest, at	0	1	8
	Item for 12 dayes and an half after harvest to the 8th instant	0	10	5
19	Item for his wages this harvest	1	12	0
	Paide to John Waites for his wages this last harvest also	1	10	0
	Paide to Thom: Eastland for six journeyes at plow, 4 before, 2 since harvest	0	4	0
20	Item for 7 dayes before harvest, at	0	5	10
	Item for threshing 5 quarters and 2 bushels of wheate of the last dressing	0	6	9
21	Item for 10 dayes work in August	0	8	4
	Item for a loade of dung before harvest upon son Nat's 3 rood in Rowze	0	1	8
23	Item for 15 dayes more in sowing, threshing seed rye and wheate, as the former 10 dayes, where for the rest of this reckoning see December following about the 6 or 7 day	0	12	6
24	Paide to Will: Fordham for his wages last harvest — junior	1	10	0

26	Paide to Mr. Thomas Malin of Elmden in Essex for 43 stone and 4 pounds of beife, mutton and lamb in harvest last, that is betwixt July 12 and August 7 at 2*d. qtr.* per pound, the sum of	5	15	4
	— We killed an hogg out of the yard — very special pork besides.			
27	Paide to George Carpenter for his yeare's wages ended at this Michaelmas, the sum of	5	0	0
28	To Marg: Cra: uxori for 5 quarters of mutton and 2 neat's tongues about 14*s.*, oisters, fish, and al other occasions this month	2	3	0
	Item for a yeare's wages of her maid Rhoda Phipp, the sum of	3	0	0
30	Item to Elizabeth Fere for her service in the time of harvest	0	12	0
	Expended in this month in al the full sum of	44	18	4
	Expended in the last six months, viz since March last, the ful summ of	256	3	1

[*p.55*]

October 1706 Receipts

	Received of Widdow Swann for her half yeare's rent ended at this September 29 last just past	1	11	6
4	Received of John Bowelsworth for a bushel of wheate August 17 and now	0	2	2
	Received of Rob: Macer for a peck	0	0	6
	Received of young G: Watson for 200 spitts	0	0	6
7	Received of Lydd: Barber for a peck of wheate	0	0	6
	Received of Rob: Macer for a peck; received of M: Waites for a peck that she had August 22 at 7	0	1	1
8	Received of Susan Carpenter for a peck	0	0	6
	Received of Edith Wedd for a peck	0	0	6
	Received of Rob: Macer for a peck	0	0	6
	Received of J: Boweles for 2 pecks	0	1	0
	Received of Robert Macer for a peck	0	0	6
14	Received of Mr. Greenhil for half a bushel of barly of 3rd dressing	0	0	10
16	Received of Mr. John Nichols of Royston for 6 quarters and 4 bushels of the first dressing, sold to him at 13*s.* 6*d.* per quarter	4	7	9
19	Received of Mrs. Westly for an acre of wheate haume neare upon, that is 2 stetches and half	0	3	6
	Received of Rob: Macer for wheate one peck at 20*d.*	0	0	5
22	Received of John Watson junior as part of his half yeare's			

	rent ended at September 29, 1705	0	17	3
	Item for a bottle of rye straw	0	0	3
24	Item fetching a score of fagots out of Northfeild in a dung cart	0	0	9
	Item for a bushel of barly May 4	0	2	0
	Item for 2 bushels and half of malt	0	6	5
26	Item for a bushel of wheate	0	2	4
	Item for 4 bushels and 3 pecks of rye	0	6	7
	Item for his half yeare's rent ended at Lady Day 1706 last	0	19	0
28	Received of Widdow Gurner for wheat a peck	0	0	5
	Item for 2 bundles of barly straw	0	0	8
	Received of J: Watson for 200 spitts	0	0	6
29	Received of Mr. John Nichols of Royston for 6 quarters and two bushels of the 2nd dressing of barly, sold to him at 3s. 6d.	4	4	4
	Received of Mr. Wedd for a peck of tithe apples or thereabouts	0	0	6
30	Item for the tithe of his Saint Foine in Dod's Close and half an acre that he bought of J: Whittingstal	0	3	6
	Item for the carriage of a loade of wooll from Sturbridge	0	2	4
31	Item for 7 acres and half of haume out of Northfeild, al stetches whereof 4 were wheate haume that were little worth	1	0	0
	Received of Rob: Macer's son for one peck of wheate, at	0	0	5
	Received of Richard Haycock for 2 pecks of the same	0	0	10
	Received this month in al	14	19	10

[p.56]

October 1706 Disbursements

	Paide to Goody Swan for the use of her horse since harvest one weeke and several other dayes, etc.	0	1	6
4	Paide to Will: Brock for his clerk's wages, due to him September 29 last	0	6	8
	Sugar candy for myself, having a great cough and wanting pectoral	0	0	4
	For a rubber to cut the mow sides	0	0	1
6	Given to a breife for the repair of St. Cuthbert's Church in Darlington in the County Palatine of Durham, the			
14	charges whereof computed at 1,705l., besides what is laide out by the inhabitants already	0	0	6
	For unloading the 6 quarters and half	0	0	3
	For a chaff sieve at Royston	0	1	3
17	Paide to Wil: Numan the coller maker for work and skins,			

	etc., laide down by Nat: about 3 weekes since — there was a new cart saddle (I think) as part of it		0	12	0
19	For a tin lanthorne for the house paide for by him at the faire		0	2	0
	Spent by him and George at the faire, going for Mr. Wedd's wool, a very wet day — that is a bushel of oates 1*s.* 6*d.*, 2*d.* toll, 4*d.* spent besides what Mr. W. spent		0	2	0
23	Spent by son Nathaniel at market going to sel wheate but could not for 5 or 6 dayes successively		0	2	0
	Paide to Widdow Gurner for a loade of dung upon my owne land		0	1	9
24	Paide to young J: Watson for half a day mending the thatch of the new barne, blowne of November last		0	0	9
	Item for 6 dayes in February and March last, lopping in close, plashing the hedge there, etc.		0	5	11
26	Item for 3 bushels of hen dung laide upon 2 of my 12 acres Barfield		0	1	6
	Item for six dayes thatching Edw: Parishes barne June 14		0	9	0
	Item for his son yelming 4 dayes		0	2	0
	Item for 5 dayes and an half thatching parsonadge old barne		0	8	3
29	Item for son yelming then, June 21		0	2	9
	Item for 300 rodds used half there and half at Edw: Parishes		0	3	0
	Item for scouring my part of the brook, about September 10		0	2	6
30	Item for a good part of a day in scouring the ditch next parsonage close		0	0	9
	Item for a day in the garden September 23		0	0	10
	Item 4 dayes proceeding upon parsonage old barne, backhouse, entry, and half a day at Goody Swann's thatching		0	6	9
31	Item for his son yelming 3 of those dayes, vide November following		0	1	6
	To Marg: Cra: uxori for mutton, tongues, etc., about 16*s.*, butter about 8*s.*, sugar about 3*s.* 6*d.*, and al other occasions, toto		2	15	0
	Expended this month in al		6	10	10

[*p.57*]

November 1706 Receipts

	Received of Rob: Macer for half a bushel of wheate		0	0	10
1	Received of Widdow Law for a peck		0	0	5
	Received of Sam Woods for a peck		0	0	5

HOUSEHOLD ACCOUNTS 1706

2	Received of Rich: Haycock for 2 pecks	0	0	10
	Received of Widdow Gooding for 2 pecks	0	0	10
4	Received of Urias Bowes of Barly for 6 quarters and 2 bushels of the 3rd dressing of barly at 13s. 6d.	4	4	4
	Received of Rob: Macer for wheate a peck	0	0	5
6	Received of Eliz: Bright for 2 pecks; received of Lyddy Barber for a bushel	0	2	6
	Received of Will Jeepes his daughter what remained due for 3 pecks of rye 1704 November 29, and deducted in part, June 7, 1705, vide disbursements June 1	0	0	3
7	Item for 100 of spitts they had now	0	0	3
	Received of Rich: Haycock for 2 pecks of wheate, as it is, at	0	0	10
9	Received of Wil: Ribshir of Hauxton for 2 bushels of the same	0	3	4
	Received of Mary Waites for a bushel	0	1	8
11	Received of Jos: Wedd for a peck	0	0	5
	Received of Widdow Swan for 2 pecks	0	0	10
	Received of Mary Brock for a bushel and half of wheate	0	2	6
	Received of Goody Hankin for a peck	0	0	5
12	Received of Mrs. Westly for 2 pecks of barly of the 5th dressing	0	0	10
13	Received of John Bowelesworth for a bushel, partly out of the same dressing, and 2 pecks before, at	0	1	8
	Item for 2 pecks of wheate now	0	0	10
14	Received of Ben: Bright for 4 pecks; received of Rich: Haycock and Widdow Good for 3 pecks more of the same; received of James Swan for 4 pecks	0	4	7
16	Received of old Wil: Thrift for 5 pecks of malt October 4, at	0	2	11
	Item for 3 pecks of rye, at	0	1	0
	Item for 7 pecks of wheate	0	3	5
18	Received of Widdow Law for a peck; received of Cass: Woods for a peck	0	0	10
20	Received of Urias Bowes of Barly for 6 quarters and one bushel of the 4th dressing of barly at 14s.	4	5	9
	Received of Rob: Macer for wheate a peck four dayes ago	0	0	5
23	Item for 2 pecks more now	0	0	10
	Received for a bushel of Rich: Haycock	0	1	8
	Received of Edw: Parish for a good loade of haume straw October 20, 1705	0	4	6
26	Item bringing 40 fagotts out of Northfield	0	1	0
	Item for what remained due for his Michaelmas rent 1705, vide February 9	0	11	0
28	Item for one his cowes 5 weeks, another 20 weekes, going in my yard last winter at 6d. per week	0	12	6

189

		£	s	d
30	Item for his smal tithes for 3 yeares ended at Michaelmas last	0	4	0
	Item for 6 pecks of malt in June last	0	3	11
	Item for 2 bushels of wheate	0	4	7
	Item for 7 bushels and a peck of rye	0	10	2
	Item for his half yeare's rent ended at Lady Day 1706	1	15	0
	Received this month in al the summ of	14	12	6

[p.58] **November 1706 Disbursements**

		£	s	d
	Paide to John Watson, my tennant, for a day of his son at haume cart	0	0	4
1	Item for his keeping my hogg about 14 dayes	0	2	0
	Item for 2 loades of dung September 21, October 10, upon son Nathaniel's hired land	0	3	4
	To old Featherstone, bottoming 2 chaires 8d., flosket 2d., basket 4d.	0	1	2
4	Spent by Nathaniel at market 6d., and unloading the six quarters and 2 bushels, being the 2nd loade to Mr. Nichols	0	0	10
	For black silk to mend my cloathes	0	0	10
6	Paide for the 2nd payment of the tax for this yeare for Hoyes	0	2	6
	Item for Cassanders to that tax	0	9	0
	Item for parsonadge to the same tax	6	18	0
8	Paide for half a yeare's duty of the window tax due at Michaelmas last	0	5	0
	Given to daughter's Hester's midwife and nurse, going away	0	10	0
	For unloading at Urias Bowes'	0	0	4
	For a skep bought at Cambridge	0	0	6
11	For a couple of skuttels then	0	0	8
	Spent by men and toll of 20 bushels of coals for myself, besides Th: Wats:	0	0	7
13	Paide to Wil Numan for half an hyde 4s. 6d., work 9d., buckles 3d., in al	0	5	6
	Item for a ridge tree to the waggon	0	2	0
	Paide to Rich. Carter for a ringe of wood last winter, vide January 7 last	0	17	0
14	Paide to James Bush for Typterye's oates and hay at Hodsdn., when son Ben went upon him, and turnepikes	0	0	9
	Item for bringing downe a barrel of oisters	0	0	9
16	Item for bringing downe an hamper of wine now, from Mr. Michel	0	1	6
	Paide to John (for a little loade of dung out of the limeyard			

	upon the stetch in Waterden, on the hither side of the 2 acres) Bowelesworth	0	2	0
18	Paide to Wil Thrift for 17 dayes work since September 8	0	14	2
	Item for hauming 9 acres and half and half a rood in Northfeild	0	12	10
19	Item for threshing 5 quarters and a bushel of the 2nd dressing of oates	0	3	6
	For unloading that 6 quarters one bushel [*c.f.* receipts, 20 November]	0	0	3
	Spent by son Nat at market the 13 instant	0	0	6
	For a letter from Mr. Michel	0	0	3
20	Spent by son Nat: at market now	0	0	4
	Paide to Wallis of Chrissol for 250 withs 1*s.* 3*d.*, and 200 spitts 5*d.*	0	1	8
26	Paide to Edw: Parish for 2 loades of dung upon Nat's land in Northfield February 18 last, al mine being sowne	0	3	4
	Item for 3 little loades in September upon my owne land in Southfield	0	4	8
28	Item for 6 dayes serving thatcher at his owne house June 14 last, vide October 29	0	6	0
29	Item for 5 dayes and half more serving about parsonadge old barne, June 21	0	5	6
	Item for 14 dayes work, hay, dung cart, etc.	0	11	8
	Item allowed for a militia tax	0	0	3
30	To Marg: Cra: uxori for linsy woolzy for herself 4*s.*, stockins, patens, waifht, taylour, about 7*s.*, candles 42 pounds 13*s.* 5*d.*, meat 5*s.*, and al other things, toto	2	16	0
	Expended this month in al the ful summ of	16	5	6

[*p.59*]

December 1706 Receipts

2	Received of Mr. Urias Bowes of Barly for 5 quarters and 3 bushels of the 5th dressing of barly at 14*s.*	4	9	3
	Received of Mary Waits for 4 pecks of the old wheate, 1*s.* 8*d.*; received of J: Bowls for 2 pecks of it	0	2	6
3	Received of Clem: Norman for 2 pecks on October 12 last at 2*s.* per bushel	0	1	0
	Item for 3 pecks of malt September 28	0	1	9
	Item in part for 6 pecks of wheate since	0	0	4
4	Received of Rob: Harvy for a bushel	0	1	8
	Received of Widdow Gooding for 2 pecks	0	0	10
	Received of J: Woodward for a peck; received of Wil: Brock in part for 2 bushels October 8 and November 7, 2*s.* 8*d.*	0	3	1

5	Received of Good: Norman for 2 pecks; received of Widdow Goode for a peck; received of Good: Hanchet for a bushel; received of R: Harvy and R: Macer for half a bushel a peice		0	4	7
6	Received of Widdow Gooding for a peck; received of Rich: Haycock for 2 pecks		0	1	3
	Item for 2 pecks October 8 and 27, 1703, at		0	1	10
	Item for a peck in part April 12, 1704		0	1	1
7	Received of Joan Woods for a peck 5*d*.; received of Good: Hankin for a peck 5*d*.		0	0	10
	Received of 2 loades of haume fetching of Thomas Watson, at		0	2	0
9	Item for fetching him half a loade of coales from Cambridge, at		0	2	6
	Item for a bushel of rye, at		0	1	4
	Item for a bushel of malt, at		0	2	4
10	Item for a bushel of wheate at 2*s*. 4*d*.; item for another bushel at 2*s*. per bushel		0	4	4
	Item for 3 bushels and a peck more at 1*s*. 8*d*.		0	5	5
	Received of Thom: Fairchild for a bushel		0	1	8
	Received of Thom: Eastland for a bushel at 2*s*. and for 2 bushels and half at 1*s*. 8*d*.		0	6	2
11	Item for 2 bushels of malt October 23		0	4	8
	Item for 2 bushels of barly of the 2nd and 3rd dressings (but he denyes one)		0	3	4
12	Received of Mr. Urias Bowes of Barly for 6 quarters and 2 bushels of the 6th dressing of barly at 14*s*.		4	7	6
	Received of Widdow Gooding for one peck of wheate, at		0	0	5
	Received of Mary Waites 3 pecks; received of Widdow Haycock for 2 pecks		0	2	1
13	Received of John Boweles for 2 pecks		0	0	10
	Received of Rob: Macer for 2 pecks		0	0	10
	Received of Rich: Haycock for 2 pecks 10*d*.; received of Good: Hankin for 2 pecks; received of Widdow Ward for a bushel		0	3	4
14	Received of Ben: Bright for a bushel; received of Widdow Haycock for 2 pecks; received of R: Haycock for 2 pecks		0	3	4
16	Received of Urias Bowes for 7 quarters and one bushel of the 7th dressing of barly at 14 shillings per quarter		4	19	9
	Received of Ralph Carpenter for 2 pecks of wheate before February 19, *1706*, at		0	1	4
	Item for a bushel of malt October 15, 1705		0	2	4
17	Item for 7 pecks more in April and June last, viz 1706, at 2*s*. 7*d*.		0	4	7
	Item for 4 bushels and half of rye before 21 of March *1706*, one at 2*s*., al the rest at 1*s*. 8*d*. per bushel		0	7	10

[p.61]	Received of Widdow Law for one bushel of wheate, at	0	1	8
	Received of Rob: Macer for 3 pecks	0	1	3
18	Received of Urias Bowes of Barly for 6 quarters and half of the 7 and 8 dressings of barly at 14s. per quarter, themselves unloaded	4	11	0
19	Received of Mary Waites for 3 pecks of wheat 1s. 3d.; received of Widdow Strett for 3 pecks	0	2	6
	Received of Wil: Brock for 2 pecks, at	0	1	0
	Received of Good: Gurner for 2 pecks rye	0	0	9
	Received of Lyd: Barber for a bushel	0	1	6
20	Received of Rob: Harvy and Cass: Woods for 2 pecks between them, at	0	0	9
	Received of Widdow Goode for 2 pecks, at	0	0	9
	Received of Widdow Ward for 2 pecks, at	0	0	9
21	Received of Mrs. Westly, the miller and Mary Waites for a bushel and 3 pecks	0	2	7
	Received of Good: Gurner and Hankin for 3 pecks of new wheate, at	0	1	6
	Received of Rob: Macer, Cass: Woods and miller for a bushel and 3 pecks of rye	0	2	8
23	Received of Ben: Bright for a bushel	0	1	6
	Received of Widdow Gooding for 2 pecks of the new wheate, at	0	1	0
	Received of Rob: Macer for 2 pecks	0	1	0
	Received of J: Waites for rye 2 bushels	0	3	0
24	Received of Rob: Harvey, Rich: Haycock, Widdow Ward, for 2 bushels more	0	3	0
	Received of Good: Gurner, Wedd, Barber, Brock, Widdow Nash, for 4 bushels more	0	6	0
	Received of Good: Hankin for wheate a peck	0	0	6
	Received of Good: Brock for a bushel	0	2	0
26	Received of Ralph Carpenter for the tithe of his cow and calf last year	0	1	0
	Received of Urias Bowes of Barly for 6 quarters 2 bushels of the 9th dressing of barly at 14s. 3d. per quarter	4	9	0
27	Item for 3 bushels of rye then	0	4	6
	Received of Edw: Godfry, Mr. Greenhil, Widdow Gurner, Mary Waits, for two bushels and 3 pecks of the same	0	4	1
	Received of Widdow Gurner for a peck of the new wheate, at	0	0	6
28	Received of Urias Bowes for 4 quarters (id est 5 bushels of the 9 and 3 quarters of the 10 dressing) of barly at 14s., carryed out with 3 quarters of son Nathaniel's, at	2	16	0
	Received of Good: Brock and Widdow Haycock for 2 pecks of barly, out of the last	0	0	10
30	Received of Good: Harvy, Mr. Greenhil, Rich: Haycock, for 7 pecks of rye	0	2	8

		£	s	d
	Received of him and Widdow Goode for 3 pecks of new wheate, at	0	1	6
31	Received of Cass: Woods and Wil: Brooks for 2 pecks of the same	0	1	0
	Received of Widdow Goode for a peck	0	0	6
	Received of her, Edw: Godfry, R: Haycock, Ben: Bright, for rye 2 bushels	0	3	0
	Received this month in all the ful sum of	32	9	2

[p.60]
December 1706 Disbursements

		£	s	d
	Paide for unloading the said 6 quarters and 3 bushels of barly	0	0	3
2	Paide to Rich: Fairechild for a constable's rate, at 1d. per pound, for parsonage 11s. 5d., and Cassanders 6d. so much as I hold of it	0	11	11
3	Paide to Rob: Harvy for 6 journeyes at haum cart and dung cart since August last, and one of them a day	0	3	6
4	Item for 3 dayes and half serving the thatcher about parsonadge old barne, backhouse entry etc., on September 26, 27, 28, etc. — see disbursements October 30	0	3	6
	Paide to Wallis of Chrissol for 1,100 spitts or sprindles for thatching	0	2	6
5	Paid to Good: Botterel for drinks for 2 of my horses and coming over	0	3	0
	Given to an ancient man in necessity	0	0	1
	Paide to Rich: Haycock for a peice of leather to the clock of the pump	0	0	3
6	Item for a day casting orchard pond in June 6, 1705, vide disbursments then	0	1	3
	Paide to Thom: Fairechild for 12 pound and an half of new iron for the mouth of the oven, vide February 14	0	2	7
7	Item for putting al the teeth into 2 dragg rakes before harvest	0	4	0
9	Item for 39 remooves of horse shooes since January 23 last, viz 10 months and 11 dayes since at 1d. a remoove	0	3	3
	Item for 77 new shooes in that time	1	5	8
	Item for a new share in that time	0	6	4
10	Item upon son Nathaniel's tally in the same space of time	1	7	0
	Paide to Thom: Watson for 11 dayes in September at 10d. per diem	0	9	2
	Item for 26 dayes and half between September 29 and November 3, at	1	2	1
11	Item for 23 dayes and an half between November 3 and December 1 at 9d.	0	17	8

	For unloading the 6th dressing, at	0	0	3
	Spent by Nathaniel at Royston	0	0	4
12	Paide to Thom: Eastland for 9 dayes work and half in October	0	7	11
	Item for hauming 8 acres and a rood	0	11	0
13	Item for threshing 49 quarters and six bushels of the first 9 dressings of barly at 9*d.* per quarter	1	17	2
	Paide for 6 ells and a quarter of doulas for myself, at	0	9	4
	To the guelder for spaying the bitch	0	0	4
14	For unloading the said 7 quarters and a bushel at Urias Bowes'	0	0	3
	Paide to Ralph Carpenter for 5 dayes serving J: Watson thatching hog-styes October 19, 1705, vide disbursements there	0	5	0
	Item for another daye's work there above	0	0	10
16	Item for killing 6 hoggs that yeare	0	2	0
	Item for 6 dayes lopping and plashing the hedge in close, with J: Watson before March 28, 1706	0	5	11
17	Item for 2 bushels of hen dung upon 2 of the 12 acres in Barfield 1*s.*; item for a loade of dung September 23 last, upon my land in Southfield	0	2	8
	Item for killing 2 hoggs November last and the 3rd of this instant	0	0	8
[*p.62*]	For a dozen of cart clouts with bradds to them at 3*d.* a peice	0	3	0
	Spent by son Nathan: at Royston	0	0	4
18	Paide to Henry Wallis of Chrissol for 100 rods 1*s.*, and 500 sprindles 1*s.* 2*d.*	0	2	2
19	Paide for 5 bushels of lime at Royston in October, laide out by Nathan: in wheate seede, and now repaide	0	2	1
20	Given to one Eliz: Browne of Wood Norton in Norfolk, loosing by fire September 21 last 352*l.*, her husband venturing in for theire cloaths being also burnt	0	0	6
	Paide to Robert Harvy for a daye's work in making a new hearth in the kitchin, and mending the floor	0	1	4
	For an hair brush for horses, at	0	1	0
21	Spent by Nathaniel at Royston Market	0	0	6
	For a couple of new sheet almanacks	0	0	3
	Spent by Nathan: at market againe	0	0	6
	Paide to Ralph Carpenter for killing the last of our bacon hoggs	0	0	4
23	Paide to Mr. Greenhil for an old sive to take out the peices of rye eares when the tasker thresheth rye, though it is too course for that use, until Eastland did mend it somewhat	0	0	6
24	Paide to George Carpenter for unloading the 11 dressing of barly at U: Bowes'	0	0	3
	Paide to Thomas Watson for six days before the 8 day of this instant	0	4	6

			s	d
26	Item for six dayes more before the fifteenth of this instant also	0	4	6
	Item for six dayes more before the 22 of this instant, al at 9*d.* per diem, and one of them also he served the mason, vide January and disbursements ibidem	0	4	6
27	Paide to Will: Morric of Harston, gelder, for cutting of 7 piggs	0	0	9
	Spent by son Nathaniel at Royston	0	0	6
	Paide for a dozen of whip sticks 4*d.*, and a plow staff 2*d.*, to Henry Wallis of Chrissol, in all	0	0	6
28	Paide to Thom: Fairchild for mending the latch to the little gate out of the greate yard into back yard	0	0	1
	Paide for an ounce of mohaire for a coate for myself, turned	0	1	0
30	For buckram and stay tape to the same, at 8*d.* and a penny tape	0	0	9
	Item for taylour's work in making of it a day and half (I suppose) about	0	1	6
	Paide to Rich: Jackson for 12 foot of glass new ledded 2*s.*, a foot of new glass 1*s.*, 14 quarryes 1*s.* 2*d.*, all in the chancell	0	4	2
31	To Marg: Cra: uxori for 37 pound of cheese 8*s.* 9*d.*, beife 7 stone one pound 16*s.* 6*d.*, mutton 2*s.* 3*d.*, a brass cock instead of a tap 1*s.* 6*d.*, of Brock, and al other occasions this month, toto	2	4	0
	Expended this month in al the full summ of	15	7	2

[*p.63*]

January 1707 Receipts

			s	d
	Received of Mr. Greenhil, Rob: Macer, Jos. Wedd, G: Law, Gurner, for four bushels and an half of rye at 1*s.* 6*d.*	0	6	9
6	Received of Thom: Watson for 6 pecks (and 2 pecks Edw: Godfry at 1*s.* 6*d.*)	0	3	0
	Item for 3 pecks of barly of the 10 dressing	0	1	3
	Item for 2 bushels and half of wheate	0	4	2
7	Received of Widdows Gooding, Reynolds and Rob: Macer for a bushel	0	2	0
	Received of Mary Waites, Lyddy Barber, Widdow Nash, for 3 bushels of rye	0	4	6
8	Received of Mary Brock, miller, etc., for six pecks of the same at	0	2	3
	Received of Edith Wedd for 2 pecks of barly	0	0	11
	Item for 2 pecks of wheate from malth[ouse]	0	1	0
9	Received of Edw: Godfry, Widdow Gurner, Mary Waites, for a bushel, at	0	2	0

HOUSEHOLD ACCOUNTS 1707

	Received of Mary Law, El: Parish, Will: Brock, Mary Waites, for 2 bushels and a peck of rye, at	0	3	5
11	Received of Liddy Barber and Widdow Carrington for a bushel and 3 pecks	0	2	7
	Item of her for a peck of wheate, at	0	0	6
	Received of Widdow Gurner for a bottle of pease straw	0	0	2
13	Received of Mart: Hankin, Eli: Parish, Cass: Woods, for a bushel of wheate, at	0	2	2
	Received of Jos: Wedd, Rob: Harvy, Ben Bright, for a bushel and half of rye	0	2	3
14	Received of Rob: Macer, Wil: Brooks, Widdow Goode, Mary Waites, Widdow Gurner, for 4 bushels and a peck more	0	6	4
16	Received of Rob: Macer, Mary Waites, Widdow Goode, for 5 pecks of wheate	0	2	11
	Received for a score of wite bricks of Mr. W.	0	0	5
17	Received of Clem: Norman what remained due for 6 pecks of wheat December 3, and a peck since 6d., in al	0	2	8
20	Item for a bushel of rye at twice; received of Mr. Greenhil for 6 pecks; received of Jos: Wedd for 4 pecks; received of Eliz: Parish for a peck	0	5	8
	Received of Lyd: Barber and W: Brock for 2 bushels of the same, at	0	3	0
22	Received of them two, Good: Hankin and Rob: Macer, for 6 pecks of wheat	0	3	6
	Item for him for 2 bottls of pease straw	0	0	5
24	Received of Mrs. Westly for a bushel of barly of the 12 and 13 dressings	0	1	10
	Item of her Widdow Nash and Rob: Harvy for 2 bushels of rye, at	0	3	0
	Item of her, Widdow Ward, Widdow Swan, Rob: Harvy, for 6 pecks of wheate	0	3	6
25	Received of Jos: Wedd for 2 pecks	0	1	2
28	Item of him, Wil: Brock, El: Jude, Eliz: Bright, Athan: Carrington, for 3 bushels and 3 pecks of rye	0	5	7
	Received of Widdow Gurner for a bushel	0	1	6
	Received of old Wil: Thrift for 2 bushels of wheat in November and December	0	3	4
29	Item for 2 bushels of malt	0	4	8
	Received of Wil: Brooks and Mr. Greenhil for a bushel of rye	0	1	6
31	Received of John Bowels for half a bushel of barly of the 13 dressing	0	0	11
	Item for half a bushel of wheate	0	1	2
	Received this month in al	4	12	0

[p.64] **January 1707 Disbursements**

		£	s	d
	Paide to Thomas Watson for 8 dayes since December 22 last at 9*d*., vide December 24 etc.	0	6	0
7	Paide to one Rich: Johnson of Royston for a pair of shack traise, at	0	2	2
	Spent by son Nathan: at Royston Market	0	0	6
8	Paide for 2 bushels of teares for the 2 half stetches under Gaines (and 2 pecks of them left)	0	5	0
9	Spent by Nathaniel that market day	0	0	6
10	Paide to Thom Fairchild for making an iron barr to the chancel window — the iron my owne, but only in many peices, a good part of it being old cart nayles, etc. He			
11	mended an iron candlestick into the bargaine, vide disbursements December 30 last	0	1	0
	Paide to Good: Soale for new ringing one of the tumbrel wheeles, at	0	9	0
13	Item for sharpening the nayles to it	0	0	6
	Item for 3 new spokes to it	0	1	6
	Item for heading a plow now	0	1	0
14	Paide to Will: Thrift for 5 journeyes at plow of his son David	0	1	8
	Item for threshing 5 quarters of pease at 1*s*. per quarter	0	5	0
16	Item for threshing 14 quarters and 2 bushels of oates at 8*d*. per quarter	0	9	6
17	Item for threshing 23 quarters and 2 bushels of the 8, 9, 10 and 11 dressings of barly, at 9*d*. per quarter	0	17	5
20	For dressing broomes, one paide to Mr. Taylour, one to Thom: Nash	0	1	0
	Spent by son Nathan at Royston	0	0	4
24	Paide to Good: Soale for two new felloes to the tumbrel wheele, the fellow to that which was new rung, vide 11 instant, at 1*s*. 6*d*. a peice, and a new spoke 6*d*.	0	3	6
27	Item for a new paire of shafts to the tumbrel, the old ones being both broken and rotten	0	5	0
28	Given to Mrs. Mary Lee, a minister's widdow, taken by the French going from Ireland to New England, between Ireland and England, her husband and son killed; 16 set on shore naked, most of them woomen	0	0	6
29	Given to Mr. Rich Cooch of Mettingham in Suffolk, an ancient family, going over to serve the Queen, taken by the French privater with 130*l*. in mony besides other things	0	1	0
31	To Marg: Cra: uxori for 26 pound of cheese 5*s*. 10*d*., six ston and 8 pounds of beife and mutton of Mr. Malin 17*s*., veale 4*s*. 6*d*., and all other occasions, toto	2	4	0
	Expended in this month in all the ful summ of	5	16	1

[p.65] **February 1707 Receipts**

		£	s	d
	Received of Will: Fordham for a couple of deale boards 11 inches, over	0	3	4
3	Received of Widdow Carrington, Rob: Macer, Jos: Wed, for 3 pecks of wheate	0	1	9
	Received of Eliz: Parish and Widdow Gurner for 2 pecks of the same	0	1	2
4	Received of Widdow Nash, Barber, Eliz: Parish, Widdow Haycock, for rye 2 bushels	0	3	0
6	Received of Mr. Urias Bowes of Barly in part for 20 quarters of the 11, 12, and 13 dressings of barly, the summ of	10	0	0
	Received of Good: Hankin for one peck of wheat, at 2s. 4d.	0	0	7
7	Received of Benj: Bright, Widdow Law, Jos: Wedd, for 6 pecks of rye	0	2	3
8	Received of Urias Bowes what remained due for 20 quarters of barly, vide the 4 instant	4	5	0
	Received of Wil: Brock and Rob: Macer for 2 bushels of rye, at	0	3	0
10	Item for a bushel of wheate, 2 pecks each, Brock, the 4 instant, at	0	2	4
12	Received for a bushel more of Rob: Harvy, Widdow Goode, Jos. Wedd	0	2	4
	Received of Rob: Harvy and Widdow Goode for a bushel of rye	0	1	6
	Received of Good: Gurner for 2 pecks	0	0	9
	Received of Urias Bowes for six quarters of the 15 dressing of barly at 14s. per quarter	4	4	0
13	Received for the calf of the brindle cow of Edw: Godfry at 5 weeks	0	11	0
	Received of Widdow Gooding, Eliz: Parish, Jos: Wedd, Will: Brock, for 2 bushels of miscellaine	0	3	0
17	Received of him, Widdow Law, Will: Brooks, for 6 pecks of wheate	0	3	6
	Received of Widdows Gurner and Goode for 2 pecks	0	1	2
18	Received of Widdow Haycock, Widdow Nash, Ben: Bright, Widdow Goode, for 2 bushells and one peck of rye or miscellaine	0	3	5
	Received of Jos: Wedd and Barber for 5 pecks	0	1	10
	Received of Rob: Macer and Jos: Wedd for each of them a peck of wheate	0	1	2
20	Received of the miller, Widdow Gooding, Mr. Greenhil, for 6 pecks of rye	0	2	3
22	Received of Widdow Nash, Jos: Wed, for 2 pecks 9d., and Wil: Brock 6 pecks, in al 3 bushels	0	4	6
	Item of him, Athan: Carrington, Macer, Jos. Wed, for 5 pecks of wheate	0	2	11

26	Received of Mrs. Westly for a peck		0	0	7
	Item of her and Widdow Gurner for rye 4 pecks		0	1	6
27	Item of her and Widdow Swan 2 pecks, and Mary Brock 3 pecks, in al 6 pecks of the 14 dressing of barly, at		0	2	9
	Received of Good: Harvy, Wed, Macer, for 2 bushels and half of rye		0	3	9
28	Received of Edw: Phip for 5 quarters of wheate at 11s. per loade		4	8	0
	Received this month in al the ful sum of		26	2	4

[p.66] February 1707 Disbursements

	Paide to Mr. Love for 40 bushels of coales at 8d. ob. per bushel	1	8	4
	Item for 2 new matts then, at	0	1	0
3	Item toll 2d., spent by George 4d., toto	0	0	6
	Paide for the 3rd quarterly payment of the land tax laide upon my parsonadge, at 4s. per pound	6	18	0
4	Item for Cassanders to that tax	0	9	0
	Item for Hoys or Edw: Parishes	0	2	6
	Spent by Nat at Royston Market	0	0	6
5	Nayles to nayle up the vines, etc.	0	0	2
	Paide to Rob: Hills the gardner for 2 dayes work, and a peice 6d.	0	3	6
7	Item for beanes 6d., other seeds 5d.	0	0	11
	Paide to young Rob: Harvey for his daye's work in helping him	0	0	10
8	Paide to Rich. Carter for stubb mony for half a ringe of wood — son Nathaniel went on purpose	0	1	0
10	Paide to Good: Bush for 12 of Mr. Lewis his 'Church Chatechismes Explained', which her husband and my son Ben paide for at London	0	5	0
12	Paide by son Nathaniel a groat, spent by George at John Kempton when he went for coales, and 3d. when they carryed the 15 dressing of barly to Urias Bowes	0	0	7
13	Spent by Nathan: at Royston	0	0	5
	To Ralph Carpenter for killing a couple of hoggs for us	0	0	8
	Paide for shooing the horse at Barly	0	0	5
17	Paide to Mr. Botterel for drinking Jolly coult for wormes	0	1	6
	Paide for a bushel of lintels for seede to show one stetch in Northfield	0	4	0
	For a new shovel with an iron tip	0	1	3
19	Spent by Nathaniel at Royston	0	0	6

HOUSEHOLD ACCOUNTS 1707

	Spent by Nat at Barly, carring in the 17 dressing of barly to Urias Bowes, meeting B: W:, at	0	0	3
20	Spent by Nathan at Ash Wednesday Faire 6*d*., and at Buntingford going with a loade of wheate 10*d*.	0	1	4
	Spent by him at Cambridge, going to buy some seede barly	0	0	6
24	Paide to Will Fordham junior for eleven journeyes at plow in barly seed, at 8*d*. per journey	0	7	4
26	To Marg: Cra: uxori for salt a bushel 6 shillings, 2 stone of beif of Mr. Malin 5*s*. 4*d*., and 2 shoulders of mutton of him			
28	about 3*s*., hemp 12 pound 6*s*., whitings 1*s*. 6*d*., cheese 17 pounds at Royston 4*s*. 4*d*. and 22 pounds of Cheshire cheese 5*s*., and al other occasions, in all	2	6	0
	Expended this month in all the summ of	12	16	0

[*p.67*]

March 1707 Receipts

1	Received of Urias Bowes of Barly for 7 quarters of the 16 dressing and 6 quarters and 3 bushels of the 17 dressing of barly, the first at 14*s*., the last att 14*s*. 3*d*., in al	9	8	9
	Received of Widdow Gooding, Ben: Bright, Jos: Wedd, Will: Brock, for 2 bushels and 3 pecks of rye	0	4	1
3	Item of him, Jos: Wedd, Rob: Macer, Widdows Gurner and Gooding, for 6 pecks of wheate, at 2*s*. 4*d*.	0	3	6
	Received of Edw: Godfry for the calf of the little red cow — the Welsh	0	13	0
4	Received of Thom: Eastland for a bushel of rye, February 13, at	0	1	6
	Item for 2 bushels of malt, at	0	4	8
5	Item for 5 bushels and a peck of wheate since December 9 last at 2*s*., 2*s*. 4*d*.	0	11	7
	Item for a young hogg, with a bushel of rye 1*s*. 6*d*., in al	0	12	0
6	Received of Rich: Haycock, miller, Will Brock, Macer, for 6 pecks of wheate at 2*s*. 4*d*.	0	3	6
7	Received of Widdow Nash, R: Macer, Thom: Barber, Widdow Gurner, Wil: Brock, for 4 bushels of rye	0	6	0
	Received of Jos: Wed and Rob: Macer for 6 pecks more, at	0	2	3
8	Item of them 2 for 3 pecks of wheate	0	1	9
	Received of Eliz: Bright, Widdow Nash, Jos: Wedd, for 6 pecks of rye	0	2	3
	Received of Ben: Wedd for 2 bottles of wheat straw carryed by Thomas East.	0	0	6
12	Received of Urias Bowes of Barly for 6 quarters and 2			

		£	s	d
	bushels of the 18 dressing of barly carryed in some time since at 14s. 6d.	4	10	6
	Received of Good: Watson for 2 pecks of it	0	0	11
13	Received of Rob: Macer for a peck of old wheate, by Rhoda	0	0	7
	Received of Widdow Gurner and Mr. Greenhil for 3 pecks of rye	0	1	2
14	Received of Rich: Haycock and his mother for 3 bushels and half of rye	0	5	3
	Received of Widdow Goode, Eliz: Bright, Edith Wedd, for 7 pecks more	0	2	7
	Item of the 2 last for 3 pecks of wheate	0	1	9
17	Received of Will: Brooks for a peice of sycamore wood to mend a wagon	0	0	6
	Received of Rich: Haycock for wheat, half a bushel to his wife	0	1	2
18	Received of Will Thrift junior for a bushel of miscellaine	0	1	6
	Received of Thom Barber, Mr. Greenhil, Rob: Harvey junior, for 2 bushels of the same	0	3	0
19	Received of him and Widdow Gurner for 2 pecks of wheate, one old and the other new wheat	0	1	2
20	Received of Rob: Macer, Widdow Swan, Jos: Wedd, for 5 pecks	0	2	11
	Received of Rob: Macer for a bushel of miscellaine	0	1	6
22	Received of Brother Wedd for a stick of elming timber	0	2	0
	Received of Will Thrift junior for a peck of rye by Elizab.	0	0	5
[p.69]	Received of Rich: Haycock for one bushel of rye or miscellaine	0	1	6
26	Received of Widdow Nash, Thom: Barber, Jos: Wedd, for 7 pecks more	0	2	7
27	Received of Robert Harvy, Widdow Haycock, Widdow Swan, for 9 bushels and half of barly at 14s. 6d. per quarter	0	17	3
	Received of Widdow Gurner for 2 pecks of rye or miscellaine	0	0	9
28	Received of Rob: Macer for wheat a peck	0	0	7
	Received of him for a peck of barly	0	0	6
	Received of Mary Brock for a peck	0	0	6
29	Received of Widdow Nash for half a bushel of wheate by Rhoda	0	1	2
	Received of Widdow Gooding for a peck	0	0	7
	Received of Widdow Goode for a peck	0	0	7
	Received of Ben: Bright for a peck	0	0	7
31	Item of him for 2 pecks of rye	0	0	9

Item received of Clement Norman for 2 pecks the 6 of February last	0	0	9
Received this month in al the full summ of	20	0	4
Received in the last six months, viz since Michaelmas last, the summ of	112	16	2

[p.68]

March 1707 Disbursements

3	Paide to one Mr. Ward at the smal bridges in Cambridge for six quarters and half of barly to sow, at 11s. 4d. per quarter	3	13	8
	Spent by men going for it, and for toll 2d., in al	0	1	0
4	Paide to Thom: Eastland for threshing 22 quarters and 4 bushels of the 12, 13 and 14 dressings of barly at 9d.	0	16	11
5	Item for threshing 5 quarters 6 bushels wanting above a peck of wheate	0	7	2
	Item for threshing 17 quarters one bushel and half of rye, viz on December 9, January 1 and 4, February 12 and March 3	1	1	5
	Item for 5 dayes work and an half on December 11, 12, 13, 16, January 31 and February 1, at 9d. per day, and so they come to	0	4	2
6	Item for 8 dayes and a good part of a day reckoned at 8d., in my closes and orchard, lopping, cutting down dead trees, sawing them out, bringing them in, fagotting, etc., viz			
7	on February 3, 4, 5, 8, 11, 13, 14, 15, 28, at	0	7	4
8	Item for ten journeyes in seed time — January 29 sowing oates, the rest barly, on February 20, 21, 22, 24, 26, 27, and 1 and 3 of this instant March, at 8d. a journey	0	6	8
	Spent by Nat at Royston Market	0	0	6
	Paide to James Swan for dressing a calve's skin with the haire on that was found dead in the morning	0	0	4
12	Spent by son Nat: at Royston Market, taking mony there	0	0	6
14	Paide to Goodman Soale for a daye's work of himself, and half a day of his son Will:, in sawing and hewing out mould boards and other plow timber, al the lime etc.	0	2	6
	Paide to Widdow Haycock for a loade of dung in February 1706 upon Nat's land	0	1	6
17	Item for a loade upon my owne land in Rowz, February 18 last, by George etc.	0	1	8
	Paide for sweeping kitchin and backhouse chimneyes to Rob: Macer	0	1	0
18	Paide to Robert Hills his nephew for strawburyes about the garden, vide April 24, 1706	0	1	0

	Given to an old man 83 yeares of age	0	0	1
19	Paide to Goodman Soale for a plow beame bought of him	0	1	6
	Item for making a new plow now, the fallow being very hard	0	1	6
20	Spent by Nathan: and George, carrying 5 quarters and half of wheat to Edw: Phipp at Buntingford this day	0	0	10
	Item spent by him at Royston last market	0	0	6
22	Paide to Rob: Harvey for 3 journeyes at plow this barly seede at 8d. per journey	0	2	0
	Item for one whole day then	0	0	10
[p.70] 26	Given to one Robert Willamite to pay his prison fees of St. Giles his parish, Cambridge	0	0	6
	Spent by Nat: at Royston taking mony	0	0	6
27	To Henry Wallis of Chrissol for fagotting up a cant of wood in the wood called Rofway 9d., and his help in loading 2d., in all	0	0	11
	For setting on a shooe at Chrissol	0	0	1
28	Paide to Thom: Holt for mending the greate tunnel, ratt trap, churne staff, etc.	0	0	6
29	To Marg: Cra: uxori for 2 calve's head and pluck about 2s. 8d., 2 quarters of mutton, one the 12 with 4 stone of beife, the other 19 instant 13s. 9d., and 3 stone and half of beife			
31	againe neare 31 instant 9s., 14 pounds and one quarter of cheese at 3d. per pound 3s. 6d. ob., a quarter of veale of Thom: Ward 1s. 8d., and al other things this month	2	10	0
	Expended this month in al the summ of	10	5	7
	Expended in the last six months, viz since Michaelmas last, the full summ of	67	1	2

[p.69]	**April 1707 Receipts**			
1	Received of Clement Norman for 2 bushels and a peck of malt now and January 30 last, at 2s. 4d.	0	5	3
2	Received of Edw: Phipp for 5 quarters and an half of wheate at 2s. 2d. per bushel	4	17	0
	Received of Widdow Gurner for a peck	0	0	7
4	Received of Widdow Law, Mr. Greenhil, Thom Barber, Widdow Goode, for rye 2 bushels	0	3	0
	Received for the calf of the great black Welsh cow of Mr. Malin	0	14	0
7	Received of him and others at home for my bull, besides the hyde	1	8	3
8	Received of Mrs. Westly for half a bushel of barly at 16s.	0	1	0

HOUSEHOLD ACCOUNTS 1707

	Item for her smal tithes this yeare	0	1	6
	Item for fetching a loade of clay	0	1	0
9	Received of Widdow Nash, Jos. Wedd, Widdow Law, for a bushel of wheate	0	2	4
	Received of Widdow Gurner, Jos. Wed, for a bushel of rye or miscellaine	0	1	6
14	Received of Rob: Macer for 4 pecks	0	1	6
	Item of him and Widdow Nash for a bushel of wheate at 7 groats	0	2	4
15	Received of Mary Brock, widdow, what remained due for 2 bushels of wheat, vide December 4 last	0	0	10
	Item for 7 bushels of wheat between November 7 and January 1 at 20*d.* and 2*s.*	0	12	0
17	Item for 6 pecks at 2*s.* 4*d.* since February 25 last	0	3	6
	Received of J: Bowles and Widdow Gooding for 3 pecks of the same	0	1	9
18	Received of Good: Gurner and Lyddy Barber for 6 pecks of rye	0	2	3
	Received of Good: Macer for barly a peck	0	0	6
19	Received of Mary Brock for fetching home her wood, three quarters of a ringe or 15 pole, from Cleave Park after her husband's death, with 2 carts and 2 good loades, 3 horses — it was worth indeed as much more	0	5	0
[*p.71*]	Received of Mr. Greenhil, Ben: Bright, Widdow Nash, Law, Goode, Rob: Harvey junior, for 8 pecks of rye	0	3	0
21	Received of Rob: Harvy junior, Widdow Gurner, Ben: Bright, Widdow Nash, Rob: Macer, Jos: Wedd, Widdow Law, Widdow Goode, for 3 bushels and 3 pecks of wheate at 7 groats	0	8	9
23	Received of James Swan, Eliz: Bright, Edw: Godfry, Good: Nash, for 7 pecks more of the same, at	0	4	1
	Received of Widdow Gurner for rye 2 pecks	0	0	9
	Received of Widdow Brock for 6 pecks	0	2	3
24	Item for 2 pecks of wheate then	0	1	2
	Item for a peck of barly	0	0	6
	Received of Mrs. Westly for 2 pecks	0	1	0
26	Item for a peck of wheate now	0	0	7
	Received of Widdow Haycock and Lyddy Barber for 6 pecks of rye	0	2	3
29	Received of Edith Wedd for 2 pecks whereof one tayle wheate	0	0	9
	Item for a peck of wheate, at	0	0	7
	Received of Widdow Gurner for a peck	0	0	7
	Received of Lydd: Barber for a peck of barly	0	0	6
30	Received of her also for a couple of little piggs that they had March 30	0	4	6
	Received of Widdow Carrington for half a bushel of barly	0	1	0

REV. JOHN CRAKANTHORP

		£	s	d
	Received of Mr. Greenhil for rye 2 pecks at 1s. 6d. per bushel	0	0	9
	Received this month in al	10	18	1

[p.70] **April 1707 Disbursements**

		£	s	d
1	Paide for the last quarter of the tax granted for the yeare 1706 at 4s. per pound, for Hoyes	0	2	6
	Item for Cassanders to that tax	0	9	0
2	Item for the same tax upon the parsonadge, by 4s. per pound	6	18	0
4	Paide to Rich: Fairchild for the half yeare's duty upon houses and windows due Lady last, by son Nathaniel, I being absent	0	5	0
	Spent by Nathan: at Royston Market	0	0	6
7	Paide for a bull bought by him at Royston the next market day	1	19	6
	For a letter by London, from R. Bridg	0	0	5
	Spent by Nathan: that market day	0	0	6
9	Paid to Widdow Brock for her husband's trimming myself in his life time, 2 quarters	0	6	0
	Item for trimming son Nathaniel	0	1	8
14	Item for the dust behind the porch door 1s. 6d., and 7 bushels of pigeon dung out of the steeple, upon my 3 stetches in Waterden Shott 2s. 4d., in al	0	3	10
16	Item for her boy Will, going with S: Sp: fetching the horse as far as Ikleton Grange or a little further	0	0	2
17	Item for clerk's wages due to him at Lady Day for the half yeare	0	3	4
	Paide for 3 wash balls at Cambridg	0	0	3
	For a coomb of coales at 8d. ob.	0	2	10
	For nayles to the pump, 3d. ones	0	0	2
19	Paide to Brother Sparh: for the interest of 20 pound due to him for a yeare ended at Christmas last, vide January 1, 1706	1	3	0
[p.72] 21	Paide to Rich: Jackson for his yearly rent, for keeping my pump in order, vide May 3, 1706	0	1	0
22	Paide to Good: Headly of Hinton for 8 bushels of lime, and brought home in his cart comeing to Royston — paide by Nathan: at Royston	0	4	0
	Spent by Nathan: at Royston Market	0	0	6
24	Paide to Rob: Harvey for 3 dayes underpinning the hog styes, backhouse, entry, kitchin, one day whereof was spent in paving the chancel, at 18d. per day	0	5	3
26	Item to him for another day, underpinning barne next			

	the street, dawbing backhouse and entry, ruff casting	0	1	6
	For a peck of haire for him	0	0	2
29	Paide to Rich: Haycock for a peice of leather to the bucket of the pump, by the hands of son Nathaniel	0	1	0
	For 5 hempen raines to the halters	0	0	10
	Spent by Nathaniel at Royston Market	0	0	2
	For a pound of nayles for carpenter mending doors and gates, etc.	0	0	4
30	To Marg: Cra: uxori for cheese 26 pound at 3d. ob., 7s., 2 quarters and a shoulder of mutton 9s. 4d., 6 ston 4 pound of beif 18s. 3d., a side of veale and an head 6s. 6d. and al other occasions in this month, in al	2	16	0
	Expended this month in al	15	7	5

[p.71] **May 1707 Receipts**

	Received of Widdow Nash for 2 pecks of rye	0	0	9
	Received of Eliz: Bright for 2 pecks of barly	0	1	0
2	Received of Ann Thrift for another peck	0	0	6
	Received of Good: Macer for 2 pecks	0	1	1
	Item for a peck of wheate	0	0	7
	Item for a bushel of rye at the same time	0	1	6
6	Received of Widdow Nash for 2 pecks of wheat at 2s. 4d.	0	1	4
	Received of Widdow Goode for a peck of barly 6d. ob., and a peck of rye 4d. ob., toto	0	0	11
7	Received of Mr. John Wells of Baldock for 6 quarters and 2 bushels of wheate at 11s. per loade, and so it comes to	5	10	0
	Received of Lyddy Barber and miller for 2 pecks of barly at 2s. 2d.	0	1	1
9	Received of Ben: Bright and Widdow Brock for a bushel of rye, at	0	1	6
	Item of her for 2 pecks of tayle wheat	0	0	9
	Item for a peck of the best, at	0	0	7
12	Received of G: Rayner for the tithe of his dovehouse in 1706, due at Michaelmas last	0	9	0
	Item for his smal tithes this yeare, not reckoning for dovehouse, but cowes and calves, etc.	0	4	6
13	Item for a yeare's quitt rents due at Lady last	0	1	4
	Received of Ann Thrift for a peck of barly — she paid an halfpenny short in the last, this therefore	0	0	7
[p.73]	Received of Rob: Harvy junior for a bushel of barley by Thom:	0	2	2
13[sic]	Received for a peck of Rhoda Carrington	0	0	6½

REV. JOHN CRAKANTHORP

	Received of Rob: Macer, Widdow Nash, Will Thrift junior, for 2 bushels and 3 pecks of rye, having 2 pecks of tayle wheate with the rye	0	4	1
14	Item of them three for 2 bushels and a peck of wheate	0	5	3
	Received of Eliz: Bright for 2 pecks	0	1	2
	Received of Widdow Gurner and Mr. Greenhil for 6 pecks of rye	0	2	3
16	Received of old Will: Thrift for a peck of barly his wife had	0	0	6
	Item for a bushel of malt April 3	0	2	4
	Item for a bushel of wheate	0	2	4
17	Item for fetching a loade of wood from Rofway worth 6s., I have	0	5	0
	Item for 4 bushels and half of rye since March 12 at several times	0	6	9
	Item for wintering 2 cowes 12 weeks until March 17 at 7d. per week each	0	14	0
19	Received of Widdow Goode and Mary Wallis for 2 pecks of wheate, at	0	1	2
	Received of Edith Wed for rye 2 pecks	0	0	9
	Received of Mary Watson for 2 pecks of barly by her daughter	0	1	1
20	Received of Clem: Norman for 2 pecks of wheat March 1	0	1	2
	Item for 4 pecks of rye February 21 and April 8, and a penny over	0	1	7
21	Received of John Watson junior, Ben Bright, Widdow Good, for 7 pecks	0	2	7
	Item of her for 8 pecks more February 28 and March 12 in *1705,* and April 2 and July 19 in 1705, at	0	3	0
	Item for a peck of barly February 20, *1705*	0	0	5
22	Item for 3 pecks of wheat March 7, *1705,* and April 2 and May 21, 1705, at	0	2	2
	Item for wintering her cow from October 30, 1705 to May 1706	0	10	7
	Received of Rob: Macer and Widdow Brock for a bushel of wheate, at	0	2	4
23	Item of her for a bushel of rye February 25 last — she had then 6 pecks	0	1	6
	Received of Widdow Gurner, Widdow Nash and Rob: Macer for 2 bushels and half	0	3	9
24	Received of Widdow Gurner, Widdow Nash, Edith Wedd, for 2 bushels of wheat	0	4	8
	Received of Edw: Godfry for a peck of barly	0	0	6½
	Received of Widdow Ward for a bushel in November 1704, at	0	1	7
	Item for a peck of pease in March 31	0	0	6

HOUSEHOLD ACCOUNTS 1707

26	Received of Thom: Barber, Cass: Woods, Widdow Ward, Good: Harvy, for 2 bushels and 3 pecks of rye	0	4	0
27	Received of Mary Wallis and Peter Nash for a bushel and half	0	2	3
	Received of young J: Watson, Good: Harvy, Mary Wallis, Peter Nash, Widdow Goode, for 3 bushels and 2 pecks of wheate	0	9	2
	Received of Will: Fordham, Widdow Brock, Rich: Haycock, miller, Steven Creak, for 2 bushels and one peck more	0	5	3
28	Received of Widdow Brock and Widdow Nash for 9 pecks of rye, at	0	3	5
	Received of Rob: Macer, Mrs. Westly, Wil: Jeeps, Wil: Thrift junior, Anne Thrift, for a bushel and half of barly	0	3	5
[p.75] 28[sic]	Received of Thom: Barber, Widdow Gooding, J: Bowles, for 5 pecks of wheate at 3 shillings per bushel	0	3	9
	Received of young Will: Thrift for a shoate, or young hogg, at	0	8	2
	Received of Widdow Law for 2 pecks of rye December 10 last	0	0	9
29	Item for a peck of wheate now	0	0	9
	Received of Athan: Carrington for his smal tithes in his father's time for 2 cowes and a calf, and for a cow and calf of his owne	0	2	4
30	Item for 2 bottles of wheate straw in his father's time, and a bottle of rye straw himself April 15 last	0	0	9
	Received of Widdow Gurner for 2 pecks of rye or miscellaine	0	1	0
	Item for a peck of wheate also	0	0	9
31	Item of her for a bushel of barly on March 28 last at 15s. 8d.	0	1	11
	Received of Mrs. Wedd and Rob: Macer for 5 pecks of barly	0	2	9
	Received this month in all the full sum of	13	13	0

[p.72]
May 1707 Disbursements

	Given to a Scotch man, blind of both his eyes by the smal pox	0	0	2
2	Paide to Mr. Wilson for tenths due at Christmas last, and 4d. an acquittance	2	19	9
6	Paide to Mr. Love for 50 bushels of coales May 10, 1705, at 10d. per bushel, and 6d. a matt then, in al	2	2	2
	Item for 50 bushels more June 25 that yeare at 9d. per bushel	1	17	6

7	Item for 60 bushels of coales June 15, 1706, at 10*d.* per bushel, 2*l.* 5*s.* 6*d.*, and 20 bushels November last at 8*d. ob.*, abateing about 14*d.*, in al	2	18	0
	For a box of Lockier's pills	0	2	0
8	For a peice of a skin to mend my black leather breeches withal	0	1	2
	Spent at Cambridge that day and to barber	0	0	11
9	Spent by men, Nat: and George, for unloading wheate 5*d.*, at Mr. Wels and spent besides then 7*d.*, setting horses, etc.	0	1	0
	For 6 quarters of oates bought at Cambridge of Mr. Howel at 9*s.* 9*d.* per quarter	2	16	0
10	Spent by men then, and toll 2*d.*	0	0	10
	Paide to John Rayner for an overseers' rate at 2*d.* per pound for what I hold of my owne	0	0	9
	Item to the same rate for parsonadge	1	2	10
12	Paide for the yeare's quitt rents for Cassanders 13*s.* 2*d.*, Hoyes 3*s.* 8*d.*, vide April 20, 1706	0	16	10
[*p.74*]	Spent by son Nathaniel at Royston Market, selling, the 7 instant	0	0	6
13	Paide to old Will: Thrift for threshing 20 quarters and 2 bushels of the 15, 16, 17th dressings of barly at 9*d.* per quarter as the rest is	0	15	2
14	Item for threshing 16 quarters more of the 18 and 19 dressings, and the last, March 10 and 28	0	12	0
	Item for a journey at plow in barly seed	0	0	8
	Item for 2 dayes in February 25 and 26 at plow and hedging in the afternoon	0	1	8
16	Item for 8 dayes in March at plow and spreading dung, lopping and hedging in the further close	0	6	8
17	Item for 22 dayes in April at wood cart, hedging against Linch Lane, against kitchin garden, and the round about, fagotting, laying up the wood, casting up the dung	0	18	4
19	Item for 11 dayes in this month, of Wednesday 14 was the last	0	9	2
	Spent by Nat: at Royston that day	0	0	2
	Paide to Clem: Norman for trimming me 2 times 6*d.*	0	0	6
20	Paide to Widdow Goode for 3 loades of dung in February *1705* — there was 8 in al, but Mrs. Nor: was paide for five, December 7, 1705 — upon son Nat's land, at	0	3	0
	Item for 5 bushels of shreds or raggs at that time, at 4*d.* per bushel	0	1	8
21	Item for 3 bushels of hen dung about March 4, *1706*, at 6*d.* per bushel	0	1	6
	Item for 3 loades of dung February 15, *1707*, upon my stetches at Hicke's hedge	0	3	0
	Given to Andr: Edwards bringing me a book for the 6th			

22	triennial visitation of Simon Lord Bishop	0	0	6
	Paide to Henry Wallis of Chrissol for 300 spitts at 2*d. ob.*, abating an *ob.*	0	0	7
23	Paide to Goodman Thomas Watson for 11 dayes and half betwixt January 4 and 19 at nine per diem or 4*s.* 6*d.* per week	0	8	8
	Item for 12 dayes more betwixt January 19 and February 2 at that rate	0	9	0
24	Item for 24 dayes between February 2 and March 2, at 10*d.* per day	1	0	0
	Item for 11 dayes between March 2 and the 16 at 10*d.* per day	0	9	2
25	Given to a breife for the inhabitants of Spilsby in Lincolnshire, viz Francis Humstance, Robert Gregory, Solomon Hutton, John Whittaker, Edw: Barker, Jane Taylour, widdow, etc., loosing by fire (happening on August 14, 1706) an 100 houses besides many goods,			
26	apparel, etc., to the value of 5,984*l.* etc.	0	1	6
	Paide to Thomas Watson for 11 dayes work betwixt March 16 and 29 at 10*d.* per diem	0	9	2
27	Item for 4 dayes and half between March 30 and April 6, at	0	3	9
	Item for 4 dayes and half between April 6 and the 13 of the same	0	3	9
[*p.* 76] 28	Item for 5 dayes between April 13 and 20 — Thursday that week he was marryed — and 5 days more before April 27 — Munday at Cambridg	0	8	4
	Item for 9 dayes between April 27 and the 11 of this instant	0	7	6
29	Item for 11 dayes and half more to the 25th day of this instant May	0	9	7
	Given to Captaine Thomas Wentworth with his son Charles, taken by a French privateer, with the loss of 1,800[*l*]. cargo	0	0	6
30	For bottoming the barne sieve, and the handle of the basket 2*d.*, and bottoming the chaff sieve in stable 6*d.*, toto	0	1	6
31	To Margar: Cra: uxori for 3 stone and 3 pound of beife and veale of Malin 9*s.* 4*d.*, 3 quarters of veale and lamb besides 4*s.* 11*d.*, in al 14*s.* 3*d.* for meate, and for a new bed tike 15 shillings, and for al other things, as pease, vinegar, sugar plums, etc., in all	2	8	0
	Expended this month in all the ful and entire sum of	25	15	5

REV. JOHN CRAKANTHORP

[p.75]   **June 1707 Receipts**

	Received of Goodman Thomas Watson for a bundle of rye straw	0	0	4
2	Item for 7 old peices of board he had of me, deale and elme	0	1	0
	Item for 2 pecks of malt, last April	0	1	2
3	Item for fetching a loade of turf from Cambridge, with 2 horses and George	0	2	6
	Item for fetching a cant of wood from Rofeway with 2 carts etc.	0	5	0
	Item for 4 bushels of barly at several times, at least 5, at	0	7	5
4	Item for 4 bushels and half of wheat	0	10	10
	Item for 7 bushels and half of rye, at	0	11	3
	Item for 2 shoats or young hoggs, one January 23 at 12*s.* one May 9 at 8*s.*	1	0	0
6	Received of Good: Hankin in January last, for a peck of wheate he had then	0	0	7
	Received of Eliz: Bright and Mary Brock for 3 pecks of rye at 2*s.*	0	1	6
	Received of one Clerk the horse courser, for the Watt horse, bought in June 1704	11	4	0
7	Received of Rhoda Car:, Good: Harvey, for 3 pecks of barly and a peck of Rob: Macer also, at	0	2	4
	Received of Good: Hankin for a peck of wheat, and 2 of Rob: Macer	0	2	2
9	Received of J: Watson junior, Ralph Carpenter, for 4 pecks of rye at 2*s.*, of Widdow Brock, Thom: Barber, Widdow Gurner, Rob: Macer, for 11 pecks more, at 1*s.* 8*d.* per bushel, in al	0	6	7
	Received of Ralph Carpenter and Eliz: Bright for a bushel more	0	1	8
10	Received for a peck now and 6 bushels May 20 last, at 1*s.* 6*d.*, of Edith Wedd	0	1	11
	Item for a jagg of rye straw, laide in January 10, *1705*, a bottle or two before, in al	0	4	6
	Item for 3 bundles of wheat straw in May 1705, at	0	0	9
11	Received of R: Carpenter, Widdow Nash, Rich: Haycock, for wheat 4 pecks	0	2	8
	Received of Widdow Goode, Edw: Godfrey, young J: Watson, for 5 pecks of barly	0	2	11
	Received of Widdow Jeeps, Rhoda Carrington, for 3 pecks more, and Lyd: Barber	0	1	9
[p.77]	Received of young J. Watson and Lyd: Barber for 7 pecks of rye, at	0	2	11
12	Received of Wil: Brooks, Rob: Macer, Rich. Haycock, for 4 pecks of wheate, at	0	2	4

HOUSEHOLD ACCOUNTS 1707

	Received of 4 pecks of James Swann, May 21 and now	0	2	4
13	Received of Edith Wedd and young J: Watson for 3 pecks of the same	0	1	9
	Item of her and Rob: Macer for 4 pecks rye	0	1	8
	Received of Widdow Brock, Widdow Ward, young J: Watson, Widdow Law, for 2 bushels and 3 pecks, whereof one peck was tayle wheate	0	4	7
14	Received of Widdow Gooding, Will: Brooks, Rich: Haycock, for 4 pecks of wheate	0	2	4
	Received of Mrs. Greenhil, Lyd: Barber, Goody Harvey, for 3 pecks of barley	0	1	9
16	Received of Joane Woods for an acre of haume, laid in October 2, 1705, and a penny due for 3 pecks of rye (besides 2d. inserted January 13 last for wheat)	0	3	7
	Received of Widdow Haycock and Will Thrift junior for 2 pecks of wheate	0	1	2
17	Received of him, R: Macer, Lyd: Barber, for 2 bushels of rye at 1s. 8d.	0	3	4
	Received of Eliz: Bright and Widdow Brock for 6 pecks of rye, with 3 of tayle wheat	0	2	6
	Received of Jos: Wedd for 3 pecks of rye	0	1	3
18	Received of Good: Harvy and Jeeps for half a bushel of barly	0	1	2
	Received of Rob: Macer for a peck	0	0	7
	Received of Mr. Greenhil and Widdow Brock for 5 pecks of rye at 1s. 8d.	0	2	1
19	Received of Widdow Nash and Rich: Haycock for 3 pecks of wheate, at	0	1	9
	Received of Wil: Brooks, Widdow Gurner, Rob: Macer, for 3 pecks more	0	1	9
20	Received of Sarah Fairechild for half a bushel of rye, at	0	0	10
	Received of Lyddy Barber for wheate a peck	0	0	7
	Received of James Bush for 2 bushels	0	4	8
23	Received of Widdow Brock for a peck of rye; received of Widdow Nash for 2 pecks 10d.	0	1	3
	Received of Robert Macer for 2 pecks	0	0	10
	Received of young Will: Thrift for a peck of barly by self, at	0	0	7
	Received of Widdow Jeeps for a peck	0	0	7
24	Received of Edward Godfry for a bushel of rye July 10, 1706, at	0	1	4
	Item what remained due for a bushel of wheate the June before, see receipts on June 17, 1706	0	2	0
26	Item in part for a bushel of wheate now at 7 groats — debt book 5th July	0	0	2
	Received of Mr. Greenhil for half a bushel of wheate yesterday	0	1	2

213

	Received of Good: Harvy for barly a peck	0	0	7
27	Received of Lyd: Barber for rye 3 pecks	0	1	3
	Received of Widdow Goodin for wheate a peck; received of young W: Thrift for a peck	0	1	2
	Received of Jon: Peck for a bushel	0	2	8
28	Received of Eliz: Bright for 2 pecks	0	1	4
	Item for a peck of rye 5*d*.; received of Widdow Law for a peck 5*d*.	0	0	10
	Received of Will: Dovy for barly a peck	0	0	7
30	Item for a bottle of wheate straw	0	0	2
	Received of Rob: Macer for wheate a peck	0	0	8
	Received this month in al the full summ of	18	14	11

[*p.76*]   **June 1707 Disbursements**

	Paide for 800 spitts or sprindls of Isaac King of Chrissol at 2*d. ob.*	0	1	8
2	Spent by Nathan: having 2 horses at Whitson Fair (and selling one, that is Watt, or Conder, that cost at the same faire, 1704, 8*l*. 4*s*. 6*d*.)	0	1	0
4	Paide to Marmaduke Feaks comeing over to the little black cow being infected with the dry garget and staling blood, for his drink and... [*a word illegible*]	0	1	0
6	Paide to Wil: Numan of Barrington for cloath and buckles a paire of bitts, and work 4*s*., some time since, and a job since that 4*d*., in al 2*s*. 7*d*., and for half a horse hyde now 4*s*. 6*d*., a paire of plow traise and al things to them 3 shillings,			
7	cloath and work now 5*d*. and 3*d*., in all I paide him	0	10	9
	For a pound of 6 and 8 penny nayles to mend the tumbrels	0	0	5
8	Given to a breife for a loss by fire, happening on the 25 of July 1706 upon John Rayner, Antony Tyrrel, and 27 other inhabitants of the town of Littleport in the Isle of Ely, to the value of 3,931*l*. 18*s*. 0*d*., etc.	0	1	6
9	For 3*d*. nayles used about tubs and payles, several of them	0	0	1
	Paide to Edith Wedd by discounte for 2 loades of dung in September 1705	0	4	0
	Given to one Johnson, a seaman from Sunderland in Durham, disbanded	0	0	1
	For half an ounce of black silk	0	0	11
10	For black galoone to edge my wastcoat, being fretted out	0	0	7
	Spent by Nathan: going to buy a colt at St. Neots Faire	0	1	0
	Spent by him againe at Royston	0	0	2

11	For another pound of 6 penny and 8 penny nayles towards the mending of the great gates next the street	0	0	4
	To James Bush for 5 pound of cherryes brought from London	0	0	10
[*p.78*]	Given to a breife for repairing of Brosely Church in Shropshire, the charges computed at 1,390*l.*, etc.	0	0	8
12	Paide to Joane Woods by the hands of her daughter Cassandra for a loade of dung in 1705, at	0	1	6
13	Item for another half loade in 1706	0	0	9
	Item for a loade laide upon my 4 acres on Fawdon Hil, on the left hand of Melbourne way, this 13th instant	0	1	6
14	Paide to Widdow Brock by her son Will:, who came for a bushel of rye, for a loade of dung the 13 of this instant upon my 4 acres aforesaide, at	0	1	8
16	Paide for a colt of John Browning out of Thorney Fen, which they cal Watt againe, instead of him sold	8	15	0
	Spent for toll for coales and horse 3*d.*, and otherwise at Cambridge	0	1	0
17	Paide to John Waites for the quarter's wages ended this Midsummer for 4 cowes as heardman at 8*d.*	0	2	8
18	For cherryes 3 pounds, at door	0	0	6
	Paide to Edw: Godfry for fleaing the little black Welsh cow that dyed	0	0	6
19	Item for 3 loades of dung the 13 of this instant upon my 4 acres on Fawdon Hil on the left hand (with 18 pence due to son Nat for a bushel of rye)	0	4	6
	Given to a poore man in want at door	0	0	1
	Item given to two other taken prisoners by the French — seamen	0	0	2
20	Paide for children since June last 1706: for son Benjamin for half a yeare's interest of 50*l.* which he oweth, for the			
23	half yeare betwixt Midsummer and Christmas last at 6*l.* per cent — for the rest of my children I have paide nothing this yeare	1	10	0
24	Given to a maimed souldier shott at the taking of Ostend last campaign	0	0	1
	For a paire of plow traise of Wil Numan, hemp and al, ready laide in and fitt for work, at	0	3	0
26	— paide at first by Nathan and allowed to him afterwards			
27	For stuff for swimming, diziness, giddiness, in my head, of Mr. John Moor, apothecary at the Pestle and Mortar in Abbchurch Lane, London	0	5	0
	For a bottle of orange flower water to take the former dropps in	0	1	6
29	Given to a breife for above 11 inhabitants of North Marston in Bucks., loosing by fire August 11, 1705, in the forenoon 3,465 pounds, the sum of	0	1	0

REV. JOHN CRAKANTHORP

30	To Marg: Cra: uxori for 9 stone and 12 pounds of beife about 1*l*. 10*s*. 0*d*., and cheese 5*s*., course linnen for table cloths and towels about 6*s*. 6*d*., 2 bushels of salt 10*s*. 8*d*., flocks for George's bed 2*s*. 4*d*., and al things else, in all	4	2	0
	Expended this month in al the full summ of	16	17	5

[*p.*79]   **July 1707 Receipts**

	Received of Widdow Gooding and Widdow Reynolds for 3 pecks of wheate	0	2	0
	Received of Edw: Parish for a peck	0	0	8
2	Received of Wil: Thrift junior for a peck May 13, 1706, at	0	0	7
	Item for a peck of barly about that time	0	0	6
	Item for 6 pecks of rye that May	0	2	6
7	Item for 2 bushels of malt July 5, 1706	0	5	2
	Received of Widdow Swan, John Watson, Widdow Jeeps, for 3 pecks of barly	0	1	9
	Received of Widdow Gurner for rye a peck	0	0	5
8	Received of Mrs. Norridge for fetching three loades of clay and one in	0	3	0
	Item for 2 deale boards 11 foote long	0	3	4
	Item for 2 yeares quitt rents due and ended at Lady last, the sum of	0	10	0
9	Received of Edw: Phipp for 15 bushels of the last dressing of wheate, at	1	17	6
	Received of Widdow Brock for rye 4 pecks	0	1	8
10	Received of Cass: Woods in part for 3 pecks of rye January 4, 1705 or 6, and April 25, 1706, at 1*s*. 3*d*.	0	0	2
12	Received of Kate Reynolds for a peck 5*d*.; received of Widdow Nash for 2 pecks 10*d*.; received of Jos: Wed for a peck 5*d*.; received of Widdow Brock for 3 pecks 1*s*. 3*d*.	0	2	11
	Received of Good: Harvy for barly a peck	0	0	7
14	Received of Clem. Norman in part for a bushel of wheate April 29 and 18	0	2	2
	Received of Good: Thomas Eastland... [*omission*], which he had in April last	0	0	4
	Item for 2 bottles of wheate straw, 7 instant	0	0	6
16	Item for 4 score white bricks on March 24 last, at	0	1	10
	Item for fetching him a cant of wood from Clavering in Essex	0	5	0
	Item for 3 bushels of rye	0	4	8
18	Item for 2 bushels of malt, at	0	5	4
	Item for 7 bushels and half of wheate since March 8 last at ten times	0	18	0

HOUSEHOLD ACCOUNTS 1707

	Item for a bushel of rye the day we reckoned — not set into the debt book	0	1	8
21	Received of Widdow Jeepes for a peck of barly	0	0	7
	Received of Rhoda for a peck	0	0	7
	Received of Widdow Strett for 2 pecks	0	1	2
23	Received of Good: Hankin for a peck of the best wheate at 2s. 8d.	0	0	8
	Received of Widdow Haycock for half a peck of barly at 7d.	0	0	3
	Received of Widdow Strett for half a bushel of barly more by self, at	0	1	2
24	Received of Mr. Greenhil for rye a peck	0	0	5
	Received of Widdow Gurner for a peck	0	0	5
	Received of Mary Thrift for a peck	0	0	5
	Received of Widdow Gurner for a peck	0	0	5
	Received of Elizab: Bright for a peck	0	0	5
	Received of Widdow Gurner for a peck	0	0	5
	Received of Mr. Greenhil for a peck	0	0	5
31	Received of Rich: Haycock his wife for a peck of the old wheate at Edw: Par.	0	0	9
	Received of Eliza: Bright for a peck of rye or miscellaine	0	0	5
	Received of Mr. Greenhil for a peck	0	0	5
	Received this month in all the summ of	6	1	2

[p.80] **July 1707 Disbursements**

2	Paide to Good: Will: Thrift junior for 8 loades of dung carryed away now, whereof 4 upon my 2 furlong stetch next the baulk, and 3 more upon Nat's, etc.	0	12	0
7	Spent at Royston the 2nd and at Buntingford this day, carrying 15 bushels of my wheate to make up his, that is son Nat's 5 quarters	0	0	6
8	Paide to Will: Numan for some leather that he brought to inlarge the last set of halters that were made to little	0	1	0
	Item for buckle 2d., and work on Becket Faire Day and now 7d., toto	0	0	9
9	Paide to Clem: Norman for mending my summer wastcoate (see uxor's account)	0	0	3
	Item for trimming me 24 of May last	0	0	2
10	For 8 paire of gloves, whereof George's cost a 1s., and 6d. allowed to Thom: Watson, who bought himself a paire against harvest at 6d. cash	0	5	6
	Spent by Nathan: going to by them and other things the same time	0	0	6
13	Given to a breife for 15 inhabitants of Towcester in the			

14	county of Northampton, looseing by fire (happening on August 18, 1705) 1,057*l.*, without the loss of such as desired no help by this breife, given I say	0	1	0
16	Paide to Goodman Thomas Eastland for threshing 25 quarters and 3 pecks of wheate at 1*s.* 3*d.*, viz on March 11, April 1 and 30, June 10 and 23	1	11	5
19	Item for threshing 17 quarters and 2 pecks of rye on March 15, April 14, and May 31 8 quarters and 2 bushels	1	1	3
	Item for 7 dayes and half in March and April, one fagotting, 3 at wood cart, one gathering stones, 2 and half serving mason, underpinning the hog styes	0	6	3
21	Item for a journey at dung cart June 10	0	0	8
22	Item for 13 dayes between June 10 and 29 in filling, spreading dung, mowing, making, getting in, the hay	0	10	10
	Item for 11 dayes and half between June 29 and July 13, filling and spreading dung, hay, and bringing in some small matter of tithe corne July 12	0	9	7
24	Item for 3 jaggs of dung carried out June 14 last upon my 4 acres on Fawdon Hil	0	4	6
29 30 31	To Margar: Cra: uxori for lamb, veale and 3 quarters of mutton about 14*s.*, for 5 stone and 8 pounds of cheese at 2*s.* 9*d.* per stone 15*s.* 7*d.*, 24 pounds of 4*d.* sugar 8*s.* Item for 4 dozen of raysins solis, some at 3*s.* 6*d.*, some at 4*s.*, some at 4*s.* 6*d.* per dozen, 16*s.* 9*d.* Item for 12 pounds of currants 6*s.* 6*d.* Item for a pound of tobacco 1*s.* 6*d.* Item 8 ounces carry: seeds 4*d.*, a quart of pepper 5*d.*, half a pound ginger 6*d.*, ryse 2 pound 8*d.*, and this lasted al harvest, that is a month, in al	3	10	0
	Expended this month in al the full summ of	8	16	2

[*p.81*]

August 1707 Receipts

	Received of old Will: Thrift for a bushel of barly June 7 and 10, July 7	0	2	4
11	Item for a bushel of malt July 8	0	3	0
	Item for 3 pecks of wheate at 7*d.* — 1*s.* 9*d.*; item for 3 pecks more at 8*d.* — 2*s.*; item for one peck May 28 at 9*d.*	0	4	6
12	Item for 4 bushels and half of rye at 6 several times between May 24 and August 4 at 20 pence	0	7	6
	Received of Mrs. Westly for a bushel of wheate May 31, June 11, July 21	0	2	8
13	Item for a bushel and half of barly June 9 and 13 last and July 4 at 2*s.* 4*d.*	0	3	6
	Received of Eliz: Bright for rye a peck	0	0	6
14	Received of Widdow Nash for a bushel	0	2	0

	Received young J: Watson for 2 pecks	0	1	0
	Received of Mr. Greenhil for a peck	0	0	6
16	Received of Rob: Macer for 2 pecks	0	1	0
	Received of Widdow Swann for 2 pecks of rye she had August 12 last, at	0	0	10
	Item for her half yeare's rent ended at Lady Day last, viz 1707	1	11	6
18	Received for a peck of rye of Eliz: Bright	0	0	6
	Received of Rob. Macer for a peck of the old wheat of 1705, at	0	0	9
19	Received of Mr. Wedd for the tithe of Mr. Latimer's wooll	1	11	0
	Item for the tithe of Mr. Taylour's wool, being three todd at 9s.	1	7	0
20	Item for the haume of the five acres in the Rowze and carring in that and his owne with 2 carts, 4 men	1	1	4
21	Received of Rob: Macer what remained due for 2 pecks of wheat ever since July 1703, vide July 24, 1703, receipts	0	1	6
23	Item what remained due for wintering his heffer in 1703, vide August 9 and 10 in 1704, receipts	0	3	6
	Item for 3 pecks of rye November 28, 1704	0	1	6
	Item for a bushel of wheat in July and August 3, 1705, at 2 times	0	3	0
26	Item in part for 6 pecks of wheate in November, and January 12, *1706*	0	1	6
	Received of Rich: Haycock's wife for half a bushel of rye, at	0	1	0
27	Received of Widdow Gurner for wintering her cow in *1707*	0	6	2
	Item for the use of my churchyard the last winter for her sheepe	0	2	6
28	Received of Wil: Jeepes for half a bushel of the new wheat at 2s. 8d.	0	1	4
	Received for a peck of Rob: Macer	0	0	8
29	Received of Will: Brooks for half a bushel of the old wheat at 3s.	0	1	6
	Received of Rob. Macer for a peck of rye or miscellaine	0	0	6
30	Received of Will: Jeeps for 2 pecks	0	1	0
	Received of Widdow Gurner for a peck, new rye	0	0	5
	Received this month in al the summ of	8	7	6

[*p.82*]   **August 1707 Disbursements**

2	Paide for nayles — 8 penny nayles — to mend the street gates, broken this harvest by the horses running against them	0	0	3

11	Given to one Mr. John Jonson, burned this Spring by the French, at St. Christopher's in West Indyes	0	0	3	
12	Paide to Thomas Davyes of Great Hormeade, one of my tithe men	1	10	0	
	Paide to his son David, another of my tithe men, in room of Ingry	1	10	0	
14	Paide to Newlin Moule of Elmden, another, in the room of Thom: Wallis	1	10	0	
	Given for largeses in the whole towne this yeare, as the yeare before	0	13	6	
16	Paide to Ben son of Jos: Wedd for his harvest this yeare, comeing ten dayes after we had begun, etc.	0	11	0	
	Paide to old Will: Thrift for fourteene dayes in May, dawbing, casting up dung	0	11	8	
19	Item for 19 days in June, carrying out, spreading dung, mowing, making, inning, hay, etc.	0	15	10	
	Item for 9 days more in July before harvest that began on the 14th	0	7	6	
20	Item for his wages this harvest	1	8	0	
	Item for 3 days since harvest	0	2	6	
	Spent at market by Nathaniel	0	0	6	
22	Given to Mrs. Blandina Cutts, widdow of a vicar belonging to the choire at Lincoln	0	0	3	
	For Roman vitriol at Royston	0	0	6	
	Given to poor at the door in want	0	0	1	
24	Given to a breife for several inhabitants in Shire Lane in Middlesex, loosing (by a fire hapning on March 16, *1706*) 3,505*l*., etc. — but 6 of the inhabitants named but attested by 30 of the Justices of the Peace	0	1	0	
25	Given to an old souldier etc. in want	0	0	1	
	Paide to Widdow Swan for the use of her horse this harvest	0	9	6	
26	Item for 3 loades of dung carryed away July 4 last at 1*s*. 6*d*. each of them	0	4	6	
27	Item for 2 loades more, one bigger than the other, and a jagg from the limeyard 2*s*., al five upon Nat's land	0	4	0	
	Paide to Rob: Harvey for 2 journys filling dung cart of Wil Ward's and partly at home	0	1	3	
28	For a pound of 8 penny nayles for the making on Good: Swann's door to her barne, by Good: Fordham	0	0	4	
29	Paide to Widdow Gurner for a loade of dung the 22 instant, on the 2 or 3rd stetch of gleabe in Short Furlong on this side the acre, at	0	1	8	
30	To Margaret Cra: uxori for stuff for a gowne and petticoat				

	bought at London by son W: 1*l*. 7*s*. 0*d*., and for al other occasions in this month 1*l*. 16*s*. 8*d*., in al	3	3	8
	Expended this month in all the summ of	13	7	10

[*p.83*]   **September 1707 Receipts**

3	Received of Will: Fordham senior for wintering 2 cowes for him 7 weekes a peice at 7*d*. per week	0	8	2
	Item for a very good bundle of wheat straw April 15 last, by Eastland	0	0	5
4	Item for a bushel of barly April 6 and June 6 and 23 at 1*s*. and 1*s*. 2*d*.	0	2	2
	Item for smal tithes for 1704 and 705, 706 and this yeare — cowes, calves	0	6	0
6	Item for 2 bushels and half of wheat at 6 several times since March 26 last at several prizes also	0	6	2
8	Item for ten bushels of rye at ten several times betwixt March 25 last and August 6 last at 1*s*. 6*d*. and 1*s*. 8*d*. per bushel	0	15	8
9	Received of Thom: Barber for 2 bundles of rye straw	0	0	4
	Item for a peck of barly in May last	0	0	6
	Item for his smal tithes this yeare — cow, one calf 6*d*., peares 1*s*. 6*d*.	0	2	10
11	Item for wintering his bullock 5 weeks last winter at 7*d*. per week	0	2	11
	Item for 2 stetches of haume on this side the 5 acres in the Rowze	0	4	6
12	Item for 6 bundles of rye at several times since 23 of September 1706	0	9	6
	Item for 4 bushels of wheat at 1*s*. 8*d*. and one peck at 8*d*. in the same time	0	6	11
	Received of Thomas Watson for 2 pecks of malt at 3*s*.	0	1	6
16	Received of him for a bushel of barly	0	2	4
	Item for 5 bushels of rye since 26 of May last at 1*s*. 8*d*. and 6 pecks at 2*s*.	0	8	10
17	Item for a half acre and stetch of haume at the top, and a little stetch on this side, 3 half acres, carriage 1*s*., in al	0	3	0
	Item for 4 bushels of wheate at 7 grots and 8 groats, and 2 bushels at 3*s*.	0	11	2
18	Received of Mr. Taylour for his smal tithes in 1706 — cowes, calves, lambs 95, etc. — the sum of	2	0	6
	Item for his smal tithes this present yeare and now al due in lik manner	2	1	0
19	Item for an hundred and half of willow sets that he had last winter	1	6	0

	Received of Thomas Eastland for half a bushel of new rye, at	0	0	10
20	Item for a peck of wheate 8*d*., and 2 bushels and 3 pecks of the old wheate besides at 3*s*. per bushel 8*s*. 3*d*., in al	0	8	11
	Item for 2 bushels of malt at 3*s*.	0	6	0
	Received of Widdow Gooding for 2 bushels of malt she had July 10, 706	0	5	2
22	Received of Susan Hankin for a peck of the new wheate at 2*s*. 8*d*.	0	0	8
	Received of Rob: Macer for a peck	0	0	8
	Received of Mrs. Norridge for half a bushel of the old wheate	0	1	6
	Received of one Nottingham for the tithe apples at Thom Wootton's	0	1	3
[*p.85*]	Received of William Brooks for 2 acres and 2 stetches of haume in Brook Shott in 1706	0	6	6
22[*sic*]	Item for 3 acres and half in Waterden and Horsted Shott since last harvest	0	8	0
	Item for little Typtery horse being growne old and lame	0	10	0
24	Received of Robert Macer for a peck of wheat this yeare, it being of the tithe of Mr. Latimer's shoud bread peice, and black	0	0	8
26	Received of young William Thrift for a peck of the old wheate that grew in the year 1705, at	0	0	9
	Received of Mrs. Norridge, now newly come to towne this last Sturbridge Faire, for 2 pecks of the same old wheate, at	0	1	6
29	Received of Robert Macer againe for a peck of the old wheate	0	0	9
30	Received of Mary Waites for half a bushel of the new wheate, being found now to be very black and soft in the spending of it, indeed not better than rye if it may be judged so good	0	1	0
	Received this month in all the summ of	12	14	7
	Received in all in the last six months, viz since March last, the full summ of	70	9	3

[*p.84*] September 1707 Disbursements

	Paide to Will: Fordham the elder for mending the stable, new barne doors, garden paths, October last	0	1	6
3	Item for 2 dayes and almost half in November and January last, mending cow racks, waggon, gates, etc.	0	3	0

4	Item for half a day mending new barne doors, street gates, etc.		0	0	9
	Item half a day making a curb for John Watson's well, April 29		0	0	9
6	Item for a day and half in June last mending the tumbrels, tressels, rakes, carts, cart ladders, posts to the manger in great stables		0	2	4
8	Item for his wages this last harvest		1	12	0
	Item for 8 dayes now at haume cart, spreading dung, etc.		0	6	8
9	Paide to Goodman Thomas Barber for his wages this last harvest		1	10	0
10	Given to a breife for Joseph Wakelin of Hartley Green in Staffordshire, loosing by fire September 26, 1705, al his goods to the value of 612*l*., etc.		0	0	6
	Paide to Goodman John Waites for his wages this last harvest, at		1	10	0
11	Paide to Rob: Macer for sweeping the kitchin and backhouse chimnys 4*d*. and 8*d*.		0	1	0
	Item for 2 loades of dung March 9, *1706*, upon my 2 acres, part of the 12 abutting south on Michel's path (see August receipts)		0	3	4
12	Item for the use of his chamber for a yeare, due the 16 of this month, to lay malt in		0	5	0
	Paide to Thomas Watson for 6 day in the month of May, at		0	5	0
13	Item for 20 dayes and half in June		0	17	1
	Item for about 12 dayes before harvest		0	9	10
	Item for his wages last harvest		1	10	0
16	Item for 34 dayes and half since the end of harvest at sowing rye, plow, filling, spreading dung, etc.		1	8	9
	For a letter from London from Mr. Mi:		0	0	3
	Spent by men going for coals at Cambridg		0	0	6
17	Paide to Mr. Taylour for an overseers' rate for parsonadge at 3*d*. per pound		1	14	3
	Item for what I hold of my owne		0	1	6
	Paide to him for the first quarterly tax granted for parsonadge		6	17	6
18	Item to the same tax for Ed: Parishes		0	2	6
	Item to the same tax for Cassanders		0	9	0
	Paide to Thomas Eastland for his wages this last harvest		1	12	0
19	Item for hauming 15 acres and half of rye and wheate stubble		1	0	8
	Item for 6 dayes since harvest threshing seed wheate, etc.		0	5	0
	Paide to Widdow Gooding for her boy keeping my hoggs about 3 weeks		0	3	2

		£	s	d
20	Item for 2 jaggs of dung July 13 last upon my 4 acres abutting north on Fawdon Hil	0	2	0
	Paide for a paire of black jersey stockins for myself — bought at Royston by son Wedd or his father	0	2	2
[p.86]	Paid to Will Brooks for half a day helping to in my reck in February 23, *1706*	0	0	5
22	Item for his wages this last harvest	1	10	0
	Item for 12 loads of dung August 25 and 26 last upon my gleabe's 4 stetches in Fincholne and 2 in Short Furlong at 1*s.* 2*d.*	0	14	0
24	Item for the use of his horse last harvest in the roome of Typtery	0	7	0
	Paide to Will: Fordham junior for his wages this last harvest	1	10	0
26	Paide to Thomas Fairchild for 3 dayes work in harvest last	0	3	0
29	Paide to Mr. Thom: Malin of Elmden in Essex for 55 stone and 9 pounds of beife, mutton and lamb that we had in harvest last, that is between July 11 and August 9, at 2*s.* and 8*d.* per ston, a calve's head	7	8	6
30	To Marg: Cra: for the wages of Eliz: Fere her maide, and for work a little before 1*s.*, 2*l.* 16*s.* 0*d.*, and al other occasions besides, toto	3	16	0
	Paide to George Carpenter for his yeare's wages now due	5	0	0
	Expended this month in all the ful summ of	43	6	11
	Expended in the last six months, viz since March last, the ful and entire summ of	123	11	2

[p.85]

October 1707 Receipts

		£	s	d
1	Received of Widdow Swann for her half yeare's rent ended at Michaelmas now immediatly past	1	11	6
	Received of young John Watson for his half yeare's rent ended at Michaelmas 1706, the sum of	0	19	0
3	Item for 2 pecks of wheate that he had November 12 last of that which was eaten by the mules or weevils	0	0	10
4	Item for 2 pecks more of the same on the 25 of last November also	0	0	10
	Item for 2 pecks of the same again on the 7 day of last December	0	0	10
6	Item for 2 pecks more that he had on the 27 day of December last, at	0	1	0
	Item for 2 pecks of the 12 dressing of barly on the 6 of January	0	0	10

HOUSEHOLD ACCOUNTS 1707

8	Item for 2 pecks of wheate he had on the 16 of January last, at		0	1	2
	Item in part of his yeare's rent ended at this Michaelmas 1707 now immediatly past		0	14	0
	Received of Widdow Goode what remained due for wintering her cow from October 30, 705, to May 706		0	0	4
9	Item in part for wintering her cow from November 12, 706, to Lady Day 1707 as part of 11 shillings		0	4	8
	Received of Clem: Norman what remained due for a bushel of wheat since July 14, vide... [*cross-reference incomplete*]		0	0	2
[*p.87*] 10	Item for 9 pecks of wheate on May 5, June 9 and 27, September 17 and the 3rd of this instant October		0	5	10
	Item for 5 pecks of rye on May 3, July 15 and September 29, at		0	2	1
14	Received of John Waites for rye a bushel, discounted for herdman's wages		0	2	0
	Received of Will: Ward for wintering a cow from November 4 to December 14 last — neare 6 weeks — at 7*d*.		0	3	0
17	Item for his smal tithes — 3 cowes, 3 calves, two lambs, a pigg, etc.		0	10	6
	Received of Will: Dovy for half a bushel of the old wheat		0	1	6
20 22	Received of son Wedd for 2 bushels and half of my seede wheate that was sowne upon his furthest 3 roods in Lank Furlong, at 2*s*. 6*d*. a bushel		0	6	3
	Received of Widdow Goode for a peck of the new black wheat		0	0	6
24	Received of Eliz: Bright for 3 pecks of the same, the last of it		0	1	6
29	Received of Mrs. Norridge for half a bushel of the old wheate, which I sel now at 3*s*. 4*d*. a bushel		0	1	8
	Received of Mary Thrift for a peck of the same wheate		0	0	10
30	Received of Edward Parish for 2 dozen of rye straw		0	0	6
	Received of Rob: Harvey junior of half a bushel of rye		0	1	0
	Received this month in all the summ of		5	12	4

[*p.86*]   **October 1707 Disbursements**

1	Paide to Widdow Swann for a loade of dung laide upon one of my stetches at Hickses hedges February 15 last, forgotten at our reckoning about August 26		0	1	8
2	Paide to Wil: Morrice for cutting the old boare		0	0	4
	Paide to Good: Brightman for trimming me one quarter, now chin		0	2	6

3	For a dozen of goose wings of Mr. Westly — other things we had are inserted in July, viz plums, sugar	0	0	5
4	Paide to young J: Watson for 3 loads of dung 14 and 15 of February last upon my stetches at Hicses hedges at 1s. 8d. per loade	0	5	0
	Item for a day in garden April 29	0	0	10
6	Item for 2 jaggs of dung September 20 last upon the 3 half acres of son Nat's in Hounsditch	0	2	6
7	Paide to John Waites by discount with his wife for his herdman's wages due at September 29 last, for 3 cows	0	2	0
8	Paide to Goodm: Will Soale for 2 dayes and half himself 3s. 9d., and as much of his son Will at 1s. per day 2s. 6d., in hewing out plow timber and cart timber, toto	0	6	3
9	Paide to Mr. Hurrel of Harston for 10 bushels and half of wheate to sow at 3s. — more was bought but the rest was sold againe	1	11	6
[p.88]	Paide to William Ward for him or his father taking a moule in my close about January last, at	0	0	2
14	Item for 4 loades of dung January 31 on Nat's land in Southfield	0	6	0
16	Item for 6 loades more carryed out June 12 upon 2 stetches of Nat's on the other side of the miller's path in Lank Furlong — let it be remembered that Nat: discounted 12s.			
20	6d. for barly straw they had of him in the winter before	0	9	0
21	Item for 6 loades more upon my stetch between baulks in Lank Furlong, that is 2 June 12 and the other 4 August 21 and 22, upon my acre in Fincholne Shot, and 2 stetches in Shott Furlong almost against it, at 1s. 6d. per load	0	9	3
22	Paide to Eliz: Bright for a jagg of dung upon Nat's 3 half acres in Hounsditch Shott in Barfield	0	1	3
	Spent and archdeacon's visitation, and to barber then	0	0	8
23	Item for procurations — they claimed 2 yeares as being due at Michaelmas, before the visitation	0	6	8
27	Paide to Rich: Fairchild for a constable's rate at 1d. per pound for what I hold of my owne	0	0	6
	Item for parsonadge to the same rate	0	11	5
31	To Marg: Cra: uxori, 24 pounds of candles 7s. 6d., salt 2 bushels 10s. 8d., 7 ston and 6 pound of beif of Sutton 17s. 2d., and mutton of him 4s. 6d., and al other things	3	4	0
	Expended this month in al the sum of	8	1	11

[p.87] **November 1707 Receipts**

1	Received of Mrs. Westly for an acre and half of haume laide in out of Waterden Shott	0	5	6

HOUSEHOLD ACCOUNTS 1707

	Item for 2 pecks of rye September 25	0	0	10
	Item for 2 pecks of wheate September 27	0	1	6
	Item for the tithe of her orchard, sold	0	1	2
3	Received of Thomas Watson for 5 pecks of rye, at	0	2	3
4	Item for a bushel of the old wheat September 24 and October 6 and 2 pecks more October 24, in al	0	4	6
8	Received of Edw: Godfry what remained due for a bushel of wheat ever since June 16 last	0	2	2
	Item in part for 2 pecks of old wheate he had now at 1s. 4d.	0	0	4
10	Received of Mary Thrift for a peck of barly, at	0	0	7½
11	Received of Urias Bowes of Barly for 8 quarters of the first dressing of barly sold to him at 1l. 0s. 8d. per quarter	8	5	4
12	Received of Mrs. Westly for a bushel of the old wheate	0	3	4
[p.89]	Received of old Will: Thrift for 3 pecks of malt September 19 at 3s. per bushel	0	2	3
	Item for a bushel of rye August 14 and 20, at	0	1	8
14	Item for 3 bushels of rye to sow out of my seed	0	6	0
	Item for 2 bushels and half of rye besides, September 3 and 22 and 30 last, at 2s., abating one peck	0	4	11
	Item for 3 pecks of my old wheat at 3 shillings	0	2	3
17	Received of Mr. Urias Bowes for 6 quarters and 2 bushels of the 2nd dressing of barly sold to him at 1l. 0s. 6d. per quarter	6	8	0
20	Received of Mrs. Norridge for half a bushel of my wheate, at	0	1	8
	Received of Mr. Greenhil for a bushel of rye paide me by son Nat out of his	0	2	0
24	Received of Widdow Gooding for 2 pecks	0	1	0
	Received of young Will: Thrift for a bushel, both out of that which was paid me by Nathaniel, what he had of me for seed	0	2	0
26	Received of Mary Waites for a peck of my old wheat at 3s. 4d. per bushel	0	0	10
27	Received of Benjamin Brightman for a bushel of rye out of my owne	0	2	0
	Received of Lydd: Barber for a peck of my old wheate at 3s. 4d. per bushel	0	0	10
29	Received of Rhoda Carrington for a peck of the 4rth dressing of barly, by self	0	0	7½
	Received this month in al the sum of	17	3	7

[p.88]   **November 1707 Disbursements**

Paide to Thom: Watson for 11 dayes and half between September 21 and October 5	0	9	7

227

3	Item for 18 dayes more betwixt that and October 26 at 5s.	0	15	0
	Item for six dayes more betwixt that and the 2nd of this instant November	0	5	0
4	For hog's rings at Royston	0	0	1
	Paide to Good: Soale for making a new plow, and putting an handle onto another 2d., in al	0	1	8
8	Paide to Edward Godfry for 2 jaggs of dung carryed October 24 last onto my acre next the South Close style, at	0	2	6
11	Paide for the 2nd quarterly tax for Cassanders at 1s. per pound	0	9	0
	Item for Hoyes or Edward Parishes	0	2	6
	Item for parsonadge to that tax	6	17	6
	Given to a souldier maimed	0	0	2
12	Paide to Rich: Carter for half a ringe of wood last winter	0	9	0
13	Spent by men carrying a loade of barly to Ur. B: of Barly	0	0	4
[p.90]	Allowed to son Nathaniel for his expences 2 market dayes at 6d. each, id est	0	1	0
16	Given to a breife for the repairing the church and tower of Orford in Suffolk, the damage valued at 1,450l. and upwards	0	0	9
	Given to a wounded souldier haveing lost his arme in the warrs	0	0	2
20	For a quarter of malt of Peter Clements	1	4	8
	For heeling my owne shoos to young Richard Haycock, his father not at home	0	0	8
24	Paide to Wil: Thrift for 14 dayes between August 13 and the 31, at plow, dung cart, threshing	0	11	8
26	Item for 21 dayes more between August 31 and September 28, threshing seed rye and wheat in malthouse, and black wheat and oats in Nat's barn, and gathering apples, threshing barly for hens	0	17	6
27	Item for 8 dayes between September 28 and October 31, and 2d. for one afternoone, gathering apples, haume cart, dung cart, threshing and dressing wheate in malthouse, etc.	0	6	10
	Item for threshing 6 quarters and five bushels of son Nat's oates which, because my horses have, I pay for threshing	0	4	6
29	Spent by Nathan: at Royston Market	0	0	6
	To Marg: Cra: uxori for six quarters of mutton of R: Sutton about 9s. 6d., and al other occasions this month, in al	1	4	0
	Spent this month in al the summ of	14	4	7

[p.89] **December 1707 Receipts**

	Received of John Woodward's wife for 3 pecks of rye July 6, 1706, at 1s. 4d.	0	1	0
1	Item for 2 pecks of wheate then, at	0	1	2
	Item for 2 bushels of wheat in November 1706	0	3	4
	Received of Ralph Carpenter, Mr. Malin and others for my bull, now killed	2	0	0
2	Received of Widdow Carrington for rye a peck	0	0	6
	Received of Widdow Haycock for a peck of it	0	0	6
	Item for a peck of my old wheate	0	0	10
3	Received of Mrs. Norridg for 2 pecks of the same	0	1	8
4	Received of Widdow Ward for rye a peck; received of Rob: Harvy for a bushel of it; received of Jos: Wedd for 2 pecks of the same; received of Thom Barber for 2 pecks	0	4	6
	Item for a peck of my old wheate at 3s. 4d.	0	0	10
6	Received of Mr. Urias Bowes of Barly for 7 quarters and 3 bushels of the 3rd dressing of barly sold to him at 1l. per quarter	7	7	6
9	Received of Mr. Latimer what remained due for his smal tithes in 1702	1	10	8
	Item for his smal tithes in 1703, the sum of	2	3	10
	Item for his smal tithes in 1704 againe	2	3	8
	Item for his smal tithes in 1705 againe	1	18	4
10	Item for 3 bushels of malt November 20 in 1704 at 2s. 3d. per bushel	0	6	9
11	Item in part of 14s. due to me for six bushels of malt that he had on September 24, 1706, at 2s. 4d. per bushel	0	6	9
	Received of Widdow Gooding for a peck of rye	0	0	6
[p.91]	Received of Widdow Gurner and Widdow Carrington for half a bushel of rye	0	1	0
12 16	Received of young J: Watson, John Waites, Rich: Fairchild, Will: Dovy, Widdow Ward, young W: Thrift and Mr. Greenhil for 6 bushels one peck al of the same rye at 2s. per bushel	0	12	6
	Received of Rob: Harvey, Widdow Ward, Will: Dovy, for 2 bushels and half	0	5	0
	Received of Mr. Norridg and Wil: Brooks for a bushel of the old wheate, at	0	3	4
17	Received of Athan: Carrington for a deale board of 11 foot long	0	1	8
22	Received of Thom Barber and Widdow Gurner for 7 pecks of rye	0	3	6
23	Received of Urias Bowes of Barly for 6 quarters and 2 bushels of the 4rth dressing of barly sold to him at 19s. 6d. per quarter	6	2	0
	Received of Thom: Barber for a peck of old wheate	0	0	10
	Received of Widdow Carrington, Mr. Greenhil, Wil:			

29	Brooks, for 3 bushels and one peck of rye		0	6	6
	Received of Goodm: Brightman for 5 pecks of the same, at		0	2	6
31	Received of Clem: Norman and Jos: Wedd for a bushel of it		0	2	0
	Received of Will: Brooks for a bushel		0	2	0
	Received of young Will Thrift for half a bushel of the same		0	1	0
	Received in this month in al the summ of		26	15	2

[p.90]   **December 1707 Disbursements**

	Paide to John Woodward's wife by discount for 2 quarters herdman's wages ended at Christmas in 1706 for 4 cowes	0	5	4
1	Paide to Ralph Carpenter by discount for 2 loads of dung carryed out October 24 last upon the hithermost of my 3 furlong stetches in Barfield, at	0	2	6
2	Item killing an hogg on March 26 last	0	0	4
	Given to a poor seaman from America — both ship and goods and all taken by French	0	0	2
3	Spent at Cambridge at the election of Mr. Bromley for knight of the shire	0	0	9
	For 2 sheet almanacks for the ensuing year	0	0	3
4	Given towards the repair of the church and steeple of St. Marye's in Dursley, Gloucester, that fell downe January 7,			
7	1698, unexpectedly and killed several that were ringing when it fell, and the damage computed at 1,995*l*. and upwards	0	0	9
9	Paide to Mr. Latimer for a churchwardens' rate at 3*d.* per pound for Cassanders in the yeare 1704	0	2	3
10	Item for another churchwardens' rate in 1705 at 3*d.* per pound, when the bells were new run, for Cassanders	0	2	3
	Item for an overseers' rate made on April 8, 1707, for Cassanders likewise	0	2	3
11	Item to the same rate for parsonadge, al by discount, together with 4*l.* 3*s.* 6*d.* due to Mr. Cotton for quit rents for the yeare ended at Lady Day 1703, being the last yeare due to him	1	14	3
[p.92]	Paide for a new padd to my saddle together with other mending of it	0	2	6
12	Paide for a new haire brush for George	0	1	0
	Spent by men at Barly carrying the 3rd dressing to Urias Bowes	0	0	4
16	For a letter from London from Mr. Mitchel paid by brother Wed	0	0	2

HOUSEHOLD ACCOUNTS 1707-1708

		£	s	d
	Paide to Urias Bowes of Barly for 4 bushels of malt of him at 3s. 2d. per bushel	0	12	8
17 23	Paide to Goodm: Ben: Brightman for one quarter for trimming me, and son Nat: once, ended at Christmas Day, by discount of mony due to me for 5 pecks of rye	0	2	6
	For a pound of 6d. and 8d. nayles when the cow racks were mended	0	0	5
30 31	To Marg: Cra: uxori for 4 quarters of mutton, 6 stone and 12 pounds of beife about 1l. 5s. 0d., and 28 pounds and half of Cheshire cheese 6s. 9d., 12 pounds of candles 3s. 10d., for 2 dozen and half of pidgeons of Mrs. Latimer in July last 3s. 9d., and 7 ducks 3s. 6d., and for butter ever since October 1706 1l. 1s. 8d., and for milk since the same time 10s. 10d., in al	4	15	0
	Expended in this month in al the summe of	7	5	8

[p. 91]   **January 1708 Receipts**

		£	s	d
1	Received of Rich Haycock what remained due for a peck of wheat he had April 12, 1704, vide receipts December 6, 1706	0	0	2
	Item for 6 pecks of rye April 20 and February 3, *1705*	0	3	6
2	Item for 4 pecks of wheat May 6, 1704, and April 5 and 9 and on June 21, 1706, at	0	2	9
	Item in part for 2 bushel of rye on March 24, *1707*	0	0	9
3	Received of Edw: Parish for a bushel he had with another that I borrowed of him, and now paid	0	2	0
	Received of Wil: Dovy for 2 pecks of my old wheate, at	0	1	8
5	Received of J: Waites for a bushel of rye December 16 last and 2 pecks of his wife had now at 1s., in al	0	3	0
	Received of Mr. Bowes of Barly for 5 quarters 3 bushels of the 5th dressing of barly at 19s. 6d.	5	4	6
6	Received of Edw: Parish by the hands of his wife for wintering his old cow in *1707*, 12 weeks at 7d.	0	7	0
	Item for his young cow 17 weekes in the same winter	0	9	0
7	Item for 2 bushels and half of wheat in November and December 1706 at 1s. 8d.	0	4	2
	Item for 2 bushels of barly on March 28 last at 15s. 6d. a quarter	0	3	10
[p. 93]	Item for 2 bushels of the old wheate, some at 3s. and some at 3s. 4d.	0	6	3
8	Item for his half yeare's rent ended on September 29, 1706	1	15	0
	Item for his half yeare's rent ended at Lady Day 1707 also	1	15	0

REV. JOHN CRAKANTHORP

9	Item in part of his half yeare's rent ended at Michaelmas 1707		0	1	2
	Received of Widdow Gurner, Mr. Greenhil, Widdow Carrington, for 7 pecks of rye		0	3	6
10	Received of Mary Waites, Leon: Parish, Widdow Gurner, Law, Richard Fairchild, for 2 bushels and a peck		0	4	6
12	Received of Lyd: Barber for 2 pecks of the old wheate, at		0	1	8
	Item for 6 pecks of rye, and 6 pecks of Rob: Macer and Rich: Fairchild		0	6	0
	Received of John Browne for the old Boy horse		1	0	0
13	Received of Mrs. Norridge for 2 of the old wheat againe — pecks		0	1	8
	Received of Goody Haycock for rye 2 pecks		0	1	0
15	Received of Wil: Dovy for a peck of the old wheate at 3s. 4d.		0	0	10
	Received of Mr. Greenhil, Widdow Gooding, Jos: Wedd, Widdow Gurner, Wil: Brooks, for 2 bushels and half of rye, at		0	5	0
16	Received of James Incarsoale for 3 quarters of the same from Great Shelford		2	6	9
17	Received of Edw: Parish by the hands of his wife in part of his half yeare's rent ended at Michaelmas last, viz 1707		1	3	0
19	Received of Mrs. Westly for a bushel of rye the 16 and 17 instant		0	2	0
	Received for 2 pecks of Eliz: Bright		0	1	0
	Received of J: Waits and Rich: Haycock for 2 bushels of the same		0	4	0
20	Received of Will: Brooks, Widdow Law, Thom: Barber, Mr. Greenhil, for 4 bushels of the same, at		0	8	0
	Received of Clem: Norman for 4 pecks		0	2	0
23	Item for two bushels of the new wheate October 15, 25, and November 3, 10		0	4	10
	Received of Lyd: Barber for a peck — old		0	0	10
	Received of Rich: Fairchild, Edw: Parish, Widdow Gurner, Edith Wedd, for 2 bushels and 3 pecks of rye		0	5	6
26	Received of Widdow Carrington for 2 pecks of it, at		0	1	0
28	Received of Widdow Westly, Rich: Haycock, Mr. Greenhil, for rye straw		0	0	10
	Received of Wil: Brooks for a bushel of rye and of Mrs. Westly and Ralph Carpenter for 2 bushels and 3 pecks more		0	7	6
31	Received of Widdow Gurner for ten weekes wintering her cow ended the 29 instant		0	5	0
	Received this month in al the ful summ of		18	16	2

[p.92] **January 1708 Disbursements**

				£	s	d
1	Paide to Rich: Haycock for 4 dayes work in scouring the river up to K: bridge, the towne joyning voluntarily, for J: Cann			0	4	0
	Item for 4 dayes of his children gathering stones for me in April last			0	1	0
	Item for soaling my shooes February *1707*			0	1	2
2	For 4*d.* nayles to nayle up my vine and peach trees, etc.			0	0	4
	For half a dozen clouts for the wagon and cart, at			0	1	6
3	Paide to Joh Waites his wife for a quarter herdman's wages for 3 cowes ended last Christmas, by discount of mony due to me for 4 pecks of rye he had on December 16 last			0	2	0
5	Paide to Edw: Parish his wife for 3 loades of dung in June 1706 upon my 4 or 5 acres in the Rowze, Southfield			0	5	0
6	Item for 2 loades of dung carryd onto Nat's land January 27 last in Southfield, one a good one, the other a smal one, valued at			0	3	0
	Item for a little time in gathering the granitings in July last			0	0	2
	Spent by Nathaniel at Royston			0	0	6
[p.94]	For a lanthorn bought for the use of the stable of Weston Frisby			0	2	0
7	Given to servants at Mr. Ben:			0	0	6
	Given to a poor souldier from Flanders, both sick and lame			0	0	1
8	Paide to Charles Winds by Nathan: for 5 bushels of malt at 3*s.*			0	15	0
9	Paide to Goodman Soale for new heading a plow now, at			0	1	0
	Paide to Ralph Carpenter for killing an hog for us — the old sow			0	0	4
12	Paide for a medicine for my cornes from Mr. Scampton's at the Angel, over against the Maremaid Taverne in Cornhil, London, at			0	1	0
	Paide to Rich: Fairchild for a constables' rate for Cassanders at 1*d.* per pound			0	0	5
13	Item for parsonadge to that rate, at			0	11	5
16	Paide to Mr. Etheridge of Buntingford for 2 yards and an half of black cloath for a riding coate for myself at 8*s.* per yarde			1	0	0
	Item for 4 yards and half of shaloone			0	7	0
	Item for 2 dozen and half of buttons			0	1	0
17	Item for half an ounce of silk for it			0	0	10
	Item one ounce of mohaire for buttonhole 8*d.*, buckram, canvas and tape 6*d.*			0	1	2

	Item to Clem: Norman for making my coate at home, at his house	0	3	6
19	Item for 5 dayes work besides, the weeke before, of himself and Thom, mending and altering my old cloaths, when			
20	also he made a new coate for son Nathaniel at our house	0	5	6
	Spent by son Nathaniel at Royston	0	0	3
	Paid for the 3rd quarterly tax for Edw: Parishes at Hoyes	0	2	6
26	Item for Cassanders to that tax	0	9	0
	Item for parsonadge to that tax	6	17	6
	Paide also to Rich: Fairchild, collect for the half yeare's duty of the window tax ended at Michaelmas last	0	5	0
27	Paide to Richard Jackson of Royston for new leading of 8 foot of glass in the chancel November 11 last, etc.	0	2	0
28	Item for 27 quarryes there and in the kitchin windowes, and repairing them, about a foot at 1*d*.	0	2	6
	Allowed to son Nathaniel for expences at Royston Market	0	0	6
29	To Marg: Cra: uxori for veale and mutton of Edw: Parish and Fidlin 12*s*. 2*d*., and 3 stone and 3 pounds of beife of			
31	Mr. Malin 9*s*. 4*d*., and for al other occasions whatever in this month besides, the sum of	1	15	0
	Expended in this month in all the full summ of	14	3	8

[*p*.95]   **February 1708 Receipts**

2	Received of Urias Bowes of Barly for 6 quarters of the 6th dressing of barly at 19*s*. 6*d*. carryed in on January 9th	5	17	6
	Received of Widdow Brock for a bushel of rye July 18 last, at	0	1	8
3	Received of Eliz: Bright for a peck	0	0	6
	Received of Widdow Gurner for 2 pecks, one January 29 and one now	0	1	0
4	Received of Rob: Macer and Mr. Greenhil for 7 pecks of the same	0	3	6
	Received of Lyddy Barber for six pecks of the same	0	3	0
6	Received of Mr. Urias Bowes of Barly for 10 bushels remaining of the sixth dressing of barly carryed in the 2nd instant, with five quarters of William Fordham's	1	5	0
9	Received of Widdow Haycock, Will Brooks, Rich Haycock, for 2 bushels of rye	0	4	0
	Received of J: Bowelesworth for 2 pecks of wheate July 7 last	0	1	4
10	Item for a peck of barly July 8 and 2 pecks now at 2*s*. 6*d*.	0	1	10

HOUSEHOLD ACCOUNTS 1708

	Received of Mrs. Norridge for 2 pecks of the new wheate, at	0	1	4
11	Received of Mr. Hughes of Royston for 6 quarters and 2 bushels of the 7 dressing of barly at 1*l*. 0*s*. 8*d*.	6	9	2
12	Received of Clem: Norman and Widdow Law for 2 bushels of rye, at	0	4	0
	Received of Wil: Brooks for wheat a peck	0	0	8
	Received of El: Bright, Waites, Gurner, for a bushel of rye	0	2	0
13	Received of her for a bottle of wheat straw	0	0	3
	Received of Joh: Watson for rye a bushel	0	2	0
	Received of Thom: Barber for wheat a peck	0	0	8
	Received of Mary Thrift for 2 pecks of rye; received of Good: Harvy and Bright for 4 pecks	0	3	0
14	Received of Rob: Macer for a peck of wheat	0	0	8
	Received of Rich: Haycock and Mr. Greenhil for a bushel of rye	0	2	0
16	Received of Betty Parish for young Will: Thrift for a bushel of the same	0	2	0
	Received of Widdow Gurner for 2 pecks	0	1	0
	Item for a peck of wheat that time	0	0	8
17	Received of Edith Wedd for half a bushel of rye or miscellaine	0	1	0
	Received of R: Kettle for a peck of wheate	0	0	8
18	Received of Mr. Urias Bowes of Barly for 1 quarter and five bushels of the 7th, and 6 quarters and 3 bushels of the 8th dressing of barly, in al 8 quarters, sold to him at 1*l*. 0*s*. 4*d*. per quarter, for	8	2	6
19	Received of Rob: Macer for rye 3 pecks	0	1	6
	Item for a peck of wheate at that time	0	0	8
	Received of Lydd: Barber for 2 pecks	0	1	4
20	Received of Mrs. Norridge for 2 pecks	0	1	4
	Received of Lyd: Barber for a bushel and half of rye; received of Will Brooks for 2 pecks	0	4	0
[*p.97*]	Received of Widdow Law for one bushel of rye or miscellaine	0	2	0
23	Received of Mary Waites for a bushel	0	2	0
	Received of Rob: Macer for a peck of wheat	0	0	8
	Received of Lyd: Barber for a peck; received of Widdow Law for 2 pecks	0	2	0
26	Received of Urias Bowes for 7 quarters of the 9th dressing of barly sold to him at 1*l*. 0*s*. 6*d*. per quarter	7	3	6
28	Received of Mr. Greenhil and Widdow Gurner for a bushel of rye between them	0	2	0
	Received of Mrs. Norridge for half a bushel of wheate, at	0	1	4
	Received this month in al the summe of	31	15	3

February 1708 Disbursements

[p.96]

		£	s	d
	Paide to Rich: Johnson of Royston for a paire of body traise and a paire of shack traise by Nathan:	0	6	6
2	Paide by him to Will: Numan for laying in the body traise and other work last time he was here	0	3	6
	Paide to Widdow Brock for 200 of rodds for thatching at 1s. per 100	0	2	0
3	Item for 25 ashen poles from Langley	0	5	0
	Given to a breife for 24 inhabitants of Southam in Warwickshire loosing 4,454l. 16s. 0d. by fire happening on the 16 of May last	0	1	0
4	Spent by Nathaniel at Royston Market	0	0	6
	Paide to Fetherstone for bottoming the chaff sieve in stable	0	0	6
	Item for mending both the fanns	0	0	9
8	Paide to John Bowelesworth for 2 loades of dung on July 4 last (one of them upon Nat's stetches together with one			
9	from the hayhouse door, see August 26 before, in al, 7) at 1s. 6d. each	0	3	0
	Paide to Thom: Fairechild for a coale shovel to the coppers in the backhouse (I think very dear), at	0	3	0
	Paide for unloading at Mr. Hughes	0	0	4
11	Paide for white wine 1s., and other things for wormes in the Watt colt and for mange, etc., in al	0	4	0
12	Paide to Mr. Redhead for a new periwig for myself, with sweet powder, I suppose about half a pound, and a glass of lemmon oyle	0	18	0
13	Paide for a pint of hony for the Watt horse for a medicine, to be added to the other things for wormes	0	1	0
14	Given to one Richard Amis of Hadenham in the Isle, haveing had great losses in cattle and sicknesses in his family	0	0	2
16	Paide to Thom: Fairechild for a new share since December 4, 1706, when I reckoned with him last, at 4d. ob.	0	5	0
	Item for 44 removes between that and December 3 last, when I reckoned but he was not paide al til now	0	3	8
	Item for 74 new shoos in that time	1	4	8
17	Item upon Nathaniel's tally in that time	1	3	0
	Given to a dumb wench at doore	0	0	2
18	Paide to Rich: Haycock for mending the heeles of my shoos and setting them more upright	0	0	2
	To Rob: Hills for a daye's work in garden	0	1	6
	Paide for 5 bushels of malt to Ch: Winds	0	15	0
19	Paide to Widdow Gurner for a loade of dung the 17 instant upon Nat's 3 roods — his part is 18 pence, but in al	0	2	0
	Spent by men carring barly to Ur: Bo:	0	0	4

	Paide to Rich: Carter by Nat at Royston, stub mony for a ringe	0	2	0
20	Given to a Welsh man taken by the Turks and now returned, a schollar	0	0	6
	Paide to Maskal the carrier for bringing downe an hamper of wine from London by brother Wedd	0	2	0
[p.98] 24	Paide for 8 quarters and an half of barly bought at Cambridg for seede at 18 shillings per quarter of one Ward, and a shillin spent, and 2d. tol, in al	7	14	2
	Spent by men carring 7 quarters to Ur: Bo:	0	0	2
26	To Marg: Cra: uxori for 40 pounds of cheese whereof 25 pounds of one Cheshire cheese 10s. 3d., mutton and veale			
28	and head and pluck about 8s. 3d., a new tap hose 1s., 12 pounds of candles 3s. 10d., and al other occasions in this month, in al	1	17	0
	Expended in this month in al the summe of	16	0	7

[p.97]
March 1708 Receipts

	Received of Rich: Haycock for 2 pecks of wheate February 20 last	0	1	4
1	Received of Thom: Watson for one bushel of the same at twice	0	2	8
	Item for 2 bushels of the old wheat December 8 and January 15 and 21 at 3s. 4d.	0	6	8
2	Item for 2 bushels of rye paide me by Nat for what he had of seede	0	4	0
3	Item for 11 bushels and half of rye more between December 8 and this time at 2s. per bushel al of it, besides 1s. 3d. for beife, and 2s. 6d. for a bushel of wheate I paid son Nat: for	1	3	0
	Received of Goodman Thomas Eastland for fetching him 2,000 of turf from Cambridge October 17	0	5	0
4	Item for a bushel of pease that he borrowed March *1707* and being unpaide until now; I reckon the prise there set			
5	downe in case of non-payment, viz as then	0	2	6
	Item for a bushel and half of the old wheate in October last at 3 shillings	0	4	6
6	Item for 4 bushels and 3 pecks of the same between November 3 and January 19 at 3s. 4d. per bushel, at	0	15	10
8	Item for a bushel and half of the new wheate at 2s. 8d. (besides 1s. 9d. for beife and 3s. 9d. for 6 pecks of Nat's wheate at 2s. 6d. not here inserted but in December)	0	4	0
	Received of Rob: Macer for a peck	0	0	8
	Received of Jonathan Peck for a bushel by son Nathaniel	0	2	8

REV. JOHN CRAKANTHORP

9	Received of Lyd: Barber for a peck	0	0	8
	Item for a peck of rye at that time	0	0	6
	Received of J: Watson for a peck of barly, at	0	0	7½
10	Received of Widdow Gurner for a peck	0	0	7½
	Item for a peck of rye then	0	0	6
	Received of Wil: Brooks for a peck of wheate 8d.; received of Clem: Norman for 2 pecks	0	2	0
	Received of Rob: Macer for a peck	0	0	8
11	Received of Edw: Parish for a peck that he had the 5 instant	0	0	8
	Item for 3 pecks of rye that he had at the same time with three more that he did not pay for	0	1	6
[p.99]	Received of Lyddi: Barber for 2 pecks of wheate at 2s. 8d.	0	1	4
12	Received of Wil: Thrift senior for a peck of the old wheat that he had January 21 last, at	0	0	10
	Item for 5 pecks of the new wheate since February 5 to this time, at	0	3	4
13	Item for 4 bushels and half of rye that he paide for, for the use of his son Wil whilest he went to [?]Camb.	0	9	0
15	Item for 4 bushels more of the same rye that he had (as I suppose) for the use and expence of his own house, besides beife inserted in December last	0	8	0
	Received of Widdow Gurner for a peck of wheate the 12 instant, at	0	0	8
16	Received of Mary Parish for a peck	0	0	8
	Item for a bushel of rye	0	2	0
	Received for 2 pecks of the seed barly of Widdow Jeeps and Jos: Wedd	0	1	3
	Received of Rob: Harvy for 3 pecks	0	1	10½
17	Received of Mrs. Westly February 6, 17, for a bushel of rye those 2 times	0	2	0
	Item for a peck of wheate the 9 instant; received of Jonath: Peck for a bushel; received of Widdow Strett for a peck	0	4	0
18	Received of Urias Bowes for 7 quarters and 3 bushels of the 10th dressing of barly at 1l. 0s. 7d. and carryed in 11 instant	7	12	0
20	Received of Clem: Norman for 2 pecks of wheat 1s. 4d.; received of Mrs. Norridge for 3 pecks	0	3	4
	Received of Rob: Macer and Thom Barber for 3 pecks more of the same	0	2	0
23	Received of Uri: Bowes for 6 quarters and 3 bushels of the 11th dressing of barly at 1l. 0s. 8d., at	6	11	6
	Received of John Cann and Wil. Jeepes for 3 bushels and a peck of barly	0	8	1½
26	Received of Edw: Parish, Rob: Macer, Widdow Gurner,			

		£	s	d
	Mrs. Norridge, for a bushel and half of wheat at 2s. 8d.	0	4	0
	Received of Clem: Norman for 2 pecks of the same, at	0	1	4
27	Received of Rob: Macer for a peck of barly, at	0	0	7½
	Received of Jonathan Peck for a bushel of wheate, at	0	2	8
29	Received of Jer: Gooding for rye 3 pecks, and one of Good: Gurner	0	2	0
	Received for 2 pecks of Rob: Macer, at	0	1	0
	Received of Edward Parish for a bushel and half	0	3	0
30	Item for 3 pecks and about a quarter of Eastland's last dressing of barly	0	2	0½
	Item for a peck of wheate	0	0	8
31	Received of Lydd: Barber that day for a peck of the same	0	0	8
	Received this month in all the full sum of	23	10	6
	Received in the last six months, viz since Michaelmas last, the summ of	123	13	0

[p.98]   **March 1708 Disbursements**

		£	s	d
	Paide to Thomas Watson for 22 dayes in November, since the 3rd, when we reckoned last, at 9d. per day	0	16	6
1	Item for sixteen dayes between November 30 and December 21 at 9d. also per day	0	12	0
2	Item for 8 dayes between December 21 and January 4 at plow, nayling up the vines and other things in garden	0	6	0
	Item for 18 dayes between January 4 and February 1 al at 9d.	0	13	6
3	Item for 21 dayes and half between February 1 and 29 at 10d., lopping in orchard and close, dung cart, garden	0	17	11
	Item for 2 journeyes at plow February 24 and 27, being gone in afternoon	0	1	4
4	Item for Munday and Tuesday the 1 and 2 of this instant, plow, and hedging next Linch Lane	0	1	8
	Paide to Charles Winds for 5 bushels of malt at 3s. per bushel	0	15	0
5	Paide to Goodman Eastland for 9 dayes work before October 20, though they were a considerable part of some of them in the afternoon about Nat's work, at plow, haume cart, etc.	0	7	6
6	Item for 6 journeyes before November 4 at plow, clunch cart, at 8d. each	0	4	0
	Item for 3 journeyes January 7, 21, 22, and 2 November, 27 and 28, at plow, sowing pease	0	3	4

			s	d
8	Item for 7 journeyes more February 16, 23, 24, 25, 26, 27, and the 3rd instant, at dung cart, sowing barly, etc.	0	4	8
	Item for 4 dayes January 28, February 17, 1 and 2 instant, sowing oates, dung cart, sowing barly, lopping in close	0	3	4
9	Item for threshing 5 quarters of the 3rd dressing of wheat at 1s. 3d.	0	6	3
	Item for threshing 23 quarters 7 bushels of the 2, 3, 4, 5 dressings of rye at 1s. 3d. per quarter	1	9	10
10	Item for threshing 49 quarters and 6 bushels of the first 7 dressings of barly at 9d. a quarter	1	17	3
	Item for a loade of dung February 18 upon the furthest of my 2 acres next Southclose Style — copyhold	0	1	6
11	Paide to Will: Numan for half an hyde 4s. 6d., 2 great buckles 4d., one little one a penny, work now of himself 10d., and his man 3d., toto	0	6	0
[p.100]	To Ralph Carpenter for killing the 2nd and 3rd hoggs	0	0	8
12	Paide to Rob: Hills the gardner for a Jinniting Hervy and a Golden Pippen tree 1s. 8d., beanes and seeds 8d., and his daye's work now 1s. 6d., toto	0	3	8
	Spent by Nathaniel at Royston Market	0	0	6
13	Paide to old Wil: Thrift for a journy at plow of David — Nathan: at Royston	0	0	6
	Item for threshing 3 quarters 4 bushels and half of pease and tares at 1s.	0	3	7
	Item for threshing 28 quarters of oats wanting but 2 pecks at 8d. a quarter	0	18	8
14	Item for threshing 21 quarters 2 bushels and half of the 8, 9, 10 dressings of barly	0	15	11
16	Given to a breife for the erecting a Protestant church and ministry in the district of Oberbarmen by the grant of the Prince Palatine of the Rhine in Germany, by petition of the Queen	0	2	6
	Paide to Rob: Harvy for a journey at plow Friday last, Nathan at Langly	0	0	7
17	Spent by Nathaniel at Royston taking mony there	0	0	6
	Paide for five quarters of mault bought of Charles Windes at 3s.	6	0	0
19	Paide to Thom: Fairchild for mending the irons at the bottom of the great copper, and 3 pound of new iron cross			
22	the mouth of the copper to hold up the other 9d., and so in al there is	0	1	6
	Paide to Rob: Harvey for his work about the copper at that time	0	1	0
23	Paide to Ralph Carpenter for killing the 4rth hogg for us	0	0	4
	Paide to Henry Wallis of Chrissol for fagotting a ringe of woode this winter in Chrissol Park	0	1	6

HOUSEHOLD ACCOUNTS 1708

24	Paide to Goodman Bright for trimming me a quarter ended on this Lady now being	0	2	6
26	Paide to Jerem: Gooding for 7 dayes going to plow, with victuals	0	1	0
	Paide to Rob: Macer for sweeping the kitchin and backhouse chimnyes just before we begun to burn wood	0	1	0
28 30	Given to a breife for 14 persons by name and several others loosing by fire, wherein 7 persons were consumed and as many houses and goods besides, as amounteth to the damage of 6,137*l*. 17*s*. 0*d*., happening on Tuesday May 13, 1707, about one in the morning, in the parish of St. Paul Shadwell, in the county of Middlesex	0	1	6
31	To Mar: Cra: uxori for 34 pounds of cheese 9*s*. 6*d*., meat 5*s*. 9*d*., and all other occasions in this month	1	12	0
	Spent this month in al the summ of	19	17	0
	Expended in the last six months, viz since Michaelmas last, the full summ of	79	13	5

[*p.101*]

April 1708 Receipts

	Received of Widdow Strett for a peck of wheate, etc., at 2*s*. 8*d*.	0	0	8
2	Received of Rob: Macer and Mary Waites for 2 pecks of the same wheat	0	1	4
	Received of Thom Barber, Mrs. Norridge and Rob: Macer for a bushel	0	2	8
6	Received of Mr. Greenhil for a peck of barly out of Thrift's dressing; received of Clem Norman for 3 pecks of it	0	2	6
8	Received of Jonathan Peck and Widdow Gurner for 5 pecks of wheate	0	3	4
	Received of Thom: Dovy for wheat straw	0	0	3
	Received of Will: Dovy for his smal tithes in 1706, 707 and this instant yeare — cowes, calves, etc.	0	4	0
9	Item for wintering his cow 12 weeks from January 12 til Easter Munday, viz 5 instant, at 7*d*.	0	7	0
12	Received of Widdow Gurner for 6 weeks wintering hers againe — or thereabout	0	3	0
	Item for a peck of rye at that time	0	0	6
14	Item for 2 pecks of barly of the last dressing and 2 pecks of Jo: Watson and Widdow Swann	0	2	6
16	Received in part of Mr. Urias Bowes of Barly at 1*l*. for 6 quarters and half of the 15 and 16 dressings	6	0	0
	Received of Mrs. Westly for her cows and calves, and this last year	0	1	10

			£	s	d
19	Item for a peck of wheate that she had on March 19 last		0	0	8
	Received for a bushel of Jonathan Peck of the same wheate		0	2	8
20	Received of Mrs. Norridge for 2 pecks		0	1	4
	Received of Widdow Swan for her half yeare's rent ended at Lady Day last past — I say received by me		1	11	6
22	Item for 2 constables' rates to Rich: Fairechild, and 2 overseers' rates to Mr. Tailor and Mr. Latim:, and 2 churchwardens' rates disbursed for her before, to him		0	3	0
23	Received of Jonathan Peck for a bushel of wheate, att		0	2	8
	Received for a peck of Widdow Gurner		0	0	8
	Received of Good: Thom: Watson for 4 bushels of rye in March last		0	8	0
24	Item for a bushel and half of wheat in the same month of March		0	4	0
	Item for 5 pecks of wheate more in this present month before 27		0	3	4
26	Item for 2 bushels and half of rye more in the same time, at		0	5	0
	Item for 2 bushels of barly of the 16 and last dressings, at		0	5	0
27	Received of William Fordham junior for a bushel of rye January 16 last		0	2	0
30	Received of Good: Harvy for a peck of barly at 2s. 6d.		0	0	7
	Received this month in al the summ of		11	0	0

[p.102] **April 1708 Disbursements**

		£	s	d
	Paide to Campian of Wittlesford the cooper for a new gett or jett	0	1	0
3	Spent by son Nat: at Royston Market	0	0	4
	Paide to Ralph Carpenter for killing the 5th hogg for us, againe	0	0	4
6	Given to a poor seaman having lost his arme, kept prisoner by the French 10 months	0	0	2
7 9 11	Given to a breife for Jane Short, John Thornburgh, Sarah Dingly, and nine others by name and several others, inhabitants of Charles Street and places adjacent in the parish of St. Margaret's, Westminster, in the county of Middlesex, loosing by fire on April 10, 1707, the sum of 3,891 pounds and upwards, 17 houses being burnt and 6 others greatly damaged, and the greatest part of the sufferers' goods, houshold stuff, implements of trade, burnt, broaken or stolen away, etc.	0	1	6

12	For unloading, and spent by men carrying in that barly 2*d*. and 4*d*.	0	0	6
	Paide to Goodman Soale for new heading a plow now	0	1	0
14	Paide for the 4rth quarterly payment of the land tax for Hoyes	0	2	6
	Item for Cassanders to that tax	0	9	0
19	Item for parsonadge to the same tax	6	17	6
	Paide for the half yeare's duty upon windowes ended at Lady Day now last past	0	5	0
20	Spent by Nat with Rich: Carter the last loade of wood, they carrying no beere with them	0	0	6
22	Paide to Fetherston for mending Eastland's fann, and a little basket	0	0	6
	Spent and given to servants at Mr. S. by self and wife, etc.	0	1	0
23	Paide to Widdow Swann for a loade of dung of October 24 last upon the hithermost of my 3 furlong stetch in Barrfield	0	1	8
	Item for another loade February 16 last upon Lamps Acre in that field	0	1	8
24	Given to some in towne at times	0	0	7
	Given to one Marshal, cast away last September, being greivously hurt by the casks in the ship, loosing 300*l*.	0	0	6
26	Paide for 20 dayes work in March to Thom: Watson, etc.	0	17	4
	Item for 24 dayes more before the 27 of this instant April	1	0	0
27	To Marg: Cra: uxori for veale about 4*s*. 2*d*., cheese 28			
30	pounds 6*s*. 4*d*., salt about 3*s*. 6*d*., and for al other occasions in this month, in al, just	2	0	0
	Expended this month in al the full summe of	12	2	7

[*p.103*] **May 1708 Receipts**

	Received of Mrs. Norridge for half a bushel of wheate	0	1	4
3	Received of Rhoda Carrington for a peck of barly	0	0	7
	Received of Mr. Greenhil for a peck	0	0	8
4	Received of Wil: Jeeps for a peck of wheate at 2*s*. 8*d*.	0	0	8
6	Received of Thom: Dovy for a bundle of wheate straw	0	0	3
	Received of Athanasius Carrington for his smal tithes whereof he had no calf	0	0	9
7	Received of John Cann for a peck of barly by Sam	0	0	8
8	Received for the Jolly horse, besides 2*s*. 6*d*. to George, sold to a Doctor of Divinity at Cambridge, they say in Lincoln	14	17	6

10	Received for a peck of rye of Widdow Gurner	0	0	6
	Received of Jonathan Peck for a bushel of wheate, to Sam	0	2	8
12	Received of Eliz: Bright for a peck	0	0	8
	Item for a peck of the tayle	0	0	4
	Received of Mrs. Norridge for half a bushel of wheate	0	1	4
13	Received of Thom: Ward for a peck of barly at 2s. 6d.	0	0	7
	Received of Mr. Taylour for 400 of spitts or sprindles	0	1	0
14	Received of Jonathan Peck for a bushel of wheate	0	2	8
	Received of Widdow Gurner for a peck of rye at 2s.	0	0	6
17	Item for a peck of wheate	0	0	8
	Received of Will: Jeeps for a peck	0	0	8
18	Received of Thomas Eastland for 2 bushels before March 25, 1708, at 2s. 8d.	0	5	4
	Item for 4 bottles of wheat straw after the saide 25 of March	0	0	9
19	Item for fetching him 2 jaggs of wood from [*place omitted*] in 2 waggons, 3 horses each of them, April 24	0	6	0
	Item for 2 bushels and 3 pecks of wheate since April 12 last	0	7	4
20	Received for a peck of Mary Waites	0	0	8
	Received for a bushel of Jon: Peck	0	2	8
	Received of old Will Thrift for 2 bushels and a peck since March 18 last at 2s. 8d. also	0	6	0
21	Item for 3 bottles of wheat straw	0	0	6
	Item for 5 pecks of barly	0	3	1
	Item for 9 bushels and half of rye since March 12	0	19	0
22	Item for wintering 2 cows since November 30 to March 22, 16 weeks, though they went 6 weeks after, at 7d.	0	18	8
	Received of Mrs. Greenhil for 4 weeks going of theire cow in winter	0	2	4
[p.105]	Item for the tithe of his cow, and calf 6d., in al	0	0	10
24	Received of Widdow Gurner for rye a peck	0	0	6
	Item for a peck of barly	0	0	8
26	Received of Mary Waites for a peck of wheate, at	0	0	8
31	Received of Mr. Urias Bowes of Barly what remained due for 10 quarters 7 bushels of the last 2 dressings of barly ever since April 14 last	4	17	6
	Received this month in al the ful summe of	23	16	6

[p.104]

May 1708 Disbursements

	Paide to Mr. Wilson at the Angel for tenths due at Christmas last, and an acquittance 4d., in al	2	19	9
4	Spent then with Mrs. West and to barber	0	1	1

HOUSEHOLD ACCOUNTS 1708

	For a new hatt for myself	0	10	9
5	Given to ringers at son Nathaniel's wedding, besides a dinner	0	5	0
	Spent by him 2 market dayes at Royston	0	0	8
	For mending the buttery lock	0	0	4
6	Paide to Richard Jackson of Royston for his yearly rent to keep my pump in order, vide April 21, 707	0	1	0
10	Item for repayreing the casements at home and several quarryes, and 2 or 3 in the chancel, in al	0	1	6
	Given to a souldier at door	0	0	1
11	Paide to Robert Hills the gardner for a daye's work 1s. 6d., and a disk of asparagus	0	2	0
	For 4 bushels of coals of Mr. Howel	0	3	0
12	Paide to Rob: Harvy for a little above half a day in emptying the felmonger's pits about my close in April,			
13	towards the beginning of that month	0	0	6
	For a letter from Mr. Mitchel	0	0	2
14	Paide to Brother Sparhawke for a yeare's interest of 20*l*. due at Christmas last from me, vide April 19, anno 1707	1	4	0
	Paide for the yeare's quitt rents due at Lady last, for Cassanders 13s. 2d., and for Hoyes 3s. 8d. (vide May 12, 1707), toto	0	16	10
17	For a pound of 6d. nayles used in orchard	0	0	5
	For a pound of binding to mend up the clay wals broaken down in the hayhouse and elsewhere	0	0	4
18	Paide to Thom: Eastland for 7 journeyes at plow in barly seede, viz March 4, 5, 9, 12, 13, 15 and 16	0	4	8
	Item for 7 journeyes more March 17 and 18, April 6, 13, 14 and May 13 and 14	0	4	8
19	Item for 10 dayes work at plow, rowles, fagotting, casting up dung and dawbing this May 17 and 18	0	8	4
20	Item for threshing 8 quarters one bushel and 3 pecks of rye of the 2 last dressings at 1s. 3d.	0	10	2
	Item for threshing 15 quarters and 5 bushels of barly March 8, 16, 22	0	11	8
	Item for threshing 14 quarters 4 bushels of the 2 last dressings of wheate at 1s. 3d. likewise	0	18	2
21	Spent by Nathaniel at market	0	0	6
	For a quire of paper of Mr. Norridge	0	0	5
	Paide to old Will: Thrift for threshing 23 quarters 5 bushels and a peck of his 5 last dressings of barly at 9d. per quarter	0	17	8
22	Item for 16 dayes between March 27 and May 2 spreading, casting up dung, hedging against the pickle at Goody Swann's and against Hockeril's Lane	0	13	4
[*p.106*]	Item for 14 dayes since May 2nd instant hedging in courtyard, casting up dung, remooving straw, etc.	0	11	8

24	For mending my shooes to R: Haycock		0	0	1
31	To Marg: Cra: uxori to weaver for 25 ells of linnen cloath 7s., cheese at 3d. ob., meate sent on April 28 against Nat brought home his bride 17s., veale and lamb this month 9s. 3d., and 4 stone and a pound of beife 14s. 3d., and al other occasions in this month, toto		3	10	0
	Expended this month in al the summe of		14	18	9

[p.105]   **June 1708 Receipts**

	Received of Jonathan Peck for a bushel of wheate, at	0	2	8
1	Received of Widdow Gurner for a peck	0	0	8
	Received of Mary Waites for a peck	0	0	8
2	Received of Mrs. Norridge for half a bushel of the same	0	1	4
	Received of Wil: Jeepes for a peck	0	0	8
	Received of Ann Thrift for a peck of tayle of the last dressing	0	0	5
4	Received of Widdow Gooding for a peck	0	0	8
	Received of Widdow Gurner for a peck of rye by self, at	0	0	6
	Received of Mary Waites for a peck of wheat — the daughter	0	0	8
7	Received of Jonathan Peck for a bushel of the same wheate	0	2	8
10	Received of Goodwife Rayner by hands of Edw: Parishes wife for a peck of the last dressing of barly	0	0	7
	Received of Widdow Strett for a peck of wheat	0	0	8
11	Received of Jonathan Peck for a bushel	0	2	8
	Received of Widdow Gurner for a peck of rye	0	0	6
14	Received of Edith Wedd for a peck of the last dressing of barly, at	0	0	8
16	Received of Mrs. Norridge for half a bushel of wheat, by Jane	0	1	4
18	Received of Clem: Norman in part for 2 pecks of wheate that he had February 16 last (a penny being paid March 10th last and 6s. 3d. being deducted for wheat that was Nat's besides at that time)	0	1	0
19	Received of Jonathan Peck for a bushel of wheate, at	0	2	8
	Received of Widdow Gurner for a peck	0	0	8
	Item for a peck of miscellaine	0	0	6
	Item for the use of the churchyard the last winter for her sheepe	0	2	6
21	Received of Edw: Godfry for what remained due for 2 pecks of wheat he had November 10 last	0	1	4

HOUSEHOLD ACCOUNTS 1708

		£	s	d
	Received of Robert Harvy for a peck	0	0	8
22	Received of Wil Brooks for half a bushel of the same	0	1	4
	Received of Mary Waites for a peck; received of Mary Wallis for a peck	0	1	6
[p.107]	Received of Thom: Barber for half a bushel of wheat, at	0	1	6
23	Received of the widdow Mrs. Norridge for half a bushel of the same	0	1	6
24	Received of Jonathan Peck for a bushel more of the same	0	3	0
	Received of Mary Wallis for a peck	0	0	9
26	Received of Mrs. Norridge for two bushels of the same	0	6	0
	Received of Will: Dovy for the tithe of Tipling's Close hay	0	2	6
28	Received of Widdow Gurner for a peck of the best wheat, at	0	0	9
	Received of Will: Jeepes by the hand of his daughter for a peck	0	0	9
	Received of Jonathan Peck for a bushel of the same wheate, at	0	3	0
29	Received of Good: Harvey for a peck	0	0	9
	Received of Edith Wedd for a peck	0	0	9
30	Received of Ralph Carpenter by the hands of his daughter Susan (by discount of mony due to him for 4 loades of dung the 19 and 21 of this instant June) for 2 pecks	0	1	6
	Received this month in al the summ of	2	12	3

[p.106]   **June 1708 Disbursements**

		£	s	d
1	Paide to Mr. John Howel of Cambridge for 40 bushels of coals at 9d. per bushel	1	10	0
	For 2 sheets of stampt Dutch paper	0	2	8
	For a pint of ink at Mr. Watson's	0	0	8
2	Spent going for coales 4d., and for toll 2d., in al	0	0	6
3	Paide then to Christopher Deane for 2 black skins for a paire of breeches for myself, at	0	18	0
	Silk and canvas for the making them	0	1	2
	Spent by Nathan at Royston	0	0	6
4	Given to a poor man and his wife in want	0	0	3
8	To Goodm: Soale for 2 new fellos 3s., and for 2 new spoakes to them 1s., for two other new spoks into the felloes 2s., into one of the tumbrels	0	6	0
	Item for fastning the boxes and mending the wagon now	0	1	0
9	For a pound of 8 penny nayles used by him, a few of them about wagon, and given to a poor man an *ob.*, toto	0	0	5

247

14	Paid for poundage of my hogs at Foxton	0	0	6
16	Paide to Mrs. Love for 120 bushels of coales I had of Mr. Love in May 10, June 24, September 26, 1707 (besides 10 bushels for Thom: Watson) at 8d. ob. per bushel	4	5	0
	Paide at the primary visitation of Dr. John Moore the 17 instant — Bishop of Ely	0	11	3
17	Paide for shaloone 1s. 6d., silk 2s., buttons 1s., bukram, for myself, etc.	0	4	8
	Spent then and to barber 6d., in al	0	0	10
18	Paid Widdow Gurner for almost 2 loades of dung upon the gleabe in Northfield, tilth this yeare	0	2	9
	Given to a poor widdow in want	0	0	1
	Paide to Edw: Godfry for a loade of dung February 18 last, upon the furthest of my copyhold acres	0	1	8
21	Item for a loade now upon the gleabe in Northfeild	0	1	6
22	Given to a breife for a loss by fire (happening upon the inhabitants of Wincanton in Somersetshire on the 13 day of May 1707 to the loss of 2,930l. and upwards, the summ of	0	1	0
[p.108]	Given to one or two of my neighbours being hard put to it	0	0	7
23	Paide to Goodman Brightman for trimming me for his quarter ended this instant June	0	2	6
24	Paide to Chrissol of Foxton for bringing my bore from Foxton haveing been there about 3 weeks	0	0	6
26	Paide for children since June last, in primis for 13 months interest ended May 23 last for son Samuel for 12 pounds			
28	he had of Mrs. Richardson, which was then paid in to Mr. St: for her	0	15	6
29	Item for a yeare's interest due from son Benjamin for 50l. ended at Christmas last, paid May 13 — for the rest I paide nothing	3	0	0
30	To Marga: Cra: uxori for salt a bushel 5s. 6d., cheese 24 pounds 6s. 4d., a paile and cup dish 2s. 2d., meate of Malin, 14 ston and 8 pounds, and veale and mutton of Fidlin 4s. 6d., toto about 2l. 4s. 0d., and al other occasions this month, toto	4	16	0
	Expended this month in al the summ of	17	5	6

[p.107]
July 1708 Receipts

	Received of Thom: Watson for fetching him a loade of turf from Cambridge the 11 of May last	0	5	0
6	Item for 2 bottles of wheat straw the 20 of that month also	0	0	6

7	Item for 2 bushels of rye that he had in that month also at 2s.	0	4	0
	Item for a bushel of wheat in that month also at 2s. 8d.	0	2	8
8	Item for a bushel of wheate more in the month of June, at	0	2	8
	Item for 2 pecks more at 3s. per bushel	0	1	6
	Item for 3 bushels of rye in that month whereof one on the last day	0	6	0
9	Received of young Will: Fordam for half a bushel of wheate	0	1	6
	Received of Will: Jeeps for a peck	0	0	9
10	Received of Widdow Gurner for a peck; received of Lyd: Barber for a peck; received of Goody Harvey for a peck	0	2	3
12	Received of youngh John Watson for 2 pecks of the last dressing at ten groats per bushel	0	1	8
14	Received for 5 quarters of the same wheate sold to Edw: Phip at Buntingford at 18s. per loade, that is 3s. 7d. qtr. a bushel wanting one farthing in five	7	4	0
17	Received of Lyddy Barber for a bushel of the same wheate	0	3	4
22	Received of Goody Harvy for a peck	0	0	10
	Received of Widdow Haycock for half a bushel of the same	0	1	8
27	Received of Widdow Gurner for a peck of rye by self	0	0	7
	Received of her againe for another peck of the same	0	0	7
[p.109]	Received of Widdow Gurner againe for another peck of rye	0	0	7
30	Received of William Chaplin for a peck of barly at 2s. 6d.	0	0	8
	Received of Good: Thom: Watson for half a bushel of the same	0	1	3
31	Received of Widdow Gurner for one peck of rye againe	0	0	7
	Received this month in al the summe of	9	2	7

[p.108] **July 1708 Disbursements**

	Paide to Mr. Warren for writing a lease between me and son Nathan	0	4	0
3	Paide to Widdow Hankin for a little parcel of dung carryed on to gleabe land June 21 last	0	0	4
4	Paide to Widdow Swan for 2 dayes, herself and daughter, makeing hay in Bridge Close, my owne and town	0	1	0
6	Given to a breife for 14 inhabitants of the towne of Bewdley in Worcester, looseing by fire May 21 and July 3, 1707, 1,384l. 4s. 0d., both in the forenoon	0	1	0

	Item to masons building brother Wed's house, by self and wife, 6d. each	0	1	0
7	Paide for 24 dayes between April 25 and May 30, most at plow, only 2 or 3 in garden, at 10d. per day	1	0	0
8	Item for 30 dayes more between May 30 and 4 of this instant, at plow, filling, spreading dung and hay, tilling and making, inning, al this to Thomas Watson	1	5	0
10	Paide to Ralph Carpenter's daughter for her mother, for 4 loades of dung carryed on June 19 and 21 on to the tilth in Northfeild gleabe, vide receipts June 30	0	6	0
12	Given to one Mr. Ste: Messail, a persecuted Protestant of the Cevennais, haveing large commendations from French ministers in England	0	0	6
	Spent by men carrying that 5 quarters	0	0	8
18 26	Given to a breife for Lisburne in Ireland in the County of Downe or Antrim, loosing by fire on April 20, 1707, in time of divine services 31,770l. etc., to the ruin of 300 families, besides the church	0	2	0
	For carriage of a letter to Dayes	0	0	2
	To carpenters building Mr. Wed's house	0	1	0
28	For mending my owne shoes and ux:	0	1	6
	For a pound of 8 penny nayles and given an *ob*. used about kitchin garden, pales being broaken	0	0	5
[p.110]	Paide to Will: Fordham junior mending kitchin garden pales etc.	0	1	6
30	Paide to Rich: Fairchild for a constable's [*rate*] for Cassanders, and 3d. to son Nat's	0	1	0
	Paide to Thom: Ward for ketching 2 moules in my close at 2d. a peice	0	0	4
	To Marg: Cra: uxori for linnen cloth for selves about 11s. 4d., 2 stone of beife 5s. 10d., lamb 8 quarters 11s. 3d., and al other occasions	2	4	0
	Expended for all things in this month the summ of	5	11	5

[p.109] **August 1708 Receipts**

20	Received of Goodman Thomas Eastland for a bundle of the wheat straw that he had of me the 21 of May last, at	0	0	4
24	Item for another bundle that he had on the 22nd of the same month of May againe	0	0	3
27	Item for 3 pecks of wheate that the miller had for him on that May 25 at 2s. 8d. per bushel	0	2	0
30	Item for 3 pecks more miller had for him on June 14 of the same wheat at the same price	0	2	0

	Item for 3 pecks more the miller had for him at 3s. per bushel of the same wheat	0	2	3
	Received this month in al but the summ of	0	6	10

[p.110] **August 1708 Disbursements**

	Given to one Mr. Robert Wentworth of Kinsale, a gentleman in want	0	0	3
6	Item to son Nat's men for a largess	0	2	0
28	Paide to Thom: Watson for threshing 7 bushels and a peck of rye in the malthouse from my copyhold stetch in Finchole Shot a day and a half	0	1	3
	Given to Mrs. Blandina Cutts, vide August 20 in 1707 — this her daughter	0	0	2
30	Given to Will: Rayner and his man bringing me 2 great loades of barly this harvest from the 2 acre at South Close style, coppyhold	1	1	0
31	For 6 and 8 penny nayls, and given	0	0	5
	To Marg: Cra: uxori for meat 2s. 8d., 5 bushels of malt 15s., and al other things in this month, toto	1	7	0
	Expended in this month in al	1	12	1

[p.109] **September 1708 Receipts**

6	Received of Goodman Thomas Eastland againe for 3 pecks of wheate miller had for him on July 26 last, at 3s. 4d.	0	2	6
7	Received of Widdow Gurner in part for a peck of rye that she had about this time at 7 groats	0	0	4
	Received of Goodman Barber for 2 pecks of wheat 29 September 1707	0	1	4
8	Item for a peck of old wheate January 19, another, new, February 9, 10d. and 8d.	0	1	6
9	Item for wintering his cow last winter 9 weeks from November 19 to January 22 at 7d. per week	0	5	2
	Received of Goodman Thom: Watson for 6 pecks of rye at 2s. per bushel	0	3	0
	Item for 2 pecks of wheate, at	0	1	6
10	Item for another bushel of the same wheate, at	0	3	4
	Item for 2 pecks of the old wheat kept ever since 1705	0	2	0
14	Item for 4 bushels of rye at 7 groats since the 23 of July last	0	9	4
	Received of Athanasius Carrington for new rye 2 pecks growing this present yeare, at	0	1	2

16	Received of Widdow Gurner what remained due for a peck of rye ever since the sixth instant	0	0	3
	Item for another peck at the deliverance of it	0	0	7
[*p.111*]	Received of old Wil: Thrift for 2 bushels and a peck of the tayle of the last dressing of wheate in May, June and July, at 20*d*.	0	3	9
17	Item for the tithe hay of Pipers	0	4	0
	Item for 2 bushels and 3 pecks of rye, at 2*s*. per bushel that is	0	5	6
18	Item for 2 bushels and half more at 2*s*. 4*d*., 3 pecks of it new	0	5	10
	Item for 3 pecks of the best wheate at 8 groats	0	2	0
	Item for 3 pecks of the same at 3*s*. and 3*s*. 4*d*.	0	2	4
	Item for a peck of the old wheate	0	1	0
20	Received of Widdow Gurner for a peck of the new rye	0	0	7
30	Received of Widdow Thompson for a yeare's quitt rents due to the rectory for the yeare compleatly ended at Lady Day 1704, vide receipts in May 11, 1703, the sum of	0	5	0
	Received this month in al the sum of	3	2	0
	Received in the last six months, viz since last March, the summe of	50	0	2

[*p.110*] **September 1708 Disbursements**

6	Given to Goodman Hanchet of Triplow, looseing al his goods by fire breaking out about 1 of the clock in the morning July 30 last, unknown how, to the value of 19*l*. etc.	0	2	6
7	Given to Good: Haycock, she and her husband both sick	0	0	6
	Paide to Goodman Eastland for a day May 21, getting straw out of barn	0	0	10
8	Given to a breife for Alconbury cum Weston in the county of Huntingdon, loosing by fire happening on April 21, 1707, 3,318 pound 10*s*. and collected the 5th instant	0	1	6
10				
11	Given to one James Le Roy, a Walloon borne at Brussels, loosing 500*l*., his vessel cast away against Hul in August last now past	0	0	3
14	Paide to Goodman Thom: Watson for 19 dayes work between July 4 and 29 at plow, dung cart, hay, keeping the field 3 or 4 dayes pro filio N:	0	15	10
16	Paide to Athan: Carrington for 2 days work, at Good: Swan's and Parishes one, the other dooing odd things at			

	home, shelf for shoos, stool in the backhouse, making 2 beetles, etc.	0	3	0
	Paide for a pound of nayles and given	0	0	5
17	Paide to old Wil: Thrift for a day in mowing orchard and carrying out grass, though he could not be a day about it, only 2 peices about June 7	0	0	10
[p.112]	Item for 18 dayes between June 2 and 27, spreading dung, mowing in close, makeing, inning hay	0	15	0
18	Item for 24 dayes more betwixt that and July 29 about the same works, only mowing pro filio Nat	1	0	0
	Item for a day and a part now, cutting hedges, mowing orchard, gathering appls	0	1	6
19	Given to a breife for John Bird of Great Yarmouth, Norfolk, loosing by fire (happening 27 December at 2 in			
27	the morning 1706) the sum of 1,228*l*. — haberdasher of hatts	0	0	9
	To Marga: Cra: uxori for meate about 1*l*. 14*s*. 0*d*. and other things	2	12	0
	Item for her yeare's wages to Jane Gurner	3	0	0
30	Paide to George Carpenter for his yeare's wages in part, son Nat: paying the rest, that is 2*l*.	3	0	0
	Expended in this month in al the summ of	11	14	11
	Expended in the last six months, viz since March last, the summe of	63	5	3

[p.111] **October 1708 Receipts**

1	Received of Widdow Swan (that she paide precisely on			
2	Michaelmas Day indeed) for her half yeare's rent now ended	1	11	6
	Received of Widdow Gurner for a peck of rye at 2*s*. 8*d*.	0	0	8
6	Received of son Nathaniel what he paide for the first			
7	payment of the tax granted for this present yeare 1708	8	5	0
	Item he paide me in mony since about the 7th instant the sum of	3	0	0
	Received of Thomas Watson for a bushel of barly neare			
26	upon that was left of last yeare's barly laide by for fowles,			
28	and abateing 2*d*. for what there wanted of a bushel	0	2	4
	Received of Widdow Gurner for another peck of rye by self, at	0	0	8
	Received this month in all	13	0	2

[p.112]   **October 1708 Disbursements**

1	Paide to Miles Manning for a key to Edw: Parishes new chamber	0	0	6
	Item for an iron to roast apples upon with a wood fire only	0	1	4
2	Paide to Wallis of Chrissol for 300 withs and an 100 spitts at 6s. and 3d.	0	1	9
4	Paide for parsonadge to the first quarterly tax for this			
6	present yeare 1708, paide by Nathaniel on September 29, the summ of	6	16	6
	Item to the same tax for Cassanders at 1s. per pound likewise	0	9	0
7	Item for Hoyes to the same tax	0	2	6
29	To Marg: Cra: uxori for 18 pound of candles 6s. 5d., for 5 quarters and a neck of mutton of Ri: Fidlin 11s. 6d., 13			
30	stone and 5 pound of beife, pork, mutton, of Mr. Malin 1l. 5s. 6d., 5 bushels of malt 15s. 0d., and al other etc., toto	4	10	0
	Expended this month in al	12	1	7

[p.111]   **November 1708 Receipts**

2	Received of young Wil Thrift's wife for 2 pecks of rye	0	1	6
6	Received of Will: Ward for wintering his cow 14 weeks, that is from November 20 to February 20, *1708*, at 7d. per week	0	8	0
13	Received for a peck of wheat (I think) delivered 3 months agoe	0	0	9
	Received of Elizabeth Bright for 2 pecks of rye now	0	1	6
16	Received of son Nathaniel al that he laide downe for tax for me and the lord's quitt rents on the tenth of this instant	8	5	0
[p.113]	Received of Goodman Thomas Barber for a bushel of rye			
19	he had of me now, having about 2 in al	0	3	0
24	Received of son Nathaniel as part of rent due to me for my parsonadge by discount of monyes laide out for my use,			
27	vide October 6 and 16 to this instant	4	15	4
	Remained due from John Watson junior as part of his			
30	yeare's rent ended at Michaelmas 1707, and which I now received, vide October 7, 1707 before	1	4	0
	Received this month in al the sum of	14	19	1

November 1708 Disbursements

[p.112]

		£	s	d
	Paide to John Bowelesworth for a daye's work just before September 9 in 1707 at plow and other things	0	0	10
3	Paide to Ben: Bright for trimming due a quarter, September 29 last	0	2	6
4	Paide to Robert Hils the gardner for a day in garden	0	1	6
8	Paide to Wil: Ward for 3 loads of dung upon my coppyhold acre at South Close style, carryed out October 24, 1707, at	0	5	0
	Given to nurse Harding when little Ben Wedd was nursed	0	2	6
[p.114]	Paid for a yard and half of dimety for capps for myself, at	0	1	10
15	Paid for the second tax for this present yeare for Hoyes	0	2	6
	Item for Cassanders to the sam tax	0	9	0
	Item to the same tax to parsonadge	6	16	6
16	Paide to Rich: Carter for a ringe of wood last winter	0	16	0
27	For a paire of black yarne stockings for myself at Royston	0	2	0
	For a box of Lockier's pills by Frank	0	2	0
30	To Marg: Cra: uxori for meate about 1*l*. 15*s*. 6*d*., candles 12 pound 4*s*. 2*d*., and al other occasions in this month in all, then there is laide out	2	14	0
	Expended this month in al the summ of	11	16	2

December 1708 Receipts

[p.113]

		£	s	d
7	Received of John Watson junior for a peck of tayle wheate that he had 5 of December 1707	0	0	6
	Item for a bushel of rye that he had 24 of February last	0	2	0
8	Item in part of his yeare's rent ended at Michaelmas last, viz 1708, vide November 30 immediately before	0	17	6
13	Received of Mr. Thomas Malin of Elmden in Essex for the calf of the red Welsh cow about February last	0	12	0
14	Received of him also for the calf of the brindled cow, either a little before Lady last or after, or there about	0	10	0
16 18 20 22	Received of Mr. Urias Bowes of Barly for 6 quarters and 2 bushels of barly which was al my cropp growing on my two acres of coppyhold land at South Close style and a 2 furlong stetch in the Rowze Shot of coppyhold also in Southfield that was of the broak cropp (excepting 3 pecks of tayle), and it was sold to him at 1*l*. 0*s*. 10*d*. per quarter for	6	10	0
23	Received of Goodman John Waites for half a bushel of			

27	wheates his wife had the 20 of April last at 2s. 8d. per				
28	bushel, towards the wages becoming due to her husband				
30	as herdman for the quarter that was then to be ended at Midsummer Day following etc. etc.		0	1	4
[p.115]	Item for another peck of the same at the same price				
31	fetched by theire daughter on April 26 last also		0	0	8
	Received this month in all the summe of		8	14	0

[p.114]   **December 1708 Disbursements**

	Paide for a militia tax for Cassanders at 1d. per pound		0	0	9
6	Item for the parsonadge to the same militia tax		0	11	5
7	Paid to the young John Watson for 3 dayes and an half thatching the end of the old barne next the pond in orchard		0	5	0
	Item for his son yelming those dayes		0	1	6
8	Paide to Robert Harvey serving him at that time, at		0	3	0
9	Item for 4 dayes and half in November last mending the stable wals, barne wals, 2 dayes and half of it about the backhouse chimny, Widdow Swan's hog sty and wals, and John Watson's hearth		0	6	0
10	Given to one Wil: Cross of Bromly in Kent loosing above				
11	200l. by merchants he traded withal, a woosted man		0	0	3
	For a couple of sheet almanacks		0	0	3
14	For half a reame of writing paper 5s. 6d., carriage J: Bush from London 4d., in al		0	5	10
24	Paide to Benjamin Brightman for trimming me for a quarter now ended at 2s. 6d. per quarter		0	2	6
27	To Marg: Cra: uxori for 13 stone and 8 pounds of beife and mutton betwixt August 12 and September 7, and a quarter of mutton 1s. 5d., at 8 groats per stone 1l. 16s. 6d., and 2 ston of beife and 2 quarters of mutton in September 10 and 26				
28	10s. 1d., and 2 quarters of mutton and 2 ston and 9 pound of beife October 3 and 24 and 9 stone and 4 pound of				
29	meate, September 7 and 21, 1l. 1s. 0d., and 5 ston December 23 in 1707 11s. 9d., and 14s. 7d. which I could not get a bil of the butcher before now, and the sum due to him for it is		4	14	6
[p.116]	Item for 4 stone and one pound now 9s. 6d., malt 5 bushels				
31	15s., and al other occasions in this month		1	15	6
	Expended this month in al the summ of		8	6	6

[p.115]	**January 1709 Receipts**			
13	Received of Goodman John Waites againe for a peck of wheate his wife had May 13 last at 2s. 8d. per bushel	0	0	8
14	Item for half a bushel of wheat she fetched on July 28 when I sold for above 3s. 7d. per bushel	0	1	8
18	Received of son Nathaniel what remained due to mak up the former summe, received in October and November, even 25 pounds, the sum of	0	14	8
19 20 21	Item received of him what he laid out for me by discount to George Carp:, Ralph Carpenter, Rob: Harvey, and allowance for Hester's board etc. 2l. 17s. 5d., and 2s. 7d. in mony, which is in all	3	0	0
22	Item he paide me in 17 guineas 18l. 5s. 6d. and in silver at the same time 3l. 19s. 6d., which make in al the summ of	22	0	0
	— which he then paid me			
24 25	Received of Clem: Norman what remained due for a peck of wheate that he had February 16, *1708*, a 1s. being paid June 16 last	0	0	4
	Item in part for 2 pecks that he had March 2 last, al by discount for work, vide uxor's reckoning	0	0	3
	Received this month in al the sum of	25	17	7

[p.116]	**January 1709 Disbursements**			
13 14	Paide to Goodman John Waites for 2 quarters wages for 4 cowes due to him as herdman at Michaelmas last — we had the milk and therefore pay though they were son N: cowes most before that — at 2s. 8d. per quarter	0	5	4
17	For 6 penny, 8 penny, 2d., nayles for mending courtyard gate and the chaires in the best chamber	0	0	7
18	Paide to Goodman Eastland for 4 dayes and half before October 26 digging up dead trees, tying up the old hedge in close, clearing out the old walnut tree, etc.	0	3	9
19	Item for threshing 13 bushels of oates in malthouse wanting a peck, of mine, carrying them into garner, emptying the malthouse of al the chaff, shack, etc.	0	2	2
20	Item for threshing 6 quarters 2 bushels of barly and 3 pecks tayle at 10d. per quarter, at	0	5	3
22	Paide to son Nathaniel for a bushel of lime left of his wheate, used at Widdow Swan's hog stye and John Watson's hearth, being broaken, vide December 8	0	0	5
24	Item given at my order to Ralph Carpenter, having his cow killed (as supposed) by our boar	0	2	0

REV. JOHN CRAKANTHORP

		£	s	d
	Paide for a letter from Mr. Mitchel	0	0	4
31	To Marg: Crack: uxori for dimety and crape, and taylour's work, and al other occasions this month	0	18	0
	Expended this month in all the summe of	1	17	10

[p.115] February 1709 Receipts

		£	s	d
1	Received of Mr. James Taylor for 5 score willow sets that he had last March at 18s. per hundred	0	14	0
3	Received of Clem: Norman by discount in part for the 2 pecks of wheat mentioned in January 25 just before	0	0	9
19	Received of son Nathaniel by discount for the 3rd quarterly tax paid January 26 for my owne, and Lord's quit rents, the summ of	8	5	0
[p.117]	Item in what he laide out for my wife at market besides etc.	1	0	10
21 24	Item for half a yeare's rent for my parsonadge which he should have paide at Lady Day 1708 only his mony did not come in until now	50	0	0
26 28	Received of Wil: Dovy for 7 pound and half of beife, part of what was thrust upon us by the butcher on August 27, 1707, an exceeding hot day, and which we feared, and so he had this and Edw: Parish 11 pound of it etc. — these are owing thus long because I could not get a bil of the butcher before this December last	0	1	2
	Received this month in al the full sum of	60	1	9

[p.116] February 1709 Disbursements

		£	s	d
2	For an ounce of Spanish liquorish for my cough, at	0	0	3
6	Given for a loss by fire (happening on November 28, 1707, at midnight, at the head of the Cannon Gate, on both sides of the street at once, in Edinburgh in Scotland, the loss amounting to the summ of 7,962l. and upwards — they were new-built stone houses, being burnt before but in the yeare 1696) the summe of	0	1	6
[p.118] 16	Paide to son Nathaniel for my owne and wive's board from the time of son Nat entring into house keeping until Christmas Day, which was 8 dayes	0	8	0
17	Given to a breife for a loss by fire in the Strand, London, in the parishes of St. Clement's Danes and St. Martin's in the Feilds on February 14, *1708*, the loss 17,880l. — 36 houses			

	burnt, 14 demolished, besides the loss in goods and merchandizes etc.	0	2	6
28	To Marg: Cra: uxori for cambrick, holland, silk, and all other occasions	1	0	0
	Expended this month in al the sum of	1	12	3

March 1709 Receipts

[p.117]

1	Received of Widdow Law, Gooding, Gurner, Barber, for a bushel of black wheat in December 1707	0	2	6
2	Received of Widdow Carrington, Widdow Brock, Mary Waits, Will: Brooks, for a bushel of the same in January 1, 3, 22, February 2	0	2	6
3	Received of Thom: Eastland for a bushel and half of the same in January 15 and 26 anno *1708*, vide March 6, 1707 or 8 receipts	0	3	9
4 7 8	Received of Clem: Norman for 2 bushels and half of the same on November 17, January 1 and 15, 23, February 3, *1708*, vide receipts June 16 last, which with 2 bushels and half sold to mil for fatting hoggs, being tayle of the same, make up 8 bushels and half which is returned to me by Nathaniel for 10 bushels at least of good wheat that he had to sow upon 4 acres one rood and half; vide 1rst dressing of wheat 1707, whereas al this very black	0	6	3
26	Received of son Nathaniel by discount for a quarter's board for self and wife now ended at this Lady Day, viz 1709	5	0	0
	Received this month in al	5	15	0
	Received in the six months since Michaelmas the ful summe of	128	7	7

March 1709 Disbursements

[p.118]

6	Given to a breife for the repaireing the church and steeple of Brenchley in Kent blowne down on November 26 in 1703, charges computed at 1,000*l.*	0	1	0
7	Paide for the 3rd payment of the land tax (discounted in receipts the last month and paid to the collectors January 26) for Hoyes	0	2	6
	Item for Cassanders to that tax	0	9	0
8	Item for parsonadge for the same tax	6	16	6

	To Goodm: Carter for stub mony for half a ringe of wood	0	1	0
9	For a paire of shooes for myself of Wilmot of Royston	0	3	6
10	Paide to Athan: Carrington and Robert Harvy for 3 dayes work in lopping, hedging, setting, trenching, in my close	0	3	3
24	Paide to Mr. Jackson for mending parsonadge windows, specially in son Nat's chamber, parlour, etc.	0	0	7
	Given to Ben: Bright for his newse	0	0	6
26	Paide to him for trimming me a quarter ended this Lady Day	0	2	6
30	Paide to son Nathaniel for one quarter's board for self and wife due unto him this Lady Day	5	0	0
31	To Marg: Cra: uxori for al occasions in this month in the whole	0	6	0
	Expended this month in al the sum of	13	6	4
	Expended in the last 6 months since Michaelmas last the ful and entire summ of	49	0	8

[p.117]

April 1709 Receipts

9	Received of Thom: Watson for the haume of a stetch of rye in Finchholn Shot, part of my coppyhold land, at	0	0	6
12	Received of Mrs. Norridge for 2 yeares quitt rents due and ended at Lady Day last	0	10	0
[p.119] 18	Received of son Nathaniel by discount for the 4 quarterly tax for self and quit rents	8	5	0
19	Item for al that remained due, even for al rent due, and ended at Lady last, viz Lady Day 1709	27	9	2
20	Received of Mr. Coxal by the hands of son Nathaniel for three score and ten willow setts he had the 11 or 12 instant	0	10	6
30	Received of James Whittingstal for 2 yeares quitt rents due to the Rectory ended at Lady Day 1696, the summ of	0	10	0
	Received this month in all the summ of	37	5	2

[p.118]

April 1709 Disbursements

9	Paide to Athanasius Carrington for 7 dayes work more, in close, in lopping, cutting downe, setting, fagotting, and to Th: W: for bringing in	0	7	6

HOUSEHOLD ACCOUNTS 1709

12	Given to Mr. John Constantine, a Neapolitan and Capuchin — out of conscience of Popish errours became Protestant — and is yet unsetld	0	1	0
[*p.120*]	Paide to Richard Carter by son Nath: for stubb mony for half a ringe of wood; vide March 8	0	1	0
18	Paide for the last quarter of the land tax for Hoyes	0	2	6
	Item for Cassanders to the same	0	9	0
	Item for the same tax for parsonadge, the last for 1708	6	16	6
19	Given to nurse at little Dorothye's christening etc.	0	5	0
24	Paide to Wallis of Chrissol for tying up a ringe of wood and 2*d*. given	0	1	8
	Given to poor at Easter	0	1	0
30	To Margaret Cra: uxori for nurse and al other occasions this month	0	19	6
	Expended in this month in al the sum of	9	4	8

[*p.119*] **May 1709 Receipts**

9	Received of James Whittingstal for four yeares quitt rents due to me for the yeares 1697 and 1698 and 1699 and ended at Lady Day 1700 at 5*s*. per annum	1	0	0
10	Item for 8 yeares more for the same quit rents due for al the yeares since Lady Day 1700 and compleatly ended at Lady			
14	Day 1708, so that there is a yeare due stil for the year ended at Lady last 1709	2	0	0
17	Received of Brother Sparhawke for Cozen Sarah his daughter what remained due until the time of her departure from us for her board — that was on October the eight last	4	10	0
24	Received of Clement Norman the taylour for what			
26	remained due for 2 pecks of wheate that he had on the 2nd of March, the latter end of 1707, paide in part February 3, *1709*, vide	0	0	4
27	Item in part for a bushel of rye that he had on January 27			
30	in *1708* also, at 2*s*. a bushel, by discount for a daye's work now due to him, vide uxo: account	0	0	6
	Received this month in al	7	10	10

[*p.120*] **May 1709 Disbursements**

3	Paide to Mr. Wilson for a yeare's tenths due at Christmas last to the Queene, and acquittance	2	19	9
	For a comb for my periwig	0	0	4

	To barber and spent etc.	0	0	10
	For 3 boxes of Lockier's pils	0	6	0
4	For a quire of paper of Mr. Norridge	0	0	7
9	Given to Mrs. Margaret Fairfax of Galloway in Ireland, merchant, sister to Mr. Ayres, captain of a ship, who, being assaulted by 2 French privateers, was killed, his brother Fairfax carried to St. Maloe being bound for his brother a trader into those parts is kept prisoner there for that debt, viz 1,000*l*. — she hath 7 children, Mrs. Mary Ayres 3, who upon this newes are deprived of al at home	0	1	0
10				
14 24	Paide to Brother Sparha: for 18 months interest of 20*l*. which I had of his, and now paid the principal also, vide May 13, 1708	1	16	0
27	Given to a Yorkshire wooman going to London side for work with her son about 8 yeare old, upon whom Widdow Eversden's wal fel as he was passing by and hurt him much	0	1	0
31	To Marg: Cra: uxori for crape 7*s*. 6*d*., holland for herself also 10*s*. 6*d*., doulas for myself 8*s*. 6*d*., and al other occasions	1	15	0
	Expended this month in al	7	0	6

[*p.119*]   **June 1709 Receipts**

	Received of John Watson for an 100 of spitts for thatching	0	0	3
4	Received of Widdow Haycock for an hundred more of the same	0	0	3
6	Received of Margaret Swann, widdow, in part of her half yeare's rent that was compleatly ended at Lady Day last past, that is at Lady Day 1709	1	7	6
[*p.121*] 7	Received of Wil: Thrift junior for a loade of haume laide in on September 26, 1705	0	4	0
	Item for wintering a cow 3 weekes in 1705 in December and January	0	1	6
8	Item for wintering a cow of Leonard Parishes in February *1709*, 2 weekes at 7*d*. per week	0	1	2
9	Item for a bundle of straw in May 10, 1708, at	0	0	3
	Item for a bushel of rye July 7, 1707, at 1*s*. 8*d*. per bushel	0	1	8
13	Received of Mrs. Westly for a year's quitt rents for the yeare ended at Lady Day 1705, left in her hand by Widdow Thompson a little before she dyed	0	5	0
14	Received likewise of Richard Fordham, yet by her hands			

20	for another yeare ended at Lady Day 1706	0	5	0
	Item for another yeare ended at Lady Day 1707 of the same quitt rents	0	5	0
22	Received also of him by the hands of the same Mrs. Westly			
24	for 2 yeares more ended at Lady Day 1708 and at Lady Day 1709, the sum of	0	10	0
	Received of Clement Norman for work done for uxor,			
24[sic]	which he discounted as due to me in part for a bushel of rye that he had in January 27, *1708*, and paid in part in May 30 last, vide ux: account in the month of July	0	0	10
	Received of John Watson junior in part of his yeare's rent that was due and ended at Michaelmas 1708, and was			
25	paid, some of it, in the beginning of December last, quem vide etc.	0	5	6
	Received of Thom: Holt for a box of Lockier's pills, at	0	2	0
27	Received of Edward Parish by discount for a peck of wheat, growing in 1705, laid up in his chamber then and			
28	measured out to him on January 21, *1708*	0	0	10
	Item for a peck of new wheate that he had 16 of February *1708*	0	0	8
28[sic]	Item for half a bushel of rye the miller had for him the fifth day of that February	0	1	0
29	Item for another whole bushel of the same rye that he had on the 16 of that February	0	2	0
	Item for a peck of wheate that he had on 24 of that February	0	0	8
	Item for 5 pecks of rye his wife had on the 26 of the same month, at	0	2	6
[p.123]	Item his wife had 2 pecks of the 9 and tenth dressings of barly on that February 27, *1708*	0	1	3
29[sic]	Item for 3 pecks of rye measured to him by son Nathaniel on March 5, *1708* — he had 3 pecks at the same time which			
29[sic]	he then paide for, vide that month	0	1	6
	Item for a peck of wheat miller had for him that March 12	0	0	8
30	Item in part for 2 bushels of barly he had out of the last dressing on April 15, 1708, at 2*s.* 6*d.* per bushel	0	0	8
	Item what remained due of his half yeare's rent ended at			
30[sic]	Michaelmas 1707, though to speak plainly I forgive him, in one thing or other, as much and more as this last summ amounts too	0	10	9
	Received this month in al the summ of	4	12	5

[p.120] **June 1709 Disbursements**

6	Given to Robert Wiliamite's wife for paying prison fees to get her husband out of prison of St. Giles his parish in Cambridg, vide March 26, 1707 or 6	0	0	6
7	Given to Mr. Thom: and John Burckett, merchant, haveing been taken by the French etc. with 2,500*l.* cargo and having one brother kept still for ransome in France, and an ancient father, etc.	0	1	0
[p.122]	For a dozen of church catechismes bought at Cambridge	0	0	10
8	Paide for putting the teeth into the iron rake by Tho: Fairechild	0	0	4
12 13	Given to a breife for a loss by fire happening in Holt Market in Norfolk on Saturday May 1, 1708, upon 11 inhabitants by name and 70 other familyes besides, consumeing theire parish church and al, the loss being 11,258*l.* etc.	0	2	0
14	Paide to old Will: Thrift for 6 dayes hedging in back yard and about horse pond between the 5 and 12	0	5	0
20	Paide to Will Thrift junior for 3 loades of dung laide upon the gleabe in Northfeild June 18, 1708, to be reckoned to son Nat because it is laide upon his 2nd tilth, Barrfield being his first	0	5	0
	For a bottom of pack thread and given *ob.*	0	0	3
24	Paide to Ben: Brightman for trimming this quarter now ended	0	2	6
26	Given to the rebuilding of the church of Harlow in Essex, burnt downe on April 28, 1708, the bels and al melted, being in good repaire before, to the damage of 2,035*l.* as in the brief	0	1	0
	Given to children since June last, to son John being here in April	1	1	9
27	Given to son Samuel, which he had in his hands ever since the yeare 1699 and 1700, but upon bond, not then knowing whither I might be able to give him it	60	0	0
27[*sic*]	Item given in 4 peices of old gold 4*l.* 16*s.* 0*d.*, and paid the weaver for his flowerd huggeback etc., 1*l.* 4*s.* in al, the summ of	6	0	0
27[*sic*]	Given to son Benjamin in like 4 peices of broad gold 4*l.* 16*s.* 0*d.*, and paide him for 18 months interest for 50*l.* ended last Midsummer 4*l.* 10*s.* 0*d.*, in al the summ	9	6	0
28	Given to daughter Hester like in 4 broade peices of gold also	4	16	0
28[*sic*]	Item in 2 deale boards for his new dresser 2*s.* 6*d.* and for her board during the time that Nathaniel kept house, and something for Mary, 1*l.* 16*s.* 0*d.*, in al	1	18	6
	Item in mony also the summ of	20	0	0

28[sic]	Given to son Nathaniel in goods, I haveing no mony for him, that is in dung and seede and wages for one man in harvest, etc., the sum of 45l. 19s. 0d., for Mrs. Norridge's			
29	land al inserted in expences since Lady Day 1704 already, and besides this paide in mony for his first rent at Lady Day 705, and town rates, the sum of	9	9	0
[p.124]	Item he had 8 quarters and 5 bushels of barly for seed at 8l. 6s. 8d. and 66 loads of dung at 2s. 6d., 5l. for his 4th tilth of Mrs. Norridge's, but forasmuch [p.124] as he paide to Mr. Westly, Wil Numan and others the sum of 6l. 5s. 1d. for me,			
29[sic]	I charge him in these but at	7	1	7
	Item he hath had 107 loades of dung already out of my yard for the gleabe of Northfild, at	8	0	0
29[sic]	Item he had 6 acres 3 roods and ½ of oates, pease and teares ready to cut at my charge	9	5	0
	Item 38 acres 2 roods and half of barly in South and Barr Feild, and 17 acres one rood and half of rye, and 14 acres of wheat ready for sickle, in al 70 acres, worth at 40 shillings	140	0	0
30	Item in cowes, hoggs, horses, traise halters, collars, wagon, carts, tumbrels, harrows, rowles, sacks, rops, cow racks, plow timber, forks, rakes, and other utensils of husbandry as much as is worth	70	17	6
30[sic]	Item given to his wife at Sturbridge Faire last, and since for Dorothy's frocks etc.	0	17	8
	To Marg: Cra: uxori going to London 13s., and al other occasions etc.	1	10	0
	Expended this month in al the summe of	350	18	5

[p.123] **July 1709 Receipts**

4	Received of Mr. James Taylor for an hundred of sprindles or spitts John Watson junior had for him, at	0	0	3
8 12	Item for 200 more of the same fetched by him also to mend up some part of his thatch before his going away at Michaelmas next	0	0	6
14 19	Received of Thomas Nash for his smal tithes in 1707, as cowes, calves, hearth and garden penny, and such like, in al	0	2	6
26	Received of Widdow Jeepes for her smal tithes in 1703 and 1704, she haveing no cow in them yeares, and paying some in kind	0	0	6
27 28 29	Received of James Whittingstal of Triplow for a yeare's quit rents for his messuage holding of the Mannor of the Rectory for the yeare compleatly ended at Lady Day 1709, vide receipts May 10	0	5	0

30	Received of Clement Norman what remained due for a bushel of rye that he had January 27, *1708*, and paide in part June 24 and on May 30, vide ibid.	0	0	8
	Received this month in all	0	9	5

[*p.124*]

July 1709 Disbursements

6	Paide to Mr. Jackson for his rent to keep my pump in order, vide May 6, 1708 — this is too late	0	1	0
	Given to poor in towne and at door	0	0	8
	For 7 boxes of Lockier's pills	0	14	0
8	Given to a breife for John Baynton of Market Rayson,			
10	Lincolnshire, July 4, 1705, and Lucy Morric of Chetton in Salop, October 11, 1707, and Thom: Wilbraem of Worleston, Cheshire, May 8, 1708, loosing theire goods and houses, amongst them 1,228*l*.	0	1	6
11	Paide to John Watson junior for 4 dayes work of himself, and 4 of his son, at 6*d*. a day, for thatching the further end			
12	of the old barne in orchard etc., done on May 17, 18, 19, 20, toto	0	8	0
	For a two hand saw of Thom Watson	0	1	6
26	For an old carpenter's axe also	0	1	0
	For a box of Lockier's pills	0	2	0
	To barber, and spent at archdeacon's visitation	0	1	1
28	For procurations then to archdeacon	0	3	4
	For a brush with a black handle	0	1	0
	For a pound of 6 penny nailes	0	0	4
29	For a glass of tincture of steele of Mr. Wright of Royston	0	1	0
	For an ordinary hammer at Royston	0	0	10
30	To Marg: Cra: uxori for all occasions this month, but	0	7	0
31	Given for rebuilding of the church in Llanvilling in county of Montgomery, become ruinous by length of time, so that they were forced to take it downe in 1707, the charges being 1,325*l*.	0	1	0
	Expended this month in al	2	5	3

[*p.125*]

August 1709 Receipts

1	Received of Richard Fordam for a fine after his mother's death, which because he sold his house to Mrs. Westly,
2	widdow, the mony was left in her hand, at the takeing of it up, and she not furnished at the instant, gave me a note

HOUSEHOLD ACCOUNTS 1709

3	under her hand, and this she paide about this day	3	0	0
4	Received then also of Mrs. Westly, she takeing up the same messuage as a purchaser for herself, one year's rent	2	0	0
22 23 24	Received of son Nathaniel which he paid for the half year's duty due upon the windows for the half yeare ended at Lady Day 1709, and must be reckoned as part of his half yeare's rent that wil be due at Michaelmas next ensuing, the summ of	0	5	0
	Received this month in all the summ of	5	5	0

[p.126] **August 1709 Disbursements**

	Paid for a militia tax laid downe by Edw: Parish for his	0	0	3
2	Item for al work done by him for me since April 15, 1708, to the 2nd of this instant August	0	7	3
4	Item for 2 loads of dung July 20 last upon my coppyhold stetch in the Rowze, at 2s. per loade	0	4	0
14	Given to the rebuilding of the church of St. Mary Redcliffe in the suburbs of the city of Bristol — 248 foot long, 122 foot broad, supported by 105 pillars — much decayed, the charge of rebuilding 4,416*l*. etc.	0	1	0
17	Paid Rich: Jackson for an elmen bucket to the pump 1s., windowes 3d.	0	1	3
	For a paire of shooes for self of Mr. Bently	0	4	0
25	For Lockier's pills 6s., tincture of steel of Mr. Wright 2s., toto	0	8	0
	For the half yeare's window tax ended at Lady Day 1709	0	5	0
29	Given to George 1s., largese to men 2s.	0	3	0
	For a book called The Weeks' Separation for the Sacrament, of Herne	0	0	8
31	To Marg: Cra: uxori buying a new scarf, gowne and other things before she went into Essex, to Sam	10	0	0
	Expended this month in al the summ of	11	14	5

[p.125] **September 1709 Receipts**

24 26	Received againe of son Nathaniel (by payment of the tax which must be accounted as part of his half yeare's rent that will be due at Michaelmas now following) the sum of	7	10	0

27	— which must be added to that which he disbursed in August for window tax, vide			
28 29 30	Item received of him againe in several things that he hath paide for me and upon my account disbursed at several times, and several of them but in little parcels which must be accounted as part of his half yeare's rent due at this Michelmas	8	1	8
	Received this month in all the summ of	15	11	8
	Received in the six months since March last the full summ of	[blank]		

[p.126] **September 1709 Disbursements**

	Paid to Rob: Harvy and his man for a daye's work at Cassanders underpinning hogg styes etc.	0	2	6
6	For shooing horne at Cambig	0	0	4
	Given to one Prier, a Londoner in want	0	0	6
7 25	Given to a breif for the Palatine's Protestants especially, whereof about 8,000 are come hither for succour, having had 2,000 of theire cityes, townes and villages burnt and wasted by frequent invasions of the French, neare the Rhine	0	2	1
26	For the first land tax for this year for Hoyes, at	0	2	6
	Item for Cassanders to the same tax	0	9	0
	Item the same tax for parsonadge	6	16	6
28	Paide for Mr. Taylour's grinstone stall and al, wanting 2 leggs	0	4	0
29	Paide Ben: Bright for trimming me for his quarter now due	0	2	6
	Item given him towards their newes	0	1	0
31	To Marg: Cra: uxori (that was most given to poor being sick, and to servants, as Jane, George, Jo)	0	14	0
	Expended this month in al	8	14	11
	Expended since March last in the preceeding 6 months in al	[blank]		

[p.127] **October 1709 Receipts**

3	Received of Margaret Swan, widdow, for what remained due as part of her half yeare's rent that was compleatly

		l	s	d
4	ended at Lady Day 1709, which was paid in part in June, vide June 6, receipts	0	4	0
6	Item received of her againe as part of her half yeare's rent ended at Michalmas now last past, partly by discount and			
7	partly in mony, the summ of	1	4	0
	Received of her againe what remained due of the same half yeare's rent, the sum of	0	7	6
10	— and this is in full, for her half yeare's rent ended on September 29 now just and immediately past			
	Received this month in al but	1	15	6

[p.128] October 1709 Disbursements

		l	s	d
	Paide for 2 quarters of dust upon my copy stetch in Barfild	0	6	8
10	For irons to the great copper new done	0	5	0
	For a quarter's board for self and wife ended on June 24 last, viz Midsummer	5	0	0
13	Item for a quarter for self ended September 29	2	10	0
	For a studying gown at London	1	0	0
	Item for 2 maps, one of Europe, and Flanders	0	3	0
24	Item 3 paire of spectacles and 2 cases	0	4	0
	Item a paper book to succeede this	0	1	2
26	Item spent by myself in a journy to London besides, going up and coming downe, given to relations and prentices and servants at son's	1	0	11
27	Paid for bookes — The Young Man's Guide, 2 of them 1s. 6d., Dr. Ken's New Yeare's Gift 1s., Practice of Piety 1s. 6d.,			
29	Christian Monitor 7d., of Hern at door	0	4	7
	Paid to Widdow Swan for a loade of dung on July 20 upon my stetch in Rowze	0	2	0
30	Given to a bill of request for Robert Hills and 17 other sufferers at Linton, loosing by fire September 30 last, 609l. etc.	0	5	0
31	To Marg: Cra: uxori, she being then at London and come from Essex	2	10	0
	Expended this month in al the sum of	13	12	4

[p.127] November 1709 Receipts

		l	s	d
10	Received of son Nathanael, by discount in paying taxes and such like and in mony, see September and August, as part of his half yeare's rent ended at Michaelmas last, the sum of	8	3	4

24	Received of James Whittingstal for a fine after his father's death, a yeare and half, abating quit rents		2	15	0
30	Item for half a yeare's quitt rent ended at Michaelmas last, vide receipts July 27 last		0	2	6
	Received this month in al		11	0	10

[p.128]   **November 1709 Disbursements**

	Paide for the 2nd land tax for Hoyes		0	2	6
	Item for the same for Cassanders		0	9	0
	Item for the same tax for parsonadge		6	16	6
10	Item for window tax ended September 29		0	5	0
13	Given to a breife for 24 familyes in Stoak next Clare in Suffolk, loosing by fire on September 13, 1708, the value of 2,463 pounds etc.		0	1	6
15	Paid Rich: Carter for a ring of wood		0	13	0
	Paide to J: Woodward for 2 dayes work in the ditch next the walk		0	2	6
30	For a riding whip of son Wedd		0	1	0
	For 2 sheet almanacks for 1710		0	0	3
	To Marg: Cra: uxori for her occasions		0	13	0
	Expended this month in al the sum of		9	4	3

[p.127]   **December 1709 Receipts**

14	Received of son Nathanael again in mony besides what is discounted in August, September and November last, as				
16	part of his half yeare's rent ended at Michaelmas last, the sum of		17	11	0
17	— together with Mr. Etheridge's bill paid by him at Buntingford 1*l*. 11*s*. 0*d*., in al				
20	Received of Edw. Parish for pease 2 pecks		0	1	4
	Received this month in al the sum of		17	12	4

[p.128]   **December 1709 Disbursements**

3	Paid for what I hold of Cassanders to a constable's rate		0	0	5
	For 2 yards and half of black cloth for a coat for self of John Etheridge		1	2	6
29	Item 4 yards of saloon for it		0	6	6

	Item 2 dozen of buttons (besides 1s. 10d. in ux: account) at 1s. per dozen	0	2	0
30	Item to Clement Norman for making	0	3	0
	Paide to Goodm: Bright for trimming me the quarter now due 25 instant	0	2	6
31	To Marg: Cra: uxori for her occasions	0	10	0
	Expended this month the sum of	2	6	11

[p.127] **January 1710 Receipts**

	Received of Edw: Parishes wife for a peck of pease	0	0	9
12	Received for 2 quarters and one bushel of barly, at 1l. 3s.			
[p.129] 6d.	[p.129] per quarter, being al my cropp that I had this			
17	yeare for my coppyhold land, saving 2 bushells and 3			
18	pecks left to sow my 2 stetches of the same land in the Northfield, that was with wheat and rye upon the tilth	2	12	0
31	Received of son Nathanael by way of discount for the 3rd land tax paid this month on the 28th day of it	7	8	0
	Received this month in al	10	0	9

[p.128] **January 1710 Disbursements**

	Paide Ben, son of Jos: Wedd for ketching a moul in close for me	0	0	2
4	For a large skepp for chips at Royston	0	0	10
14	For cloth for handkercheifes and bands for myself, 3s. 6d. and 2s. 6d.	0	6	0
[p.130]	For nayles used about the bearer of the old trunk etc.	0	0	7
20	Paide for the 3rd land tax — Hoyes	0	2	6
28	Item for Cassanders to the same tax	0	9	0
	Item to the same tax upon parsonadge	6	16	6
31	To Marg: Cra: uxori for stuff of Brother Wedd's 8s. 8d., saffron sent into Essex 5s. 6d., hony sent to London and carriadge 6d., toto 3s. 6d., and all other occasions this month	1	8	0
	Expended this month in al	9	3	7

[p.129] **February 1710 Receipts**

23	Received of William Thrift junior for half a bushel of the best pease at 8 groats per bushel, and so they come to	0	1	4

		£	s	d
24	Item for a bushel of tayle of the same pease which he had in November last — he paide me in part for them then also	0	1	0
27 28	Received of Edward Parish also for one peck of the same pease when he threshed them — they al were growing upon my coppyhold acre next South Close style etc.	0	0	8
	Received this month in al but	0	3	0

[p.130]

February 1710 Disbursements

		£	s	d
8	For 11 pounds of nayles used in the malthouse when it was drawn up, being run an end — 6, 8, 10 penny ones	0	3	11
12	Given to a breife for about 300 Protestants, cheifly Brittains setled in Mittau, the metropolis of Courland, towards the setling a ministry among them	0	1	0
18	Paide Athan: Carring: for 5 dayes work in drawing up the malthouse with Wil Fordham, his son, and Th: Holt	0	7	6
22	Paid to Rich: Carter stub mony for a ringe of wood by son Nathanael	0	2	0
27	Paide to Robert Harvy for a day in filling the well under the woodstack	0	0	10
28	To Marg: Cra: uxori whereof about 4s. given to poor etc., in al	0	10	0
	Expended this month in al	1	5	3

[p.129]

March 1710 Receipts

		£	s	d
24	Received of Edward Parishes wife for a peck of pease, theire price being advanced	0	0	10
25 27	Received of her againe for half a peck of tayle wheat and rye together, the tayle wheat I reckon as rye, that is about 5s. per bushel, and good wheat was sold this yeare at 10s. in Royston and 12s. at Hitchin	0	0	8
28	Item for 3 pecks of tayle barly that was but indifferent in its kind neither, at	0	1	6
29	Item for a peck of wheate he had on Apprill 10 in the year 1708	0	0	8
30	Item for another peck of the same wheate that he had by			

31	Thom: Eastland (myself from home) on the same April 23, vide receipts in April 1710		0	0	8
	Received this month in all		0	4	4
	Received in the last 6 months, viz since the feast of St. Michael last, the summ of			*[blank]*	

March 1710 Disbursements

[*p.130*]

6	Paide to Wallis of Chrishol for 600 sprindles 1s. 3d., 300 withs 1s. 6d.		0	2	9
	Item for tying me up a ringe of wood		0	1	6
7	Paide for 4 bunches of reeds used about malthouse chamber		0	3	0
14	Paide Athan: Carrington for 4 dayes and half lopping in close, fagot		0	4	6
22	Paide to Thom: Dovy for 3 dayes serving thatcher over malthouse chamber and a day in yard		0	3	10
23	For 2 pound of 6d. nailes to nayle up the reeds in that place		0	0	8
	For a little red ink		0	0	3
25	Paide to Ben: Bright for his quarter trimming me ended this day, at		0	2	6
30 31	To Marg: Cra: uxori for her occasions (whereof for 31 pounds of beife and bread to it given to the fast that was the 15 instant, and given besides 6d., that is 9s. 4d.), in all to her		1	4	0
	Expended this month in al		2	3	0
	Expended in the last six months, viz since Michaelmas last, the summ of			*[blank]*	

April 1710: vide librum 7

GLOSSARY

The intention of the glossary is to provide a convenient explanation of obscure, archaic or specialised words used in the accounts and of more familiar words where Crakanthorp's spelling may puzzle or mislead the modern, especially urban, reader. Except where stated, definitions are derived from the *Oxford English Dictionary*. References to the accounts are of examples only and not necessarily to all mentions of the word in the text.

acreman, 136. Cultivator of the soil, ploughman.
baite, 181. Bait, refreshment on a journey.
Bartholomew Day, 61. St. Bartholomew's Day, 24 August.
bayle, 158. Bail, hoop handle of a bucket, etc.
Becket Fair, 217. The annual Royston Fair on 6–7 July, the latter the feast of the translation of St. Thomas (Becket) of Canterbury. cf. A. Kingston, *A History of Royston* (1906), p. 257.
beetle, 253. A ramming or crushing implement.
bill of request, 132. An authorisation for a charitable collection within a county, granted by the justices in Quarter Sessions.
breife, 159. Brief, a government letter authorising a charitable collection.
broad gold, 264. Twenty-shilling piece.
broak(e), brock, broke, 61, 83. The year after the tilth (*q.v.*) in three-year course rotation, so called, perhaps, because the land was already broken, whereas in the preceding year a tilth had to be made from the fallow. See *Introduction,* p. 12.
brock, 124. Inferior (of animals).
bottle, 125. Quantity (of straw).
bottom, 264. Skein or ball (of thread).
bugle cuff, 171. Cuff with tubular decoration.
bullimong, 42. Mixture of oats and pease.
(to) bush up, 168. Protect with bushes.
calash, 135. A light carriage with low wheels.
callimanco, 132. Calamanco, woollen stuff of Flanders.
cant, 173. Quantity (of wood).
cortex peruviana, 146. Bark of various trees used medicinally, esp. for ague.
courser, 212. Corser, (horse-) dealer.
crape, 258. Crepe.
crooper, 135. Crupper, leathern strap buckled to back of saddle.
dimity, 258. Stout cotton cloth.
disk, 245. Collection of florets.

doulas, 133. Dowlas, coarse kind of linen.
dow, 133. Perhaps dowlas, *q.v.*
each, eddish, 98, 111. Stubble or aftergrowth. cf. broak.
elming, 202. Elmen, pertaining to elm.
fan, 243. Basket of special form for winnowing.
(to) fan up, 87. Winnow, separate corn from chaff.
fannful, 153. Quantity of corn that a fan will hold.
fatt, 175. Fat, cask or barrel for dry things.
felloe, 171. Exterior rim of a wheel.
ferrent, 132. Ferret, stout cotton tape.
(to) flea, 215. Flay, to skin.
flosket, 190. Flasket, long shallow basket.
(to) foe, 130, 135. Not in *O.E.D.* Apparently to clean out.
fustian, 133. Coarse cloth of cotton and flax.
galoon, 163. Ribbon or braid used for trimming.
garget, dry, 214. Inflammation of the udder (Primrose McConnell, *The Agricultural Notebook,* 9th edn., 1919. p. 412).
gavel, 90. Sheaf (of corn).
gett or jett, 242. Jet, jett, a large ladle used esp. in brewing.
ghenting, 132. Kind of linen made in Ghent.
graniting, 233. Pomegranate.
groat, grote, 105. Fourpence.
grunsel, 181. Groundsel, timber foundation.
haum(e), 137. Haulm, stems of corn as left after gathering, for litter or thatch.
headstall, 135. Part of bridle which fits round head.
holland, 132. Linen fabric.
horse tree, 163. Swingletree, cross-bar to which traces fastened in plough, etc.
huggeback, 264. Huckaback, stout linen fabric, esp. for towelling.
jack, 162. Contrivance, esp. for turning spit in roasting.
jagg, 244. Jag, a small cartload.
(to) in, 76. Gather into barn, stackyard, etc.
Kentish wheat, 61. See **red wheat.**
knap, 36. A small hill, a knoll.
Lady, Lady Day, 207. Feast of the Annunciation, 25 March.
lainer, 172. Thong, strap.
lime, 203. Limb or limber, in either case the plough shaft or member.
linsy woolzy, 191. Linsey-woolsey, coarse dress material.
Lockier's, Lockyer's pills, 138. A purgative drug, said 'to work gently by stool and vomit' (W. Lewis, *The New Dispensatory,* 1753).
lop, 147. Branch lopped off tree.
lyeletts, 179. Water conduit for washing water.
malagoes, 179. Malaga raisins.
malling, 132. Malines lace.
mow, 39. Division of a barn.
(to) neale, 162. Anneal, to enamel.

ob., 162. Obelus (*Latin*), halfpenny.
offcorn, 36. Waste corn produced by winnowing, often used as poultry feed.
orange flower water, 215. Unidentified medicine which according to Lewis in 1753 (*op cit.*) 'is now laide aside'.
pack thread, 264. Pack thread, strong twine for tying packs, etc.
paving, 162. Tile.
pectoral, 126. Medicine for the chest.
pickle, 245. Pightle, a small field or close.
pitchbrand, 162, 166. Not in *O.E.D.* Apparently a fertilizer.
(to) plash, 188. Make or renew (a hedge, etc.) by interlacing stems.
put, 167. Option of delivering a specified amount of stock at a certain price within a certain period.
quit(t) rent, 260. Annual rent for copyhold property, payable to Lord of the Manor (as rector Crakanthorp was Lord of the Rectory Manor).
red wheat, 36. A variety of wheat which produced a reddish grain and consequently a darker flour than that produced by white wheats. One of three wheats originating in Kent (red, yellow and downy) enumerated in J. Percival, *Wheat in Great Britain,* hence also referred to as Kentish wheat by Crakanthorp.
reeke, 143. Rick.
(to) reive, 170. To twist or thread.
ridgetree, 176. Horizontal timber at ridge of a roof.
ringe(e), 190, 270. Not in *O.E.D.* but apparently a quantity, presumably a bundle (of wood).
Roman vitriol, 220. Blue vitriol, i.e. sulphate of copper, used as a coagulant for haemorrhages.
(to) rowel, 179. Cause a discharge by inserting a rowel between skin and flesh of a horse.
rowles, 245. Rolls, roller for levelling, etc., soil.
(to) ruff-cast, 207. Coat with a mixture of lime and gravel.
Saint Foine, 129. Sainfoin, a perennial herb grown for forage.
saloon, shaloun, 132, 270. Shalloon, closely woven woollen material, esp. for lining.
seed lepp, 159. Seed-lip, basket for sowing.
shack traise, 198. Shackle trace, trace with coupling.
shaling, 45. Shale *or* shellings, husks.
shoate, 209. Shoat, pig under one year.
shott, 63. A division of the open field. See *Introduction,* pp. 14-15.
skep, 190. Bin, skip.
(to) skreen, 137. Screen, to sift or sieve.
staddle, 161. Stump of tree left when it has been felled.
stale, 172. Handle.
(to) stale, 214. Pass (blood in urine).
stetch, 36. Strip of arable in the open field, a selion. See *Introduction,* pp. 15-16.
strake, 123. Iron rim of a cart wheel.
stub(b) money, 260, 272. Not in *O.E.D.* Apparently money for rooting out a tree stump.

stud, 170. Lathes used as uprights in partition walls.
sweet, 127. Sweet cordial.
Swithin, 61. Feast of St. Swithin, 15 July.
tasker, 91. Labourer paid by the task or piece.
tayle, 36. Second-rate corn, see *Introduction,* p. 20.
tilth(e), 35. Tilth, year after the fallow in crop rotation. See *Introduction,* p. 12.
tike, tyke, 211. Tick, mattress case.
tod, 176. Todd, weight of wool, usually 28 pounds.
uxor, 257. Wife (*Latin*).
waifht, 191. Waft, a kind of braid, *or* weft, yarn.
wanty, 176. Band or rope used to fasten pack to horse, body band of a shaft horse.
wash ball, 206. Soap.
wennel, 174. Weanel, an animal newly weaned.
white wheat, 35. A variety of wheat which in later times at least commanded a better price on account of the whiter flour that could be obtained. cf. **red wheat.**
wipple tree, 163. See **horse tree.**
woosted, 132. Worsted.
(to) yelm, 188. To separate and straighten corn for thatching.

INDEX OF PERSONS AND PLACES

The accounts pose a number of problems, the foremost of which is identification, especially where the term of civility 'Goodman', 'Goodwife' or 'Goody' (itself usually abbreviated ambiguously to Good: or G:) is used as an alternative to a Christan name. 'Mr', 'Wid:', etc. present the same problem, and in lists of names (especially grain sales in the household accounts) it is occasionally uncertain whether such a term refers to one or more of the surnames that follow. Such references have been identified in this index according to the balance of probability in each instance. 'Good:' has usually been read as an abbreviation for Goodman, but it must be recognised that Crakanthorp did use it for Goodwife of Goody also, and that he even used this interchangeably with 'Widow' (see for example the references to Goody or Widow Swann on pp. 187, 220, 252, 256).

The abundance of references to a small number of persons and places has made a measure of abbreviation unavoidable. Consequently names with occurrences on more than twenty pages or runs of consecutive pages have been reduced to covering references for those less than five pages apart, preceded by *passim. For example:*

Barber, Lydia, 124-5, 128, 130, 149, 153, 156, 163-4 *etc. has been reduced to*

Barber, Lydia, *passim* 124-30, 149-56, 163-4 *etc.*

It should also be noted that there may be more than one reference to a person on a page. A dash is shown for those persons who are not given Christian names or titles by Crakanthorp or who are described as 'one Cook' etc.

Identification has been much assisted by the transcript of the Fowlmere parish registers by T.P.R. Layng. Information from this and other sources is shown in square brackets.

Aberdeen, 126
Alconbury Weston, Hunts, 252
Aldred, Edward, 11
Allen, Samuel, 148
America, 230
Amis, Richard, 236
Ardeley, Herts, Hare Street in, *q.v.*
Arkle, A.H., 3
Arnold, John, 167
Ashampstead (Hempstead Ash), Berks, 155
Ashwell, (Aswell), Herts, 21, 25, 38, 48-50, 56, 66-7, 71, 73, 78, 80-1, 88, 93, 99-100, 107-9, 118
Aspinall (Aspenal), John, 84, 93, 173-4, 177
Aswell *see* Ashwell

Ayres, Mr., Mrs. Mary, 262

Babraham (Babran), Cambs, 112
Bainton, John, 145
Baker, W.P., 4
Baldock, Herts, 24, 56-7, 84-5, 99, 118, 172, 175, 207
Balleson, John, 156
Barber, Goody *or* Lydia (Lyddy), *passim* 124-30, 149-56, 163-4, 170-3, 182-97, 205-14, 227-39, 249
    Goodman *or* Thomas, *passim,* 101-2, 108, 117-28, 140-61, 168-81, 193, 199-212, 221, 223, 229-41, 247-59
Barker, Edward, 211

INDEX OF PERSONS AND PLACES

Barkway (Barkeway, Berkway), Herts, 24–5, 81, 96, 107, 109; *see also* Shaftenhoe End
Barley (Barly), Herts, *passim* 24–5, 38–41, 64, 102, 116, 148–68, 189–93, 199–201, 227–35, 241–4, 255
Barrington, Cambs, 123, 214
Bartholomew, —, 167
Basford, Notts, 176
Bassingbourn, Cambs, 33, 155
Bayes, Widow, 52, 70, 80
Baynton, John, 265
Beaman, Simon, 93, 114
Beldam, Valentine, 116
Bell, Jacob, 184
Ben or Pen, Mr., 121
Benning, Mr., 127
Bennington, Herts, 6
Bentley, Mr., 267
Berkway *see* Barkway
Beverley, Yorks, St. John's church, 162
Bewdley, Worcs, 249
Biggleswade, Beds, 166
Bird, John, 253
Birkenhead, Ches, 3
Bishop, Vincent, 184
Body, Mr., 161, 163, 174, 176
Boman, *see* Bowman
Bonnis, Ambrose, 56–8
Bonnis, J:, 42
Boreham, Goodman, 43
Botterel, Goodman *or* William, 98, 145, 194, 200
'Bourne' [Bourn, Cambs or perhaps Bassingbourn, which is closer], 71
Bourne, Widow [Joyce], 114
  Mr. [Thomas], 44, 50, 57–8, 71
Bowes, Mr. Urias, *passim* 71–3, 86, 95–6, 101–3, 116, 134, 140, 148–68, 189–201, 227–44, 255
Bowles (Boweles) *or* Bowlesworth (Bowellsworth), John, *passim* 10, 123, 140–7, 154–72, 182–97, 205–9, 234–6, 255
Bowman (Boman), Goodman, 175
Bradmore, Notts, 158
Braintree, Essex, 181
Brand, Gregory, 117
  John, 73, 117
Brenchley, Kent, 259
Brian, Mr. 36
Bridgar, *see* Bridger
Bridge (Bridg), R, 206
Bridger (Bridgar), John, 184
Bright *or* Brightman, Goodman *or* Benjamin, *passim* 124, 130–7, 144–208, 225–35, 241, 248, 255–73

Goody or Elizabeth, *passim* 123–37, 154–63, 169–89, 197–219, 225–6, 232–5, 244, 254
Bristol, Glos, St. Mary Redcliffe church, 267
Brock, Goody, Widow *or* Mary, 134, 164, 172, 174, 189, 196, 202, 205–9, 212–13, 215–16, 234, 236, 259
  Goodman *or* William, *passim* 46–9, 71, 87, 122–32, 139–74, 183–205
  William, younger, 206, 215
Bromley, Kent, 256
Bromley, Mr., 230
Brooks (Brook), Mary, 169, 182
  Widow [Mercy], 83
  Goodman *or* William, *passim* 6, 50, 65–7, 125, 137–8, 146, 151–3, 164–82, 194–202, 212–13, 219–38, 247, 259
Broseley, Salop, 215
Browne, Elizabeth, of Wigan, 178
  Elizabeth, of Wood Norton, 195
  John, 232
Browning, John, 171, 215
Brussels, [Belgium], 252
Bunning, Goodman [Thomas], 77
Buntingford, Herts, *passim* 24, 85, 91, 102, 124–48, 155, 161–4, 174–6, 201–4, 217, 233, 249, 270
  George Inn, 166
Burckett, John, Thomas, 264
Burton, Mary, 163
Bush, Goodman *or* James, 43, 126, 190, 200, 213, 215, 256

C:, J:, *see* Casbourne
C:, Mr., of Cambridge, 86; *see also* Cook
C:, Mr., of Gidding, 159
Cambridge, *passim* 18, 25, 36, 44–5, 67, 74, 86, 126, 131–8, 144–62, 171–2, 190–2, 201–15, 223, 230, 237–48, 264–8
  Christ's College, 11, 15
  King's College, 138
  Midsummer Fair, 131, 169
  Sidney Sussex College, 7
  Small Bridges, 203
  Sturbridge Fair, 112, 140–1, 143, 187, 222, 265
  Trinity College, 5
Cambridge Antiquarian Society, 5
Camden, William, 23
Campian, —, 242
Cann, John, 233, 238, 243
Carpenter, George ('George'), *passim* 126, 136, 143, 151, 158, 166, 172, 178, 186–8, 195–204, 210–17, 224, 230, 243, 253–6, 268

279

Ralph, 33, 124-6, 128, 130, 137-8, 146, 174, 192-3, 195, 200, 212, 229-30, 232-3, 240, 242, 247, 250, 256
Susan, 186
Carrington, Widow [Anne], 177, 179-80, 182, 197, 199, 205, 229, 232, 243, 259
   Goodman *or* Athanasius, *passim* 40-2, 54-7, 106, 122-37, 149-78, 197-9, 209, 229, 243, 251-2, 260, 272-3
   Athanasius, junior, 143
   Rhoda, 207, 212, 227
   Widow [Susanna], 45-6, 77
Carter, Goodman *or* Richard, 33, 148, 155, 190, 200, 228, 237, 243, 255, 260-1, 270, 272
Casbourne, John (J: C:, J: Casb:, Casbourn) *passim* 35-118
Caucutt, —, 70
Cevennais, France, 250
Chaplin, William, 249
Chapman, Thomas, 38
   William *or* Mr., 81, 88, 93, 118
Charles, Goodman, 50
Chartres, J., 30
Chatteris, Cambs, 132
Cherryhinton (Hinton), Cambs, 206
Cheshunt, Herts, 4
Chetton, Salop, 266
Chilterne, Widow [Elizabeth], 77
   Ralph, 55, 64
Chishill, Cambs, formerly Essex, 24, 72, 94, 96, 107-8
Chishill, Great (Upper), 74, 80-2
Chrishall (Chrissol, Crissol), 98, 101, 116, 124, 126, 145, 155, 162, 164, 184-5, 194-6, 204, 211, 214, 240, 254, 261, 273
   Chiswick Hall in, 65
   Chrishall Grange *or* Park in 14, 18, 82, 90, 96, 114, 142, 155, 162, 182, 240
Chrissol, —, 248
Clavering, Essex, 216
Clements, Peter, 228
Clerk, Mr., 171, 212
Clerk, William, 159
Clothall, Herts, 4
Cock, Goodman, 123
Cole, Rev. William, 9
Coleman, Robert, 85
Constantine, Mr. John, 261
Cooch, Mr. Richard, 198
Cook, —, of Cambridge, 36; *see also* Mr. C:
Cook, Esquire *or* Mr., 73-4, 94, 96, 107-8
Cornhill, Widow, 46, 55-6
Cornwel, —, 112
Cotton, Mr., 230
   Sir G:, 166
Course, Goodman, 80

Coxal, Mr., 260
Crakanthorp, Benjamin, 7-8, 132, 171, 176, 185, 190, 200, 215, 248, 257, 264
   Hester, (*later* Wedd), 5, 7-8, 28, 126, 131, 143, 159, 170, 176, 181, 190, 264
   John the elder, 5
   John (1642-1719), 1, 5-10, *passim*
   John (b.1672), 264
   Margaret, *passim* 6-8, 123-32, 139-273
   Mary, 138-9
   Nathaniel, *passim* 8, 32, 123-220, 226-72
   Richard, 5
   Samuel, 7-8, 132, 138, 142, 176, 181, 185, 243-4, 248, 264, 267
   Samuel (fl. 1795), 5
   William, 7
Creak, Steven, 209
Crigglestone in Sandal Magna, Yorks, 171
Crissol, *see* Chrishall
Cross, William, 256
Cutts, Mrs. Blandina, 220, 251

Danvers, Mr., 155
Darlington, Co. Durham, St. Cuthbert's church, 187
Davies (Davyes), Goodman 86, 179
   David, 180, 220
   Richard, 159
   Thomas, 138, 180-1, 220
Daye, —, 250
Deane, Christopher, 247
Defoe, Daniel, 18
Derby, John, 167
Dewe, M., 45
'Dick the tailor', 178
Digswell, Herts, 99
Dingle, Thomas, 175
Dingly, Sarah, 242
Dodgen, John, 184
Douglas, James, 18
Dovy, Widow [Elizabeth], 39, 46, 52, 58, 87, 100, 114
   Thomas, 127-8, 130, 241, 243, 273
   William *or* Goodman, 88, 94-6, 107, 143-4, 146, 153, 160-1, 177, 214, 225, 229, 231-2, 241, 247, 258
Drayton, Dry, Cambs, 162
Dun:, Widow, 45
Durham, 148
Dursley, Glos, 230
Duxford (Duxworth), Cambs, 131, 171

Eastland, Thomas *or* Goodman, *passim* 115, 133-42, 150-1, 158-67, 173-5, 183-5, 192-5, 201-3, 216-23, 237-50, 257-9, 273
   Edinburgh, Canongate, 258

INDEX OF PERSONS AND PLACES

Edwards, Andrew, 210
Ellis, Thomas, 128
Elmdon, Essex, 186, 220, 224, 255
Ely, Bishops of:
  John Moore, 248
  Simon Patrick, 211
England, New, 198
Enniskillen (Inniskilling), Co. Fermanagh, 170
Essex, 9, 160
Etheridge, Mr. John, 233, 270
Evans, G. Ewart, 19
Everet, —, 99-100
Eversden, Thomas *or* Goodman, 16, 69, 145
  Widow, 153, 262

Fairchild, (Fairechild), John, 17, 29-30
  Richard, 29, 42, 88, 123, 153, 155, 175-6, 180, 194, 206, 226, 229, 232-3, 242, 250
  Sarah, 213
  Thomas, 131, 150, 156, 158, 170, 192, 194, 196, 198, 224, 236, 240, 264
Fairfax, Mrs. Margaret, 262
Feaks, Marmaduke, 179, 214
Featherstone, Old, 145, 190, 236, 243
Fere, Elizabeth, 143, 180, 186, 224
  John, 97
Fidlin, Richard, 172, 179, 234, 248, 254
Finch, Mr., 138
Finchingfield, Essex, 139
Finckel, W., 16
Flanders, 233
Fliton, Mr., 72
Fordham, (Fordam, Fordhams), Henry, 40, 49, 136
  Widow [Mary], 58
  Richard, 262, 266
  William or William senior, *passim* 16, 37, 49, 95, 122, 130-42, 150, 167, 179-85, 199, 209, 220-2, 234, 272
  William junior, 143, 163, 181, 201, 224, 242, 249-50, 272
Fowlmere, Cambs, advowson, 5
  Cassanders, *passim* 123, 141, 147-55, 167, 185-94, 200, 206-10, 223, 228-34, 243-61, 268-71
  Chrishall Grange: see Chrishall
  Church, 8, 151, 166, 171, 196, 198, 206, 219, 234, 245-6
  Fowlmere Manor, 5, 15
  Heath, 12, 14, 50-1, 58, 89, 103, 110
  Heslartons Manor, 11
  Hoyes, *passim* 123, 141, 147, 167, 185-90, 200, 206-10, 228-34, 243-5, 254-61, 268-71
  Inns, 11

Lordship, 61, 104, 111
Minor field-names and localities:
  Branditch Shot, 14, 45, 70, 92; Bridge Close 140, 249; Brook Shot, 43, 53, 179, 222; the Buts, 173; Cross Shot, 14; Dods Close, 30, 51, 129, 134, 172, 187; Duck Acre Shot, 15, 36, 85; Fawdon Hill, 59, 69-70, 215, 218, 224; Fincholne Shot, 91-2, 224, 226, 251, 260; Finchway Shot, 60; Gaines, 198; Haberlow, Harborow Hill, 51, 162, 166; Hickses Hedges, 105, 210, 225-6; Horsted Shot, 35, 74, 105-6, 222; Houndsditch (Shot), 226; Lamps Acre, 44, 59, 70, 92, 113, 243; Lamps Stetch, 70; Lank Furlong, 225-6; London Way Shot 51, 63; Lower Shot, 54; Middle Shot, 54; Moor (More) Shot, 41, 53-4, 59; Northmoor Shot, 16, 41, 54; Pipers, 252; Rofeway, 148, 204, 208, 212; Rowze, 36, 40, 63, 84-5, 105-6, 119, 185, 203, 219, 221, 233, 255, 267, 269; Short Furlong, 15, 36, 45, 85, 92, 219, 224, 226; Short Shot, 14, 70, 113; Short Spots, 44-5, 91-2, 153, 180; South Close Style, 228, 240, 251, 255, 272; Tipling's Close, 247; Waterden Shot, 15, 40-1, 48, 51, 63, 88, 105-6, 117, 140, 206, 222, 226; White Acres, 70, 113
Open fields 12, 14-16, 28-30, *passim*
Moor, 10, 14, 69
Roads and byeways: Finch Way, 45, 70; Hockeril's Lane, 245; Linch Lane, 176, 239; London Way, 36, 105-6; Melbourne Way, 45, 215; Mitchell's (Michell's) Path, 45, 162, 166, 223; Mill Path, 15; Norwich Road, 17, 83; Royston Road, 75; Shepreth Way, 14; Walden Way, 89
Fox, Sir Cyril, 3
Foxton, Cambs, 23, 51, 69, 78, 138, 175, 180, 248
France, 179
Frank, —, 255
French, John, 52, 58, 114
Frinkle, William, 170
Frisby, Weston, 233
Fuller, William, 162

Gar:, Goodman, 39
'Gadwell End' (*unidentified*), 94
Gayton, Staffs, Hartley Green in, *q.v.*
George, Widow, 106, 114, 118
Gibson, Thomas, 184
Gidding, Hunts, 159
Gladwell, Goodman, 84

Glenister (Glen:), Mr., 25, 47, 55–7, 59, 65, 72–3, 79–80, 101, 117
Godfrey (Godfry), —, (of Ashwell), 73
   Edward or Goodman, *passim* 42, 62, 75–83, 87, 112, 122–31, 137, 144–9, 154–83, 193–215, 227–8, 246–8
   Thomas, 51, 59, 83
Goode (Good), Widow [Elizabeth], *passim* 125–33, 144–64, 172–80, 189–212, 225
   William or Goodman, 130, 160, 164, 169, 172
Gooding, Ann or Goody, 164, 174, 176
   Jeremy or Goodman, 118, 161, 164, 239, 241
   Widow, *passim* 129–30, 147–63, 183, 189–216, 222–32, 246, 259
Grant, Audrey, 4
Greenhill (Greenhil), Mr., *passim* 146–57, 164–86, 193–219, 227–35, 241, 243
   Mrs., 161, 213, 244
Gregory, Robert, 211
Grey, Mr., 83, 97
Groves, Capt. Timothy, 167
Gurner, Widow [Elizabeth], *passim* 124–38, 146–220, 229–53, 259
   Jane, 253
   John or Goodman, 50, 74, 124, 127–30, 133, 144, 160, 170, 193, 199, 205
   Rose, 138

Haddenham, Cambs, 236
Haggar, Steven, 82, 108
Hales, Francis or Goodman, 164
Hall, Rev. Henry, 29
Hanchet, Goodman or Mr., 141, 192, 252
Hankin, Goodman, 124, 128–9, 146–8, 152, 157, 159–60, 164, 169, 173–4, 177, 179–80, 192–3, 197, 199, 212, 217
   Goody, 129, 133, 172–3, 189
   Susan, 222
   Widow, 249
Hanson, George, 171
Harding, Nurse, 255
Hare Street in Ardeley, Herts, 24, 79, 86, 138
Harlow, Essex, church, 264
Harper, Samuel, 156
Harris, Mr., 159
Harrison, Richard, 156
Harston, Cambs, 196, 226
Hartley Green in Gayton, Staffs, 223
Harvey (Harvy), Dr., 123, 159
   Goody, 213, 249
   Robert or Goodman, *passim* 58, 67, 121–5, 144–70, 191–206, 214–20, 229, 235–47, 256–60, 268–72

   Robert, junior, 205, 207, 225
Hauxton, Cambs, 126, 189
Hawker, Steven, 72
Haycock, Goody, 128, 161, 166, 232, 252
   J:, 48, 56, 115
   Richard or Goodman, *passim* 55, 83–7, 107, 122–8, 138, 144, 152–64, 170–3, 181, 187–94, 201–19, 231–7, 246
   Richard, junior, 228
   Widow, *passim* 124–34, 143, 165–79, 192–3, 199–205, 213–17, 229–34, 249, 262
Haydon, *see* Heydon
Headly, Goodman, 206
Hempstead Ash, *see* Ashampstead
Herne, —, 267, 269
Hertford, 98
Heydon, Cambs, formerly Essex, 59, 109
Hills, Robert, 167, 171, 200, 203, 236, 240, 245, 255, 269
Hinton, *see* Cherryhinton
Hinxton, Cambs, 171
Hitch, Captain [Christopher], 135, 181
   Goodman, 40, 42
Hitchin, Herts, 24, 113, 272
Hoddesdon (Hodsdn.), 190
Hodge, Mr., 164
Holt, Norfolk, 264
Holt, Thomas, 170, 204, 263, 272
Holton, Suffolk, 46
Honywood (Hony:), Mr., 89
   Sir Thomas, 5
Hope, Thomas, 166
Hormead, Great, Herts, 112, 180, 220
Howel, Mr., 210, 245, 247
Hughes, Mr., 109, 235–6
Hull, Yorks, 252
Humstance, Francis, 211
Hunt, Christopher, 171
Hurrel, Mr. William, 122, 226
Hutton, Solomon, 211

Ickleton (Ikleton), Cambs, 136, 179
   Fair, 178
   Grange, 206
Ilet, Goodman, 65
Incarsoale, James, 232
Ingrey, Benjamin, 138, 180, 220
   Widow [Elizabeth], 106, 114
   William, 85–6
Inniskilling, *see* Enniskillen
Ireland, 158

Jackson, Richard or Mr., 81, 115, 117, 132, 151, 170–1, 196, 206, 234, 245, 260, 266–7
Jelgava, *see* Mittau

Johnson (Jonson), —, 214
  John, 220
  Richard, 198, 236
  William, 184
Jeeps (Jeepes), Widow, 172, 212-13, 216-17, 238, 265
  William *or* Goodman, 128, 130, 169, 189, 209, 213, 219, 238, 243-4, 246-7, 249
Jude, —, 56
  Elizabeth, 125, 129-30, 197

'K: Bridge' (*unidentified*), 233
Kefford, Goodman, 48-9, 65-7, 87-8, 94, 104
Keitley (Keitly, Keytly), Mr., 82, 94, 109
Kempton, John, 200
Ken, Dr. [Thomas], 269
Kettle, Richard, 159, 235
Keytly *see* Keitley
Killenbeck, Mr., 10, 37, 42-3, 74, 83
King, Rev. Ezekias, 5
  Gregory, 7
  Isaac, 214
  John, 171
Kinsale, Co. Cork, 251
Kirton in Lindsey, Lincs, 145
Kneesworth, Cambs, 33, 155

Lambert, Sir Anthony, 4
  Rev. Frederick Fox, 4
Lamott, John, 5
Landy, Mr., 154
Langley, Essex, 173, 236, 240
Latimer, John, 123
  Mr., 129-30, 165, 182, 219, 222, 229-30, 242
  Mrs., 231
Laurence, —, 81
Law, Henry *or* Goodman, 42, 54, 67, 78, 100, 107, 118, 129, 164-5, 169, 172, 196, 232
  J:, 56
  Mary *or* Goody *or* Widow, *passim* 122-30, 146-72, 188-93, 197-214, 232-5, 259
  Mat: [i.e. Matty *or* Martha, cf.register], 91
Lea, river, 32
Lee, Mrs., Mary, 198
  Mr. Oliver *or* Valentine, 99
Le Roy, James, 252
Lewis, Mr., 200
Lidlington *see* Litlington
Lincoln, cathedral, 220, 243
Linton, Cambs, 167, 269
Lisburn, Co. Antrim, 250

Litlington (Lidlington, Lydlington), Cambs, 65, 72, 79-80, 88, 106, 113
Littleport, Cambs, 214
Llanfyllin (Llanvillig), Montgomery, 266
London, 4, 10-11, 24-5, 88, 126, 131-2, 141, 148, 151, 178, 206, 221, 223, 230, 237, 265, 269, 271
  Abchurch Lane, 215
  Mermaid Tavern, Cornhill, 233
  Strand, 258
  see also Shadwell Southwark Westminster
Love, Mr., 162, 200, 209, 248
  Mrs., 248
Lunn, Edward, 138

Macer, Goody, 160
  Robert *or* Goodman, *passim* 93, 107, 114, 124-213, 219-23, 232-41
Macfarlane, Alan, 9
Mackrer, Mr., 48
Malin, Thomas *or* Mr., 123, 125, 143, 163, 167, 170, 172, 186, 198, 201, 204, 211, 224, 229, 234, 248, 253, 255
Malton (Molton) in Orwell, Cambs, 133
Manning, John, 5
  Miles, 162, 170, 254
Markshall, Essex, 5
Marshal, —, 243
Marston, North, Bucks, 215
Martin, Mr., 139
Maskal, —, 237
Matthias, Peter, 24
Maulden, Mr., 40, 51, 53, 55
Meldreth, Cambs, 123
Meriden, Warw, 156
Messail, Mr. Ste:, 250
Metcalfe, Rev. William, 28
Mettingham, Suffolk, 198
Michel, *see* Mitchell
Miles, Mr., 89
  James, 6, 180
Miller, George, 85
  Lady, 164
Minshull (Minshal), Church, Ches, 131
Mitchell (Michel, Mitchel), Mr. James, 11, 148, 166, 190-1, 223, 230, 245, 258
Mittau in Courland (now Jelgava in Latvia, U.S.S.R.), 272
Molton *see* Malton
Moor, John, 215
Morden, Rev. John, 5, 15
Morley, —, 81
Morris (Morric, Morrice), Lucy, 266
  William, 131, 171, 196, 225
Morse, Mr., 118
Moule, Newlin, 220

N:, J:, *see* Norris, John
Namur, [Belgium], siege of, 147
Nash, Widow [Elizabeth] *passim* 127-30, 147-65, 174-84, 193-218
  Peter, 129, 134, 147, 149-50, 152, 154, 159-60, 173-4, 209
  Thomas *or* Goodman, 10, 16-17, 42, 51, 110-11, 124, 129, 165, 172, 175, 183, 198, 205, 265
Newling, William, 16
Newman (Numan), Widow, 106
  William, 123, 135, 138, 146, 151, 155, 158, 162, 166, 187, 190, 214-15, 217, 236, 240, 265
Newton, [near Cambridge], 167
Nichols, John *or* Mr., 121, 123, 137, 186, 190
Norman, Clement *or* Goodman, *passim* 170, 177-9, 191-210, 216-17, 225-46, 257-71
Norridge (Norridg), Mr., 124, 229, 245, 262
  Mrs., *passim* 126, 136, 147-50, 159, 170-3, 210, 216, 222-47, 260-5
Norris, John (J:N:), 38-41, 72-3
Norton, Wood, Norfolk, 195
Norwich, Norfolk, 33
Nottingham, —, 222
Numan, *see* Newman
Nuthampstead, Herts, 67

Orford, Suffolk, church, 228
Orwell, Cambs, 123; *see also* Malton
Ostend [Belgium], 215
Ostlar (Ostler), John, 50, 56
Oundle, Northants,Fair, 158
Overton, M., 29
Oxford, All Saints church, 148

P:, Goodman, *see* Payne
Page, Thomas, 46
Palatinate [West Germany], 240, 268
Pale, John, 167
Parish, Anthony *or* Goodman, 74, 79, 95, 101-2, 104, 108, 118, 124, 128-30, 137-8, 160-1, 165
  Edward *or* Ned, *passim* 122-34, 140-57, 168-80, 188-91, 200, 216-39, 246, 252-72
  Elizabeth, 197, 199, 235
  Leonard, 232, 262
  Mary, 157, 176, 182, 197, 238
Payne, Edward, 114
  J:, 57
  Matthew (M:P:) *or* Goodman, 50, 53, 55-8, 62, 65-8, 74, 87-8, 90, 93-6, 101-3, 107-10, 113, 115, 117
Peck, Edward, 83

Jonathan *or* Goodman, 167, 173, 214, 237-9, 241-2, 244, 246-7
  Mrs., 121, 123-5, 128-30, 133, 167
Peirson, —, 58
Pen or Ben, Mr., 121
Pepys, Samuel, 9
Phage, Anthony, 57
Phipp, Edward, 124-6, 130, 139, 141, 143, 147, 153, 200, 204, 216, 249
  Rhoda, 143, 186, 217
Pigg, Mr., 101, 116
Pinnock, Thomas, 123
Pitty, James, 80-1
Porter, —, 136
Postgate, M.R., 15, 18
Potter, Rev. Simon, 6
Preston, Widow, 36
Priest (Prist), —, 113
Prior (Prier), —, 268
Puckeridge, Herts, 77, 121

R: Mrs., *see* Richardson
Rainer, *see* Rayner
Rand, John, 15
  Widow [Mary], 100, 106, 115, 159
  William, ,93, 115
Rasen, Market, Lincs, 266
Rayner (Rainer), John *or* Goodman, 16-17, 30, 104-5, 122, 169, 171, 207, 210
  John (of Littleport), 214
  Goodwife [Mary], 246
  Thomas, 45
  William, 251
Reach, Cambs, Fair, 170
Reading, Berks, 5
Redhead, Mr., 236
Reynolds (Reinold, Reinolds), Widow [Elizabeth], 62, 77, 114, 134, 159, 196, 216
  Kate, 125, 127-8, 147, 149, 154, 156-8, 216
  Thomas, 46, 51, 53, 58-9, 66, 68, 83, 144, 153, 160, 183
Ribshir, William, 189
Richardson (R:, Rich:), —, 166
  Mrs., 132, 176, 248
Riesman, —, 9
Roberts, J:, 57
Rolleston, Staffs, 148
Rowley, Mr., 65
Royston, Herts (part formerly Cambs), *passim* 3-11, 18-25, 36-187, 195-247, 255-60, 266-72
  Ash Wednesday Fair, 201
  Becket Fair, 135, 217
  Whitsun Fair, 214
Rule, George, 148

INDEX OF PERSONS AND PLACES

Rumbald, Mr., 47, 79, 86

S:, Mr., *see* Stevenson
St. Christopher, West Indies, 220
St. Malo, France, 262
St. Neots, Hunts, 169
    Fair, 171, 214
Samford, Great (Old), Essex, 181
Sandal Magna, Yorks, Crigglestone in, *q.v.*
Sandon, Herts, 57
Scampton, Mr., 233
Scott-Fox, Commander C., 3
Selby, Mr., 144, 146
Sell, Christopher, 18
    Richard *or* Goodman, 40-1, 43, 59, 73, 81, 87-8, 94-5, 101-4, 108, 111, 116
    William, 80
Shadwell St. Paul, Middx, 241
Shaftenhoe (Shafnoe) End in Barley, Herts, 24, 71, 101, 108
Shelford, Great, Cambs, 138, 232
Shepreth, Cambs, 23, 71, 78, 115, 178
Sherwin, William, 6
Shingay (Shingey), Cambs, 93, 114
Shire Lane, *see* Westminster
Short, Jane, 242
Sleigh, James, 148
Smith, —, 80
Soale, William *or* Goodman, 148, 163, 166, 171, 182, 184, 198, 203-4, 226, 228, 233, 243, 247
    William, junior, 226
Southam, Warw, 236
Southwark St. Olave, Surrey, 184
Southwark St. Saviour, Surrey, Bankside in, 159
Sparhawk (Sp:, Spa:), Edward (Ned) *or* Brother, 141, 154, 206, 245, 261-2
    Sarah, 206, 261
Spilman, George *or* Goodman, 42, 49, 51
Spilsby, Lincs, 211
Spufford, H. Margaret (née Clarke), 9-11, 17-18
Stanford Dingley, Berks, 5
Stevenson, (S:, Stevens:), Mr., 126, 167, 243, 248
Stiffen, William, 5
Stoke by Clare, Suffolk, 270
Stort, river, 32
Stortford Bishop's (Stortford), Herts, 105
    Fair, 131
Strett, Widow [Jane], 157, 165, 177, 193, 217, 238, 241, 246
    William, 150
Sturbridge Fair, *see* Cambridge
Sunderland, Durh, 214
Sutton, R:, 226, 228

Swann, James (J:) *or* Goodman, 111, 115, 118, 122, 130, 172-3, 189, 203, 205, 213
    Goody *or* Widow (Margaret), *passim* 128-37, 144-71, 177-89, 197-202, 216-25, 241-69
Symonds, William *or* Goodman, 43, 59, 100, 105

T:, W:, *see* Thrift
Taylor (Taylour), James *or* Mr., 124, 126, 132-3, 165, 167-8, 174, 198, 219, 221, 223, 242, 244, 258, 265, 268
    Jane, 211
Thake (Thak), J:, 67
Thompson, John, 140, 142, 152-3, 155, 162
    Goodman, 42, 169
    Henry, 10, 104, 112
    Widow, 252, 262
Thornburgh, John, 242
Thorne, J:, 64
Thorney, Cambs, Thorney Fen in, 215
Thorowgood, William, 71
Thrift, Ann, 121, 124-5, 128-9, 207, 209, 246
    David, 198, 240
    Mary, 174, 217, 225, 227, 235
    William, *passim* 20, 38-119, 126-167, 174-7, 183-91, 197-8, 208-10, 218-20, 227-8, 238-45, 252-3, 260-4
    William, junior, 160-1, 164, 168-70, 173-4, 202, 208-9, 213-14, 216-17, 222, 227, 229-30, 235, 254, 262, 264, 271
Thriplow (Triplow), Cambs, 21-2, 40, 41, 48-51, 59, 73, 81, 87-9, 94, 101-2, 104, 116, 125-6, 167, 252, 265
Thurgood, Miles, 25, 37-41, 47-50, 57, 65-6, 72, 79, 86, 95, 101, 107-9, 115-16, 118; *see also* Thorowgood
Torrington, Great, Devon, 175
Towcester, Northants, 217
Triplow, *see* Thriplow
Turpentine, Mr., 24, 102-3
Tyrrel, Anthony, 214

Vancouver, Charles, 18

W., Mr., Mrs., *see* Wedd
Wadnam, John, 116
Waites (Wates, Weights), John *or* Goodman, *passim* 127, 143-54, 160, 168-74, 185, 193, 215, 223-35, 255-7
    Mary, *passim* 122-38, 148-9, 157-8, 164-97, 222-35, 241-7, 259
    Widow, 160
Wakelin, Joseph, 223
Walden, Saffron (Walden), Essex, 11, 18, 123, 162
Wallington, Herts, 6

Wallis, Henry (of Chrishall), 59, 83, 104, 124, 126, 150, 155, 164, 170, 184–5, 191, 194–6, 204, 211, 240, 254, 261, 273
    Mary, 122, 128, 208–9, 247
    Richard, 51, 59, 162
    Thomas, 138, 143, 180, 219
    Widow, 59
    William, 124, 126, 143
Ward, Betty, 178
    Widow [Mary], 125, 127, 152–3, 157, 174, 178, 184, 192–3, 197, 208–9, 213, 229
    Mr., 203, 237
    Robert, 58
    Thomas, 92, 100, 106, 108, 139, 142, 147, 204, 244, 250
    William, 139, 183, 185, 220, 225–6, 254–5
Ware, Herts, 18, 87
Warner, Goodman, 99
Warren, Mr., 249
Watson, Widow [Anne], 100, 114
    John *or* Goodman, *passim* 56, 64, 80–3, 111, 136, 144, 164, 187–95, 216–23, 235–41, 254–6, 262
    John, junior, 145, 165, 186, 208–9, 212–13, 224–6, 229, 249, 255–6, 263, 265–6
    Mary *or* Goody, 126, 208
    Thomas *or* Mr., *passim* 20, 124–7, 137–42, 151–5, 167, 182–4, 190–8, 211–27, 237–53, 260, 266
Wedd, Benjamin, 4, 7–8, 132, 159, 181, 201, 220–1, 224–5, 233, 270–1
    Benjamin, junior, 255
    Dorothy, 261, 265
    Edith, 129, 134, 144, 163, 170, 186, 196, 202, 205, 208, 212–14, 232, 235, 246–7
    Hester, *see* Crakanthorp
    Josiah *or* Goodman, *passim* 110–11, 122–37, 144–83, 189–205, 213–16, 229–32, 238
    Mr. *or* Brother, 126, 129, 131, 149, 164, 175, 187–8, 197, 202, 219, 230, 237, 250
    Mrs., 172, 209
Weights, *see* Waites
Welch (Welsh), Henry, 83
    John, 16
Wells, John, 172, 207, 210
Welsh, *see* Welch
Wentworth, Mr. Robert, 250
    Capt. Thomas, 211
West, (*perhaps for* Westly), Mr., 244
Westly, Mr., 226, 265
    Mrs. [Elizabeth], *passim* 122–8, 144–54, 160, 169–209, 218, 226–32, 238–41, 262–7
    Widow, 232
Westminster, Middx, Charles Street, 242
    Shire Lane (*now lost*), 220
Weston, —, 114
Whit:, J., 58
Whitby, Widow [Elizabeth], 88, 106
    Thomas, 45, 65–6
    William, 46, 52, 57–8, 63, 114
Whitehead, G:, 37
    Widow, 44, 100
Whitham, Mr., 82, 86
Whittaker, John, 211
Whittingstall, James, 187, 260–1, 265, 270
Whittlesford (Wittlesford), Cambs, 23, 100, 105, 242
Wigan, Lancs, 178
Wilbraem, Thomas, 266
Willamite, Robert, 204, 264
Wilmot, —, 260
Wilson, Mr., 126, 170, 209, 244, 261
Wiltshire, 25
Wincanton, Som, 248
Winchester, Hants, 25
Winds (Windes), Charles, 233, 236, 239–40
Winstanleye, Mrs., 185
Wittlesford, *see* Whittlesford
Wolfe, J., 45
Wood, Robert, 45
Woodforde, Rev. James, 8
Woods, Cassandra, 146–8, 152–4, 189, 193–4, 197, 209, 215–16
    Goodman, 144, 165
    Joan, 114, 127–8, 144, 149, 154, 157, 168, 170, 180, 184, 192, 213, 215
    Samuel, 122–3, 126, 128, 130, 138–9, 141, 188
    Widow, 106
Woodward, J[ohn] *or* Goodman, 122, 129–31, 153, 155, 157, 160, 164–5, 168–9, 172–5, 177, 179, 191, 229–30, 270
    Widow, 127
Woollard (Woolward), Robert *or* Goodman, 46, 61, 64
    Widow, 70, 76–7, 100, 106
Wootton, Thomas, 131, 222
Worleston, Ches, 266
Wright, M:, 130
    Mr., 146, 166, 266–7

Yarmouth, Great, Norfolk, 253
Yorke, Rev. A.C., 3

INDEX OF SUBJECTS

Most subjects are arranged in groups, the principal groups being under the headings: Cloth, Crops, Ecclesiastical matters, Farm buildings, Farm equipment, Farming, Food and drink, Fuel and light, Horses, Household articles, Livestock, Medicine, Occupations, Stationery, Tithes, Tools

Agriculture, *see* Crops; Farm buildings; Farm equipment; Farming; Horses; Labourers; Livestock; Manure; Pests; Tools
Animals, *see* Livestock; Pests

Bills of request, *see* Briefs, etc.
Board and lodging, 132, 143, 257-8, 260-1
Books, 8, 200, 210, 267, 269
    catachisms, 170, 264
    see also Stationery
Briefs, bills of request, etc, for churches, 131, 148, 162, 176, 187, 215, 228, 230, 259, 264, 266-7
    for losses by fire, 132, 145, 148, 155-6, 158-9, 162, 167, 170, 175, 184, 195, 211, 214-15, 217-18, 220, 223, 236, 241-2, 248-50, 252, 258, 264, 270
Building, house, 250
    walls, 245
    materials, bricks, 166, 197, 216
        clay, 173, 205, 216
        clunch, carting, 239
        lime, 195, 206, 257
        pitch, 167
        stud, 170
        tar, 151
    see also Thatching
Bullimong, 42, 274

Chaff, 153, 160, 182, 257
Chimney sweeping, 203, 223, 241
Churches, *see* Briefs; Ecclesiastical matters; *Persons and Places Index* under particular places, esp. Fowlmere
Churchyard, grazing in, 168, 246
Clergy, income of, 7
    widow, distressed, 198, 220
Cloth
    broadcloth, 181
    buckram, 133, 181, 196, 233
    calamanco, 132, 274

calico, 132
cambric, 259
canvas, 181, 233, 247
cotton, 132
crepe, 262, 274
damask, 139
dimity, 132, 177, 255, 274
dowlas, 133, 195, 262, 274
ferret, 132, 275
fustian, 133, 275
galoon, 163, 214, 275
ghenting, 132, 275
holland, 132, 259, 262, 275
huckaback, 123, 264, 275
lace, 132
linen, 132, 172, 216, 246, 250
linsey-woolsey, 191, 275
malling, 132, 275
mohair, 133, 181, 196, 233
muslin, 132
plush, 132
shalloon, 132, 163, 181, 233, 276
silk, 132-3, 145, 163, 177, 181, 190, 214, 233, 247, 259
tape, stay, 196, 233
thread, silver, 132
worsted, 135, 151, 277
Clothes, 195
    aprons, 132, 168
    breeches, leather, 133, 210, 247
    coats, 133, 177, 181, 196, 270
    coat, riding, 233
    cuffs, 171
    frock, 133
    gloves, 177-8
    gowns, 220, 267
    gown, studying, 269
    handkerchiefs, 132-3, 172
    hood, riding, 156
    scarf, 267
    shifts, 172
    shirts, 133, 172
    stockings, 132, 191, 224, 255

undercoat, 145
waistcoat, 132, 181, 214
wedding clothes, 132
see also Footwear; Headwear
Colts, *see* Horses
Copyhold land, 28, 240, 248, 251, 255, 260, 271-2
Cordage, binding, 170
  cart rope, 178
  pack thread, 156, 264
  plough cord, 176
  whip cord, 123, 171
Crops, sale of, 21-8, 121-273 *passim*
  sample bag, 159
  seed corn, 58, 64, 66, 70, 73-4, 89, 95, 97, 109, 118, 145, 171, 176, 195, 200-1, 210, 227, 240
  tithes of, *q.v.*
  yields, 21-31
  types of:
    apples, 142, 228, 253
    barley, 17-18, 38-272 *passim*
    beans, 200, 240
    hay, 133
    hemp, 201
    lentils, 40, 42, 48, 52, 58, 61, 90, 200
    oats, 42-257 *passim*
    pears, 137, 142
    pease, 42-119 *passim*, 134, 151, 208, 237, 240, 270-2
    saffron, 18, 271
    sainfoin, 17, 129, 134, 172, 187
    tares, 17, 198, 240, 265
    turnips, 17
    wheat, 35-273 *passim*
  see also Vines

Debtors, imprisoned, 171, 178, 204, 264
Diary-keeping, 9
Disabled persons, blind, 146, 209
  crippled, 176, 178, 181
  deaf and dumb, 171
  dumb, 236
  distracted, 162
  see also Seamen; Soldiers
Dovehouses, tithes of, *q.v.*
Dung, *see* Manure
Dutch, 148

Ecclesiastical matters, gifts to foreign Protestants, 250, 261
  procurations, 131, 171, 226, 266
  register bill, 133
  tenths, 126, 170, 209, 244, 261
  visitation, archdeacon's, 131, 171, 226, 266
    episcopal, 211, 248

see also Briefs; Churches
Elections, 126, 230

Fairs, 112, 131, 135, 140-1, 143, 158, 169-71, 184, 187, 201, 214, 217, 222, 265 *or see Persons and Places Index* under particular places
Farm buildings, etc
  barns, 38-118 *passim*
  hire of, 126
  repairs to, 136, 145, 150, 206-7, 222-3
  thatching, 188, 256, 266
  chaff house, 64, 78
  garner, granary, 39-118 *passim*
  gates, 150, 167, 196, 207, 215, 219, 222-3
  hayhouse, 89, 96, 181
  hogsties, 142, 206, 218, 256, 268
  malthouse, 35-118 *passim*, 139, 150, 196, 228, 251, 257, 272-3
  shed, 58, 73
  stable, 59-110 *passim* 123, 143-4, 162, 178, 181, 222-3
  strawhouse, 147
  see also, Gardens; Orchard; Ponds
Farm equipment, 265
  bins, 136
  cart, etc., 136, 181-2, 223, 226, 233
  cheese press, 123
  cow bells, 144
  dung cart, 187
  hog's rings, 228
  ladders, 136, 181
  nails, 135-257 *passim*
  pails, 148, 158
  racks, 136, 150, 179, 181, 222, 231
  red lead, 135
  skip, 190, 271
  tubs, 214
  tumbril, 163, 171, 184, 198, 214, 223, 247
  wheelbarrow, 167, 181
  wagon, 123, 150, 163, 171, 181, 190, 222, 233, 244, 247
  see also Clothing, gloves; Cordage; Horses, tackle; Household articles; Pump; Tools
Farming
  dunging 126-253 *passim*
  field-keeping, 252
  fruit picking, 142, 228, 233, 253
  gathering crops, 138
  haulming, haulm carting, 142, 154, 183, 190-1, 194-5, 223, 228, 239
  hay making, 142, 176, 191, 218, 220, 249, 253
  hedging, plashing, 127, 135-6, 176, 188, 195, 210, 239, 253, 260, 264
  lopping, 127, 240, 260, 273

288

INDEX OF SUBJECTS

mowing, 136, 176, 178, 218, 220, 253
ploughing, 123, 127, 131, 142, 163, 167, 176, 185, 198, 201, 210, 223, 228, 239-40, 245, 248, 252
reaping, 136, 179
rick inning, 150, 154, 162, 176, 181, 218, 224
rick thatching, 138
rolling, 123, 175, 245
scouring brook, etc., 188, 233
setting trees, 163
sowing, 136, 142, 154, 156, 158, 166-7, 185, 203, 223, 239-40
stone gathering, 123, 126, 166, 218, 233
threshing, 19-20, 35-119 *passim*, 135-272 *passim*
see also Farm buildings; Farm equipment; Fuel, wood, carting, fagotting, gathering chips; Labourers; Livestock; Manure; Pests, Tools
Fields, open, 10, 12-16, 28
see also Person and Places Index under Fowlmere
Financial matters, see Loans; Rents; Stationery, account books
Fires, 98, 112
see also Briefs
Food and drink
apples, 254
asparagus, 245
beer, 243
bread, 273
butter, 139, 188, 231
caraway seeds, 139, 218
cheese, 127-248 *passim* (at the end of most months' disbursements)
cheese, Cheshire, 231
cherries, 215
currants, 139, 179, 218
ducks, 231
ginger, 218
goose wings, 226
honey, 236, 271
malaga raisins, 136, 139, 179
meat, 123-273 *passim* (at the end of most months' disbursements)
milk, 231, 257
oranges, 126
oysters, 143, 190
pease, 211
pepper, 218
pidgeons, 142, 231
plums, 226
raisins, 218
rice, 139, 179, 218
salad, 127

salt, 137, 156, 159, 179, 201, 216, 226, 248
strawberries, 203
sugar, 136, 139, 179, 188, 218, 226
sugar-plums, 211
vinegar, 211
whitings, 201
wine, 143, 184, 190
see also; Crops; Inns; Farming, fruit picking; Medicine; Refreshment
Footwear
boots, 132
pattens, 123, 191
shoes, 132, 138, 163, 177, 181, 228, 233, 236, 246, 250, 260, 267
French, the, 148, 167, 179, 198, 211, 215, 220, 230, 262, 264
Fuel and light
candles, 137, 148, 158, 191, 226, 231, 254-5
coal, 131, 153, 162, 190, 192, 200, 206, 209-10, 215, 223, 245, 247-8
oil, 127
turf, 140, 161, 212, 237, 248
wood, 134, 190, 200, 228, 241, 243-4, 255, 270, 272-3
from Chrishall Park, 182
from Cleave Park, 205
from Clavering 216
from Essex, 160-1
from Langley, 173
from 'Rofeway', 133, 148, 204, 208, 212
carting wood, 127, 175, 210, 218
fagotting, 136, 210, 218, 240, 245, 260, 273
gathering chips, 176
stub money, 237, 260-1
see also Timber
see also Household articles, lantern
woodstack, 272
Furniture, 8
bedding, 156, 211, 216
chairs, 190, 257
doors, 156
dresser, 264
stall, 268
table, 123, 133
Garden, 167-8, 171, 188, 203, 236, 239, 255
Garden, kitchen, 210, 250
Gentleman, distressed, 250
Glazing, 151, 156, 171, 196, 234, 245, 260
Glebe 8, 28-9, 45, 48, 60-1, 63, 70, 76, 83, 90, 92, 105, 112, 220, 224, 248-50, 264

Haulm, 137-262 *passim*
Headwear, caps, 255
hats, 133, 171, 178, 245
wig, 236, 261

Herdsman, 131, 142, 155, 175, 215, 225, 230, 233, 256
Hides, skins, bull, 159
  bullock, 135
  calfskin, 132, 135, 146, 166, 203
  horse, 146, 166, 214
  shammy skin, 102
  unspecified, 187, 190, 240
  see also Clothes, breeches
Horses, colts
  Ball, 162, 169
  Bonny, 131, 169
  Boy, Buoy, 166, 232
  Buck, 128
  Duke, 135, 145
  Jolly, 171, 200, 243
  Typtery, 126, 179, 190, 224
  Watt, 212, 215, 236
  breakages by, 172, 219
  hire of, 138, 179, 187, 220, 224
  oats for, 41–119, *passim*
  purchases of, 131, 158, 214–15
  sales of, 128, 135, 171, 212
  shoeing of, 126, 155, 166, 194, 200, 204, 236
  sickness of, 131, 145, 162, 179, 194, 200, 236
  tackle for:
    bits, 151, 214
    brush, 195, 230
    collar, 155
    crupper, 135, 274
    currycomb, 155
    halters, 146, 151, 171, 207, 217
    hames, 155, 162
    lainers, 172, 275
    reins, 207
    saddle, 181, 230
    trace, 155, 236
    wantys, 159, 176, 277
    whips, 141, 155, 196, 270
  see also Farm equipment, cart etc.; Occupations, collarmaker, farrier; Servants, horsekeeper; Tools, ploughs
House, bakehouse, 170, 188, 203, 206–7, 223, 241, 256
  buttery, 156, 245
  kitchen, 195, 203, 206, 233
  study, 51–2, 117
Household articles
  baskets, 190, 243
  brass cock, 196
  cheese press, 17, 123
  clock, 162, 170, 194
  copper, 166, 170, 240, 269
  dish, 248
  flasket, 190, 275
  hearth, 256
  jack, 159, 162
  lantern, 188, 233
  locks and keys, 123, 138, 156, 162, 245, 254
  mats, 200, 209
  ovens, 136, 158–9, 162, 166, 194
  pot, 172
  quern, 8, 172, 175
  rat trap, 204
  roasting iron, 254
  scuttles, 190
  shoe-horn, 268
  tablecloths, 216
  towels, 216
  wash-balls, 206, 277
  see also Chimney-sweeping; Farm equipment; Furniture; Pump; Tools; Well
Housekeeping, 78, 257–60

Implements, *see* Tools
Injuries, 142, 151, 166
Inns, Angel (*not located*), 244
  George, Buntingford, 166
  Mermaid, London, 233
  White Lion (*not located*), 185
Irish, 158

Labourers, acremen, 97, 104, 111, 134, 274
  farm hands, 108
  harvest men, 177
  taskers, 75, 77–8, 84, 91, 116, 195, 277
  see also Servants; tithe men
Largesses, 220, 243, 251, 267
Leather, *see* Hides
Livestock, 16–17, 265
  bitch, spayed, 195
  boar, brought from Foxton, 248; cow killed by, 257; gelded, 171, 225
  bull, bought, 206; slaughtered, 75; sold, 204, 229; stray, 33, 155
  bullocks, wintering of, 158, 221
  cows, calves, heifers, found dead, 203; hurt, 178; killed by boar, 257; losses of, 236; sick, 179, 214; sold, 124–5, 170, 199, 201, 204; wintering of, 137–262 *passim*
  hogs, pigs, shoats, sows: churchyard spoilt by, 168; corn for, 37–119 *passim*, 259; gelded, 171, 196; kept, 190, 223; pounded, 138, 248; slaughtered, 186, 195, 200, 230, 233, 240, 242; sold, 201, 205, 209, 212
  poultry (fowls, hens), barley for, 39, 56, 59, 66, 87–9, 93, 103, 107, 109, 115–18, 228, 253

INDEX OF SUBJECTS

sheep, wintered in churchyard, 219
wennel, wintering of, 174
see also Herdsmen; Hides; Tithes
Letters, 136, 148, 151, 166, 179, 206, 223, 230, 250
Loans, interest on, 122, 132, 141, 154, 176, 206, 215, 245, 248, 262

Malt, exchanged for barley, 117
  purchases of, 231, 233, 236, 239-40, 251, 254, 256
  sales of, 88, 130-229 *passim*
Malting, 24, 40-108 *passim*
Manors, see Copyhold land; Rents, quit; *Persons and Places Index* under Fowlmere
Manure, dung, 76, 127-269 *passim*
  dust, 269
  hen dung, 162, 166, 188, 195, 210
  pidgeon dung, 206
  pitchbrands, 162, 166, 276
Maps, 269
Materials, building, *q.v.*
  clothing, see Cloth
  thatching, *q.v.*
Medicine, bleeding, 123, 141
  candy, 126, 187
  cortex peruviana, 146, 274
  liquorice, 142, 258
  Lockier's pills, 138, 158, 210, 255, 262-3, 266-7, 275
  orange flower water, 215, 276
  pectoral, 126, 187, 276
  Roman vitriol, 220, 276
  tincture of steel, 266-7
  white wine, 236
  unspecified, 159, 166, 215, 233
  see also Nursing
Medicine, veterinary, see Horses; Livestock

Nursing, 143, 190, 255

Occupations, apothecary, see Moor, John,
  [attorney] see Warren, Mr.
  bakers, of Cambridge, see Brian, Mr.
    of Royston, see Whitehead, G.
  barbers, at Braintree, 181
    at Cambridge, 131, 146, 171, 226
    unspecified, 245, 248, 262, 266
    see also Trimming
  [blacksmith], see Fairchild, Thomas
  butcher, see Fidlin, Richard, Malin, Thomas; Peck, Ed:
    unspecified, 256, 258
  carpenters, at Royston, 115
    unspecified, 207
  chapman, see Bowes, Urias
  [cobbler], see Haycock, Richard
  collarmaker, see Newman, William
  cooper, see Campian, —
  currier, unspecified, 132, 159
  farrier, see Thompson, John; Horses, shoeing of
  felmongers, unspecifed, 245
  gardener, see Hills, Robert
  gelder, see Morris, William
  glazier, see Jackson, Richard
  haberdasher, see Bird, John
  horse corser, see Clerk, —
  [malster], see Kefford, Goodman; Sell, Richard
  mason, unspecified, 163, 196, 218, 250
  mealman, see Cook, —
  miller, of Buntingford, see Body, Mr.
    of Fowlmere, unnamed, 149, 170, 172-4, 177, 179-80, 184, 193, 201, 207, 250-1, 263, see also Hales, Francis
    of Herts, see Weston, —
    of Hormead, see Cornwel, —
    of Royston, see Whitehead, G:
    of Shepreth, unnamed, 71, 115
    of Whittlesford, see Symonds, William
  physicians, see Medicine
  tailor, see Dick
    unspecified, 151, 181, 191, 196, 258
  thatcher, unspecfied, 138, 191, 194-5
    see also Watson, J:
  weaver, unspecified, 168, 246, 264
  worsted man, see Cross, William
  see also Herdman; Labourers; Parish Officers; Seamen; Servants; Soldiers
Old people, 194, 204
Orchard, 52, 67, 175, 203, 239, 245, 253

Painting, daubing, 220, 245
Parish Officers,
  clerk, 145, 187, 206
  see also Rates
Pests, mice, rats, 62, 64, 70, 77, 84-5, 91
  moles, 226, 250, 271
  vermin, 78
  weevils, 105, 224
Poor and needy, 37, 45-6, 63, 70, 78, 85, 100, 111, 131-2, 135, 142, 145, 155, 176
  see also Briefs; Clergy, widows of; Debtors, imprisoned; Disabled persons; Gentlemen, distressed; Largesses; Old people; Scholars, distressed; Seamen, distressed; Soldiers, distressed
Population, 10-11
Pump, 132, 156, 170, 194, 206-7, 266-7

Rags, shreds or, 210
Rates, church, 230, 242
   constable's, 155, 194, 226, 233, 242, 250, 270
   poor (overseers'), 126, 167, 175, 210, 223, 230, 242, 265
Refreshment (bait), 181, 274
Rents, Parish's, 157, 189-90, 231-2, 263
   Swann's, 143, 163, 186, 219, 224, 242, 253, 262, 268-9
   Watson's, 144, 224-5, 254-5, 263
   for parsonage, 254, 257, 260, 266-70
Rents, quit, paid, 123, 167, 210, 230, 245
   received, 124, 216, 252, 254, 258, 260-3, 265, 270
Roads, 161
   see also Tolls; *Persons and Places Index* under Fowlmere

Scholars, distressed, 139, 171
Scotch, 209
Seaman, castaway, 243, 252
   disabled, distressed, 126, 179, 185, 214-15, 230, 242
Seed, see Crops
Servants, gifts to, 143, 175, 186, 224, 233, 268
   wages of, 143, 186, 224, 233, 253
   types of:
      bailiff, 164
      horsekeeper, 68, 90, 92, 159, 171
      maids, 143, 154, 175, 186, 224
      midwife, nurse, 190, 261
   see also Labourers; Nursing; Parish officers, clerk
Shoes, see Footwear
Soldiers, disabled, distressed, 135, 138-9, 147, 166-7, 185, 215, 220, 228, 233
Spectacles, 185, 269
Stationery, account books, 126, 135, 213
   almanacs, 151, 195, 230, 256, 270
   ink, 247, 273
   paper, 245, 247, 256, 262
   parchment, 133
   sealing wax, 126
Straw, barley, 125, 187
   pease, 174, 197
   rye, 122-225 *passim*
   wheat, 128-250 *passim*
   unspecified, 124-262 *passim*
   see also Haulm

Tailoring, 151, 191, 196
Taxes, hearth, 10-11
   land, 141, 147, 155, 167, 185, 190, 200, 206, 223, 228, 234, 253-4, 259, 261, 267-71

militia, 171, 191, 256, 266
window, 147, 166, 190, 206, 234, 243, 267-8, 270
Thatching, of barns, 188, 256, 266
   of malthouse, 273
   of stables, 145
   of strawhouse, 146
   materials for (reeds, rods, spits, withs), purchased, 126, 146, 150, 153, 155, 170, 184-5, 191, 194-5, 211, 214, 236, 244, 254, 273; sold, 172, 183, 187, 262, 265
   see also Yelming
Timber, 226
   ash, 236
   deal, 174, 199, 212, 216, 229, 264
   elm, 202, 212
   oak, 161, 182
   sycamore, 202
   willow, 166
Tithes, of
   apples, orchard, 169, 187, 222, 227
   barley, 30, 48
   cattle, 139-40, 169, 193, 207, 209, 221, 225, 241, 244, 265
   corn, 218
   dovehouse, pigeons, 122, 169, 207
   grass, hay, 33, 129, 140, 247, 252
   lambs, 139, 221, 225
   oats, 82, 89, 96, 104, 118
   pease, 30, 51, 67, 74, 97, 111
   pigs, 164, 225
   sainfoin, 33, 129, 172, 187
   wheat, 30, 105
   wool, 33, 129, 176, 219
   unspecified small tithes, 124-5, 138, 165, 180, 190, 205, 221, 229, 243, 265
   claimed by Mr. Miles, 89
Tithemen, 30, 69, 138, 180, 220
Tobacco, 139, 181, 218
Tolls, 131, 158, 190, 203, 210, 215, 247
Tools
   axe, 266
   beetles, 253
   brush, 266
   churnstaff, 204
   dressing brooms, 146, 158, 198
   fans, 236, 243
   flail, 19
   forks, pitchforks, 136, 172
   hammer, 266
   knives, 148, 171
   ploughs, making new, 163, 166, 204, 228
   repairs to, heading, 148, 171, 184, 198, 228, 233, 243
   beam, timber, 203-4, 226
   share, 236

INDEX OF SUBJECTS

trace, etc., 138, 176, 214-15
ploughstaff, 196
rakes, 181, 194, 264
rubber, 187
saw, 266
seed-lip, 159, 276
shovels, 200, 236
sieve, chaff sieve, 145, 187, 195, 211, 236
see also Farm equipment
Transport, calash, 135, 181
   coach, 135
   see also Farm equipment, Tolls
Travellers, 126, 148, 158, 262, 268
Trees, dead, 257
   peach, 233
   walnut, 257
   see also Timber; Willows

Trimming, 142, 185, 206, 210, 225, 231, 241, 248, 255-6, 260, 264, 268, 271, 273

Vehicles, *see* Farm equipment; Transport
Vines, 200, 233, 239

Weather, 35, 44, 48, 60, 76, 83, 90, 97-8, 104, 111-12
Wedding, 245
Weeds, 69
Well, 223, 272
Willows, 122, 134, 147, 168, 173, 221, 258, 260
Wool, 129, 132-3, 136, 176, 187-8, 219
   tithes of, *q.v.*

Yelming, 150, 188, 256

CAMBRIDGESHIRE RECORDS SOCIETY
Hon. General Editor: P.C. Saunders, B.A., D.Phil.

The Cambridge Antiquarian Records Society was founded in 1972 as the result of a decision by the Cambridge Antiquarian Society to form a separate society for publishing documentary sources relating to the history of Cambridgeshire and neighbouring areas. In 1987 its name was changed to the Cambridgeshire Records Society.

Membership is open to all interested persons, and to libraries, schools and other institutions. Members receive one free copy of each volume when published and can purchase back volumes at a special price. Further details and application forms for membership can be obtained from the Hon. Secretary, Cambridgeshire Records Society, County Record Office, Shire Hall, Cambridge, CB3 0AP.

Volumes published so far and still available are:

1. *Letters to William Frend*, edited by Frida Knight
2. *John Norden's Survey of Barley*, edited by J.C. Wilkerson
3. *The West Fields of Cambridge*, edited by Catherine P. Hall and J.R. Ravensdale
4. *A Cambridgeshire Gaol Delivery Roll 1332–1334*, edited by Elisabeth G. Kimball
5. *The King's School Ely*, edited by Dorothy M. Owen and Dorothea Thurley
6. *The Church Book of the Independent Church (now Pound Lane Baptist) Isleham, 1693–1805*, edited by Kenneth A.C. Parsons
7. *Catalogue of the Portraits in Christ's, Clare and Sidney Sussex Colleges*, by J.W. Goodison
8. *Accounts of the Reverend John Crakanthorp of Fowlmere 1682–1705*, edited by Paul Brassley, Anthony Lambert and Philip Saunders

The following volumes are in active preparation:

1. *Court Rolls of the Manor of Downham 1310–1327*, edited by Clare Coleman
2. Masters' *Short Account of Waterbeach (1795)* and Denson's *A Peasant's Voice to Landowners (1830)*, edited by J.R. Ravensdale and Denis Cheason
3. *The Ely Coucher Book, 1251, for Cambs, Isle of Ely and Hunts.*, edited by Edward Miller
4. *Edmund Pettis's Survey of St. Ives, 1732*, edited by Mary Carter
5. *The Sawtry Abbey Cartulary*, edited by K.J. Stringer
6. *The Diary of Joseph Romilly, University Registrary, 1842–1848*, edited by Patrick Bury